DOCUMENT ANALYSIS AND RECOGNITION WITH WAVELET AND FRACTAL THEORIES

SERIES IN MACHINE PERCEPTION AND ARTIFICIAL INTELLIGENCE*

Vol. 63: Computational Intelligence in Software Quality Assurance
(*S. Dick and A. Kandel*)

Vol. 64: The Dissimilarity Representation for Pattern Recognition: Foundations and Applications
(*Elżbieta Pękalska and Robert P. W. Duin*)

Vol. 65: Fighting Terror in Cyberspace
(*Eds. M. Last and A. Kandel*)

Vol. 66: Formal Models, Languages and Applications
(Eds. *K. G. Subramanian, K. Rangarajan and M. Mukund*)

Vol. 67: Image Pattern Recognition: Synthesis and Analysis in Biometrics
(Eds. *S. N. Yanushkevich, P. S. P. Wang, M. L. Gavrilova and S. N. Srihari*)

Vol. 68: Bridging the Gap Between Graph Edit Distance and Kernel Machines
(*M. Neuhaus and H. Bunke*)

Vol. 69: Data Mining with Decision Trees: Theory and Applications
(*L. Rokach and O. Maimon*)

Vol. 70: Personalization Techniques and Recommender Systems
(Eds. *G. Uchyigit and M. Ma*)

Vol. 71: Recognition of Whiteboard Notes: Online, Offline and Combination
(Eds. *H. Bunke and M. Liwicki*)

Vol. 72: Kernels for Structured Data
(*T Gärtner*)

Vol. 73: Progress in Computer Vision and Image Analysis
(Eds. *H. Bunke, J. J. Villanueva, G. Sánchez and X. Otazu*)

Vol. 74: Wavelet Theory Approach to Pattern Recognition (2nd Edition)
(*Y. Y. Tang*)

Vol. 75: Pattern Classification Using Ensemble Methods
(*L. Rokach*)

Vol. 76: Automated Database Applications Testing: Specification Representation for Automated Reasoning
(*R. F. Mikhail, D. Berndt and A. Kandel*)

Vol. 77: Graph Classification and Clustering Based on Vector Space Embedding
(*K. Riesen and H. Bunke*)

Vol. 78: Integration of Swarm Intelligence and Artificial Neural Network
(Eds. *S. Dehuri, S. Ghosh and S.-B. Cho*)

Vol. 79 Document Analysis and Recognition with Wavelet and Fractal Theories
(*Y. Y. Tang*)

*For the complete list of titles in this series, please write to the Publisher.

Series in Machine Perception and Artificial Intelligence – Vol. 79

DOCUMENT ANALYSIS AND RECOGNITION WITH WAVELET AND FRACTAL THEORIES

Yuan Yan Tang
University of Macau, Macau

NEW JERSEY • LONDON • SINGAPORE • BEIJING • SHANGHAI • HONG KONG • TAIPEI • CHENNAI

Published by

World Scientific Publishing Co. Pte. Ltd.
5 Toh Tuck Link, Singapore 596224
USA office: 27 Warren Street, Suite 401-402, Hackensack, NJ 07601
UK office: 57 Shelton Street, Covent Garden, London WC2H 9HE

British Library Cataloguing-in-Publication Data
A catalogue record for this book is available from the British Library.

Series in Machine Perception and Artificial Intelligence — Vol. 79
DOCUMENT ANALYSIS AND RECOGNITION WITH WAVELET AND FRACTAL THEORIES

ISBN-13 978-981-4401-00-5
ISBN-10 981-4401-00-5

Printed in Singapore.

I dedicate this book to my family and friends

Preface

The mathematics of fractal geometry and wavelet analysis has recently generated much excitement within the community of information science and engineering among those seeking broad new and more realistic classes of models for wide-ranging applications. In particular, fractal geometry and wavelet analysis, which have a common property of multiscale signal representation, have provided solutions to important science and engineering problems in the areas of document analysis and understanding (or recognition). This optimism has arisen out of the perspective that many phenomena around the research in document analysis and understanding are much better described through the powerful multiscale signal representations than by the traditional ways.

From this perspective, the recent emergence of powerful multiscale signal representations in general and fractal/wavelet basis representations in particular has been particularly timely. Indeed, out of these theories arise highly natural and extremely useful representations for a variety of important phenomena in document analysis and understanding.

This book presents both the development of these new approaches as well as their application to a number of fundamental problems of interest to the scientists and engineers in document analysis and understanding. This book is organized into three parts, namely:

Part 1: Chapter 1 presents a brief survey of the basic concepts of document analysis and understanding, including the main theories of this area as well as the major techniques used in document analysis and understanding;

Part 2: Chapters 2 and 3 are employed to present the basic knowledge of fractal geometry and wavelet analysis;

Part 3: Chapter 4 through Chapter 8 deal with the detailed presentation of the applications of the fractal and wavelet theories to document analysis and understanding, which is the core of this book.

Considering the major readers of this book are scientists and engineers, thus, in some chapters/sections, we give up the exactness in mathematics temporarily.

Initially, in Chapter 1, a brief description of basic concepts of document analysis and understanding is introduced. Document analysis and understanding are two phases of the document processing. A basic model of document processing is developed. Some important conceptions, such as document structure, document analysis, document understanding, form document processing, character recognition, document image processing, are described in this chapter. This chapter reviews established major techniques, which are widely employed in document processing. Some new theories, new methodologies as well as new algorithms, which have been developed by the author, are also presented. Chapter 1 works as a brief survey, the review is not detailed, since this book concentrates on the novel research results developed by the author. However, the references are cited if additional details are desired.

Throughout Chapters 2 and 3, both the fractal geometry and wavelet analysis are of critical concerns, which formulate the basic theory of this book. In Chapter 2, the general theory of the fractal geometry is addressed. The theory of the fractal geometry contains many conceptions, however, in this book only the fractal dimension is discussed, because we apply only the fractal dimension to document analysis and understanding. Hausdorff dimension, box computing dimension and Minkowski dimension are introduced in details in this chapter. Chapter 3 presents the basic concepts of wavelet theory, which considers continuous wavelet transforms and multiresolution analysis (MRA). In this chapter the following main ideas are discussed: the general theory of continuous wavelet transforms, the continuous wavelet transform as a filter, description of regularity of signal by wavelets, some examples of basic wavelets, basic concept of multiresolution analysis (MRA), construction of MRAs and biorthonormal wavelet bases. As an important algorithm for implementing the wavelet transform, Mallat algorithm is introduced.

By formulating the above fractal and wavelet theories, the third part of this book (Chapters 4 - 8) demonstrates more detailed applications, which become the core chapters. All of these applications were made by our research group.

Chapter 4 aims at exploring an approach using the fractal dimensions to the document analysis for automatic knowledge acquisition. A document is considered to have two structures: geometric structure and logical structure. Extraction of the geometric structure from a document is defined as document analysis. In this way, a document image is broken down into several blocks, which represent coherent components of a document, such as text lines, headlines, graphics, etc. Traditionally, the top-down and bottom-up approaches have been used in document analysis. However, both approaches have their weaknesses. This chapter presents an approach based on *Modified Fractal Signature (MFS)* or *Modified Fractal Feature (MFF)* for document page segmentation. The MFS does not need iterative breaking or merging, and can divide a document into blocks in only one step. It is anticipated that this approach can be widely used to process various types of documents including even some with high geometrical complexity.

In Chapter 5, a method for text segmentation is proposed using wavelet decomposition of *pseudo-motion image*. The aim of this approach is to detect the position of a special kind of objects in a document image. This method works on a single image using a motion of eyeshot: the translation of the function. The translation of one document image provides a sequence of moving images, which are called *pseudo-motion document images* and the translation to be a *pseudo-motion*. When a pseudo-motion of a document image occurs, the wavelet decomposition coefficients oscillate. Meantime, the amplitudes of the oscillation give the information about the frequency of the document components. Thus we can use this significant property to segment the areas of different frequency of the document components.

The problem of invariant pattern recognition (including document recognition) is considered to be a highly complex and difficult one. The technique of rotation invariance is one of the most important branches of pattern recognition. Chapter 6 is concerned in particular with the feature extraction of document related to rotation invariance, which combines central projection transform and fractal theory. Central projection refers to mapping a multi-contour pattern such as a Chinese character into a unique contour image. One of the most powerful aspects of fractal geometry is its ability to express complex pattern as a few parameters with rotation invariant. This chapter provides an approach to invariant character recognition using fractal theory.

Documents printed in different languages need to be identified and processed in an international environment. Chapter 7 proposes two methods to identify different scripts: wavelet-based method and fractal-based method.

(1) In wavelet-based approach, a document image is decomposed into several sub-images in the frequency domain by wavelet transform, and thereafter, two feature vectors are extracted from these sub-images, namely, WED - wavelet energy distribution, and WEDP - wavelet energy distribution proportion.

(2) In fractal-based method, a document image is divided into several non-overlapping sub-images in the time domain, each of which is then mapped onto a gray-level function. The δ parallel body technique, thereafter, is applied to compute the fractal signature.

Document images used in this chapter are in six languages, i.e., Chinese, English, Japanese, Korean, Russian, and Devanagari (a language used in Indian) with various fonts.

Finally, Chapter 8 deals with writer identification. In this Chapter, we propose a method based on hidden Markov model (WD-HMM) in wavelet domain for writer identification. An introduction of the hidden markov model (abbreviated as HMM) and relevant statistical knowledge is presented. As the key point, the HMM in wavelet-domain(WD-HMM), which is a joint statistical model of the wavelet coefficients is then discussed in this chapter.

The main components of this book are the achievements in our research group with visiting research scholars. I would like to express my deep gratitude to the professors from various universities and the graduate students who have made contributions to this book.

A specific international journal called "International Journal on Wavelets, Multiresolution, and Information Processing (IJWMIP)" was founded by myself in 2003. In Chapter 1, some papers from IJWMIP are quoted. I would like to record my appreciation to the authors for their contributions to this book.

Many research projects involved in this book were supported by the research grants received from several communities in Mainland of China, Hong Kong and Macao.

Contents

Chapter 1

Basic Concepts of Document Analysis and Understanding

The basic concepts and underlying techniques of document processing are presented in this chapter. A basic model for document processing is described. In this model, document processing can be divided into two phases: document analysis and document understanding. A document has two structures: geometric (layout) structure and logical structure. Extraction of the geometric structure from a document refers to document analysis; mapping the geometric structure into the logical structure deals with document understanding. Both types of document structures and the two areas of document processing are discussed in this chapter. Two categories of methods have been used in document analysis, namely, (1) hierarchical methods including top-down and bottom-up approaches, (2) no-hierarchical methods including modified fractal signature and order stochastic filtering. Tree transform, formatting knowledge and description language approaches have been used in document understanding. All the above approaches are presented in this chapter. A particular case - form document processing is discussed. Form description and form registration approaches are presented. A form processing system is also introduced. Finally, many techniques, such as skew detection, Hough transform, Gabor filters, projection, crossing counts, form definition language, wavelet transform, etc. which have been used in these approaches are discussed in this chapter.

1.1 Introduction

Documents contain knowledge. Precisely, they are medium for transferring knowledge. In fact, much knowledge is acquired from documents such as technical reports, government files, newspapers, books, journals, magazines, letters, bank cheques, to name a few. The acquisition of knowledge from such documents by an information system can involve an extensive amount of hand-crafting. Such hand-crafting is time-consuming and can severely limit the application of information systems. Actually, it is a bottleneck of information systems. Thus, automatic knowledge acquisition from documents has become an important subject. Since the 1960's, much research on document processing has been done based on Optical Character Recognition (OCR) [Ascher et al., 1971; Nagy, 1968]. Some OCR machines which are used in specific domains have appeared in the commercial market. Surveys of the underlying techniques are made by several researchers [Davis and Lyall, 1986; Govindan and Shivaprasad, 1990; Hilderbrandt and Liu, 1993; Impedovo et al., 1991; Mantas, 1986; Mori et al., 1992; Mori et al., 1984; Sabourin, 1992; Suen et al., 1980]. The study of automatic text segmentation and discrimination started about two decades ago [Johnston, 1974; Nagy, 1968]. With rapid development of modern computers and the increasing need to acquire large volumes of data, automatic text segmentation and discrimination have been widely studied since the early 1980's [Abele et al., 1981; Toyoda et al., 1982; Wong et al., 1982]. To date, a lot of methods have been proposed, and many document processing systems have been described [Casey et al., 1992; Ciardiello et al., 1988; Esposito et al., 1990; Masuda et al., 1985; Nagy, 1990; Nakano et al., 1986; Tang et al., 2000; Tang, 2009; Tsujimoto and Asada, 1990]. Thousands papers have been presented in the *International Conferences on Document Analysis and Recognition* ICDAR'91 - ICDAR'09 [ICDAR'91, 1991; ICDAR'09, 2009]. Several books which deal with these topics have been published [Bunke et al., 1994; O'Gorman and Kasturi, 1995; Spitz and Dengel, 1995; Tang et al., 2000; Tang, 2009].

What is document processing ? Different definitions have caused a bit of confusion. In this chapter, the definition is chosen from a *basic document processing model* proposed by [Tang et al., 1994; Tang et al., 1991b]. And its principal ideas will be seen throughout the entire chapter. According to this model, this chapter is organized into the following sections:

(i) Basic Model for Document Processing
(ii) Document Structure

  (i) Strength of Structure
  (ii) Geometric Structure

   (iii) Logical Structure

(iii) Document Analysis

   (i) Hierarchical Methods
      - Top-down Approach
      - Bottom-up Approach
   (ii) No-hierarchical Methods
      - Modified Fractal Signature
      - Order Stochastic Filtering
   (iii) Web Document Analysis

(iv) Document Understanding

   (i) Tree Transform Approach
   (ii) Formatting Knowledge Approach
   (iii) Description Language Approach

(v) Form Document Processing

   (i) Characteristics of Form Documents
   (ii) Wavelet Transform Approach
   (iii) Form Description Language Approach
   (iv) Form Registration Approach
   (v) A Form Document Processing System

(vi) Character Recognition and Document Image Processing

   (i) Handwritten and Printed Character Recognition
   (ii) Document Image Analysis based on Multiresolution Hadamard Representation (MHR)

(vii) Major Techniques

   (i) Hough Transform,
   (ii) Skew Detection,
   (iii) Projection Profile Cuts,
   (iv) Run-Length Smoothing Algorithm (RLSA),
   (v) Neighborhood Line Density (NLD),
   (vi) Connected Components Algorithm,
   (vii) Crossing Counting,
   (viii) Form Description Language (FDL),

(ix) Texture Analysis,
(x) Wavelet Transform,
(xi) Other Segmentation Techniques.

1.2 Basic Model of Document Processing

A basic model of processing the concrete document was first proposed in our early work, which was presented at the *First International Conference on Document Analysis and Recognition* [Tang et al., 1991b], and also appeared in the *Handbook of Pattern Recognition and Computer Vision* [Tang et al., 1993b]. Its graphic illustration can be found in Figure 1.1, where the relationships among the geometric structure, logical structure, document analysis and document understanding are depicted.

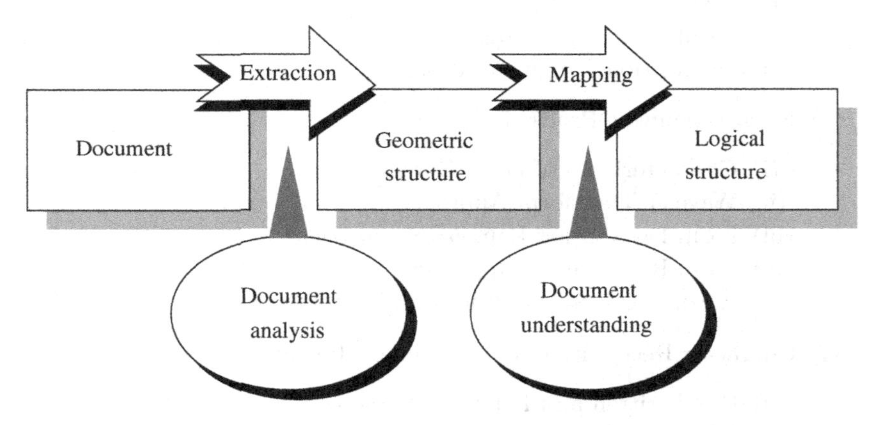

Fig. 1.1 Basic model of document processing.

The following principal concepts were proposed in this model:

- A concrete document is considered to have two structures: the *geometric (or layout) structure* and the *logical structure*.
- Document processing is divided into two phases: *document analysis* and *document understanding*.
- Extraction of the geometric structure from a document is defined as document analysis; mapping the geometric structure into the logical structure is defined as document understanding. Once the logical structure has been captured, its meaning can be decoded by

AI or other techniques.

- But in some cases, the boundary between the two phases just described is not clear. For example, the logical structure of bank cheques may also be found during an analysis by knowledge rules.

The basic model of document processing can be formally described below [Tang et al., 1993b]:

Definition 1.1 A document Ω is specified by a quintuple

$$\Omega = (\Im, \Phi, \delta, \alpha, \beta) \quad (1.1)$$

such that

$$\Im = \{\Theta^1, \Theta^2, ..., \Theta^i, ..., \Theta^m\} \quad (1.2)$$

where

$$\Theta^i = \{\Theta^i_j\}^*$$

and

$$\begin{aligned} \Phi &= \{\varphi_l, \varphi_r\} \\ \delta &= \Im \times \Phi \rightarrow 2^{\Im} \\ \alpha &= \{\alpha^1, \alpha^2, ..., \alpha^p\} \subseteq \Im \\ \beta &= \{\beta^1, \beta^2, ..., \beta^q\} \subseteq \Im \subseteq \Im \end{aligned} \quad (1.3)$$

where

- $\Im$ is a finite set of document objects which are sets of blocks Θ^i $(i = 1, 2, ..., m)$.
- $\{\Theta^i_j\}^*$ denotes repeated sub-division.
- Φ is a finite set of linking factors. φ_l and φ_r stand for leading linking and repetition linking respectively.
- δ is a finite set of logical linking functions, which indicate logical linking of the document objects.
- α is a finite set of heading objects.
- β is a finite set of ending objects.

Definition 1.2 Document processing is a process to construct the quintuple represented by Eqs. (1.1) - (1.3). *Document analysis* refers to extracting elements $\Im$, Θ^i and Θ^i_j in Eq. (1.2), i.e. extraction of the geometric structure of Ω. *Document understanding* deals with finding Φ, δ, α, and β in Eq. (1.3), considering the logical structure of Ω.

Example - A simple example is illustrated in Figure 1.2, we have

$$\Im = \{\Theta^1, \Theta^2, \Theta^3, \Theta^4, \Theta^5\}$$
$$\Theta^4 = \{\Theta^4_j\}^* = \{\Theta^4_1, \Theta^4_2\}$$
$$\Theta^5 = \{\Theta^5_j\}^* = \{\Theta^5_1, \Theta^5_2, \Theta^5_3\}$$
$$\alpha = \{\Theta^1, \Theta^2\}$$
$$\beta = \{\Theta^4, \Theta^5\}$$
$$\delta = \Im \times \Phi \to 2^{\Im} : \delta \begin{pmatrix} (\Theta^1, \varphi_l) \\ (\Theta^2, \varphi_l) \\ (\Theta^3, \varphi_l) \\ (\Theta^4, \varphi_r) \\ (\Theta^5, \varphi_r) \end{pmatrix} = \begin{pmatrix} \Theta^3 \\ \Theta^4 \\ \Theta^5 \\ \Theta^{4*}_i \\ \Theta^{5*}_i \end{pmatrix}$$

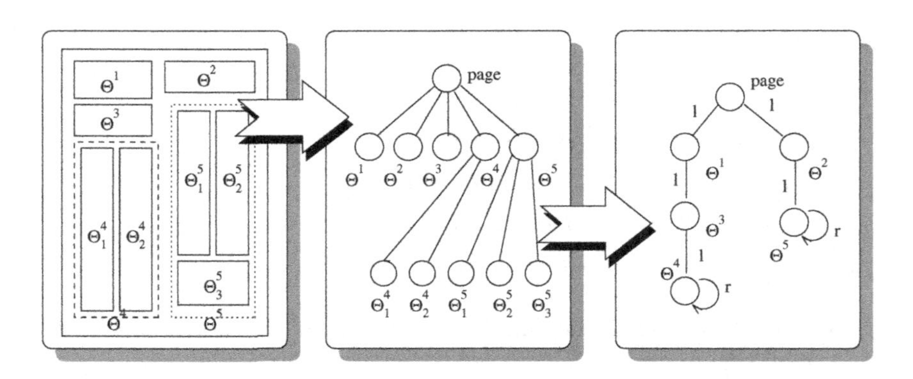

Fig. 1.2 A simple example of document processing described by the basic model.

From the above definition, it is obvious that there is a nondeterministic mapping from the geometric structure into the logical structure. However, as the geometric structure is extracted, a deterministic mapping can also be achieved. It can be formally described below:

Theorem 1.1 *Let Ω be a document defined by a quintuple $(\Im, \Phi, \delta, \alpha, \beta)$ having nondeterministic mapping from geometric structure into logical structure, then there exists a quintuple $(\Im', \Phi', \delta', \alpha', \beta')$ which contains a deter-*

ministic mapping from the geometric structure of Ω into the logical structure.

Proof $\Omega = (\Im, \Phi, \delta, \alpha, \beta)$ has nondeterministic mapping from geometric structure into logical structure. We define $\Omega' = (\Im', \Phi', \delta', \alpha', \beta')$ such that

$$\Im' \subseteq \Im, \quad \Im' = 2^{\Im},$$

$$\beta' = \{\beta'^1, \beta'^2, ..., \beta'^i, ..., \beta'^s\}, \quad \exists_i(\beta'^i \in \beta)$$

An element of $\Im'$ will be denoted by $[\theta^1, \theta^2, ..., \theta^i]$, where θ^1, θ^2, ..., θ^i are in Ω. Observe that $[\theta^1, \theta^2, ..., \theta^i]$ is a single block of Ω' corresponding to a set of blocks of Ω. Note that $\alpha' = [\alpha]$.

We define that

$$\delta(\{\theta^1, \theta^2, ..., \theta^i\}, \varphi) = \{\vartheta^1, \vartheta^2, ..., \vartheta^j\}$$

$$\overset{\text{iff}}{\Longrightarrow} \quad \delta'([\theta^1, \theta^2, ..., \theta^i], \varphi) = [\vartheta^1, \vartheta^2, ..., \vartheta^j].$$

Which means that δ' applied to an element $[\theta^1, \theta^2, ..., \theta^i]$ of Ω' is computed by applying δ to each block of Ω represented by $[\theta^1, \theta^2, ..., \theta^i]$. On applying δ to each of θ^1, θ^2, ..., θ^i and taking the union, we get some new set of blocks ϑ^1, ϑ^2, ..., ϑ^j. This new set of blocks has a representative, $[\vartheta^1, \vartheta^2, ..., \vartheta^j]$ in Ω', and that element is the value of $\delta'([\theta^1, \theta^2, ..., \theta^i], \varphi)$.

We will prove the following equation is true:

$$\delta(\theta^0, \varphi) = \{\vartheta^1, \vartheta^2, ..., \vartheta^i\} \quad \overset{\text{iff}}{\Longrightarrow} \quad \delta'(\theta'^0, \varphi) = [\vartheta^1, \vartheta^2, ..., \vartheta^i]$$

where, $|\varphi| = N$.

The proof of induction on N will be used.

Basis:

$$N = 0 \quad \Longrightarrow \quad \theta'^0 = [\theta^0] \quad \Longrightarrow \quad |\Im| = 0.$$

Inductive hypothesis: Suppose that the hypothesis is true for $|\varphi| \leq m$.

$$\delta(\theta^0, \varphi) = \{\vartheta^1, \vartheta^2, ..., \vartheta^j\} \quad \overset{\text{iff}}{\Longrightarrow} \quad \delta'(\theta'^0, \varphi) = [\vartheta^1, \vartheta^2, ..., \vartheta^j].$$

Induction: Let $|\varphi\varphi_i| = m + 1$, then

$$\delta'(\theta'^0, \varphi\varphi_i) = \delta'(\delta'(\theta'^0, \varphi), \varphi_i).$$

By the inductive hypothesis and definition of δ'

$$\delta(\{\vartheta^1, \vartheta^2, ..., \vartheta^j\}, \varphi) = \{\mu^1, \mu^2, ..., \mu^k\} \quad \overset{\text{iff}}{\Longrightarrow}$$

$$\delta'([\vartheta^1, \vartheta^2, ..., \vartheta^j], \varphi) = [\mu^1, \mu^2, ..., \mu^k].$$

Therefore

$$\delta(\theta^0, \varphi\varphi_i) = \{\mu^1, \mu^2, ..., \mu^k\} \quad \overset{\text{iff}}{\Longrightarrow} \quad \delta'(\theta'^0, \varphi\varphi_i) = [\mu^1, \mu^2, ..., \mu^k].$$

To complete the proof, the item of $\delta'(\theta'^0, \varphi)$ has to be added, Thus

$$(\Im, \Phi, \delta, \alpha, \beta) \equiv (\Im', \Phi', \delta', \alpha', \beta').$$

■

1.3 Document Structures

The key concept in document processing is that of structure. Document structure is the division and repeated subdivision of the content of a document into increasingly smaller parts, which are called *objects*. An object, which can not be subdivided into smaller objects, is called a *basic object*. All other objects are called *composite objects*. Structure can be realized as a geometric (layout) structure in terms of its geometric characteristics, and a logical structure due to its semantic properties.

1.3.1 *Strength of Structure*

To measure a document structure, a *Strength of Structure* S_s has been introduced [Watanabe, 1985].

Definition 1.3 Suppose a document is divided into n objects associated with n variables. H_i stands for the partial entropy of the i-th variable, and H for the entropy of the whole document. The strength of structure is

$$S_s = \sum_{i=1}^{n} H_i - H. \tag{1.4}$$

For instance, if the entire document consists of four composite objects associated with the variables x_1 - x_4, the strength will be

$$S_s = -\sum_{i=1}^{4}\sum_{j=1}^{n} p_j(x_i)\ log\ p_j(x_i) + \sum_{j=1}^{n} p_j(x_1,x_2,x_3,x_4)\ log\ p_j(x_1,x_2,x_3,x_4). \tag{1.5}$$

1.3.2 *Geometric Structure*

Geometric structure represents the objects of a document based on the presentation, and connection among these objects. According to the International Standard ISO 8613-1:1989(E) [ISO, 1989], the geometric or layout structure can be defined below:

Definition 1.4 Geometric or layout structure is the result of dividing and subdividing the content of a document into increasingly smaller parts, on the basis of the presentation.

Geometric (Layout) Object is an element of the specific geometric structure of a document. The following types of geometric objects are defined:

- *Block* is a basic geometric object corresponding to a rectangular area on the presentation medium containing a portion of the document content;
- *Frame* is a composite geometric object corresponding to a rectangular area on the presentation medium containing either one or more blocks or other frames;
- *Page* is a basic or composite geometric object corresponding to a rectangular area, if it is a composite object, containing either one or more frames or one or more blocks;
- *Page Set* is a set of one or more pages;
- *Document Geometric (Layout) Root* is the object at the highest level in the hierarchy of the specific geometric structure. The root node in the above example represents a page.

The geometric structure can be formally described by the following definition according to the basic model given by Eqs. (1.1) and (1.2).

Definition 1.5 The geometric structure is described by the element $\Im$ in the document space $\Omega = (\Im, \Phi, \delta, \alpha, \beta)$ shown in Eqs. (1.1) - (1.2) and β_U which is a set of operations performed on $\Im$ such that

$$\begin{aligned} &\Im = \{\Im_B, \Im_C\} \\ &\beta_U = \{\cup, \cap\} \\ &\forall_{i \neq j}(\Im_i \cup \Im_j) \subseteq \Omega) \\ &\forall_{i \neq j}(\Im_i \cap \Im_j) = \phi) \end{aligned} \tag{1.6}$$

where $\Im_B$ represents a set of *Basic objects*, and $\Im_C$ stands for a set of *Composite objects.*

$$\begin{aligned} \Im_C &= \{\Theta^1, \Theta^2, ..., \Theta^m\} \\ \Im_B &= \{\Theta^i_j | \Theta^i \in \Im_C\} \end{aligned}$$

A page of newspaper The blocks that construct the page of newspaper; The graphical description of the geometric structure; The graphical representation of logical structure.

This is a general definition of the geometric structure. Different types of specific documents have their specific forms. For example, for a specific document shown in Figure 1.3(a) which is a page extracted from newspaper, it is divided into several blocks illustrated in Figure 1.3(b). According to the above general model, its specific document geometric structure can be presented graphically in Figure 1.3(c). In this page, a document is divided into several composite objects - text areas and graphic areas, which are broken into headline blocks, text line blocks, graphic blocks, etc.

1.3.2.1 *Geometric Complexity*

The geometric complexity of a document can be measured by a *complexity function* μ which is defined below:

Definition 1.6 Let $|\Im_T|$ and $|\Im_G|$ be the number of elements in sets $\Im_T$ and $\Im_G$ respectively. Complexity function μ can be presented as

$$\mu = |\Im_T| + |\Im_G|. \tag{1.7}$$

In terms of the complexity, documents can be classified into four categories:

- Documents without graphics (e.g. editorials): $\Im_G = \phi$;

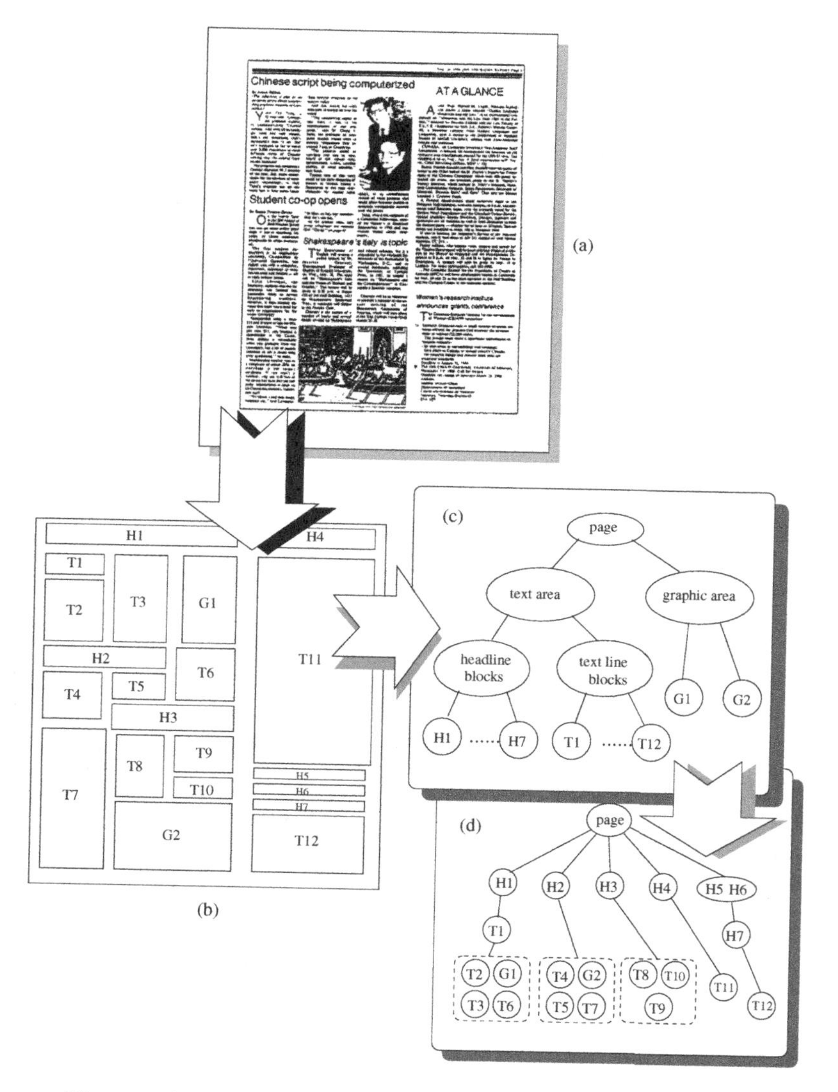

Fig. 1.3 Geometric and logical structures of a page in a newspaper.

- Document forms (e.g. bank cheques and other business forms): $\Im_G = \{\Theta_i^G\}$;
- Documents with graphics (e.g. general newspaper articles): $\Im_G \neq \phi$;

- Documents with graphics as the main elements (e.g. advertisements, front page of magazine): $|\Im_T| \leq |\Im_G|$.

1.3.3 *Logical Structure*

Document understanding emphasizes the finding of logical relations between the objects of a document. To facilitate this process, a logical structure and its model have been developed in our early work [Tang et al., 1991b] which can be summarized below.

Logical structure represents the objects of a document based on the human-perceptible meaning, and connection among these objects. According to the International Standard ISO 8613-1:1989(E), the logical structure can be defined as follows [ISO, 1989]:

Definition 1.7 Logical structure is the result of dividing and subdividing the content of a document into increasingly smaller parts, on the basis of the *human-perceptible meaning* of the content, for example, into chapters, sections, subsections, and paragraphs.

Logical Object is an element of the specific logical structure of a document. For logical object, no classification other than *Basic logical object, Composite logical object* and *Document logical root* is defined. Logical object categories such as *Chapter, Section* and *Paragraph* are application-dependent and can be defined using the *Object class* mechanism [ISO, 1989]. The document understanding process finds the logical relations between the objects of a document. According to the basic model represented by Eqs. (1.1) and (1.2), a formal description of the logical structure is presented as follows:

Definition 1.8 The logical structure is described by the elements Φ, δ, α, and β in the document space $\Omega = (\Im, \Phi, \delta, \alpha, \beta)$ in Eqs. (1.1) - (1.2), such that

$$\begin{bmatrix} \Phi \\ \alpha \\ \beta \\ \delta \end{bmatrix} = \begin{bmatrix} \{\varphi_l, \varphi_r\} \\ \{\alpha^1, \alpha^2, ..., \alpha^p\} \\ \{\beta^1, \beta^2, ..., \beta^q\} \\ \Im \times \Phi \rightarrow 2^{\Im} \end{bmatrix} \tag{1.8}$$

For a specific document shown in Figure 1.3, its logical structure can be represented graphically by Figure 1.3(d).

1.4 Document Analysis

Document analysis is defined as the extraction of geometric structure of a document. In this way, a document image is broken down into several blocks, which represent coherent components of a document, such as text lines, headlines, graphics, etc. with or without the knowledge regarding the specific format [Bagdanov and Worring, 2002; Spitz, 2002; Tang et al., 1991b; Tsujimoto and Asada, 1990]. This structure can be represented as a geometric tree shown in Figure 1.3(c). To build a such tree, there are many methods which can be classified into two categories:

- Hierarchical Methods: When we break a page of document into blocks, we conceder the geometric relationship among the blocks. In this way, we have two approaches, i.e.
  - Top-down approach
  - Bottom-up approach
- No-hierarchical Methods: When we are break a page of document into blocks, we do not consider the geometric relationship among the blocks. In this way, we have two approaches, i.e.
  - Modified Fractal Signature
  - Order Stochastic Filtering

1.4.1 *Hierarchical Methods*

In the hierarchical Methods, we have two ways: (1) from parents to children, or (2) from children to parents. Corresponding to these two ways, there are two approaches: top-down and bottom-up approaches.

These two approaches have been used in document analysis. Each has its advantages and disadvantages. The top-down approach is fast and very effective for processing documents that have a specific format. On the other hand, the bottom-up approach is time consuming. But it is possible to develop algorithms which are applicable to a variety of documents. A better result may be achieved by combining the two approaches [Nagy et al., 1988].

1.4.1.1 Top-Down Approach

Top-down (knowledge based) approach proceeds with an expectation of the nature of the document. It divides the document into major regions which are further divided into sub-regions, etc. [Bourgeois et al., 2001; Fujisawa and Nakano, 1990; Higashino et al., 1986; Ingold and Armangil, 1991; Kreich et al., 1988; Krishnamoorthy et al., 1993; Kubota et al., 1984; Lau and Leung, 1994; Nakano et al., 1986; Niyogi and Srihari, 1986].

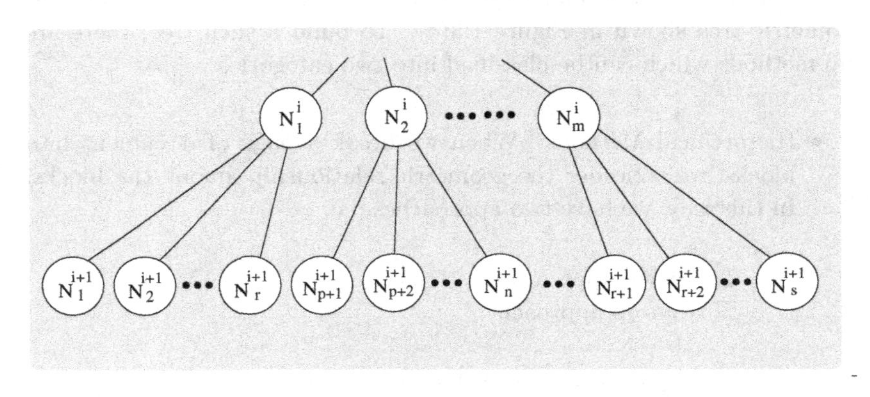

Fig. 1.4 The i-th and (i+1)-th level of a structure tree.

The geometric structure of a document can be represented by a tree. Suppose this tree contains K levels. Figure 1.4 indicates the i-th and $(i+1)$-th levels. Suppose the upper layer has nodes N_1^i, N_2^i, ..., N_m^i; and the lower layer has nodes N_1^{i+1}, N_2^{i+1}, ..., N_n^{i+1}. The relations between these two layers are expressed by edges between the nodes. They can also be represented in the form of

$$\begin{bmatrix} N_1^i \\ N_2^i \\ \cdot \\ \cdot \\ \cdot \\ N_m^i \end{bmatrix} \Longleftrightarrow \begin{bmatrix} 1 & 1 & \ldots & 1 & 0 & 0 & \ldots & 0 & 0 & 0 & \ldots & 0 \\ 0 & 0 & \ldots & 0 & 1 & 1 & \ldots & 1 & 0 & 0 & \ldots & 0 \\ & & \ldots & & & & \ldots & & & & & \\ & & \ldots & & & & \ldots & & & & & \\ & & \ldots & & & & \ldots & & & & & \\ 0 & 0 & \ldots & 0 & 0 & 0 & \ldots & 0 & 1 & 1 & \ldots & 1 \end{bmatrix} \tag{1.9}$$

Values 1's in Eq. (1.9) correspond to the edges in Figure 1.4 meaning that

$$N_1^i \Longleftrightarrow (N_1^{i+1}, N_2^{i+1}, ..., N_r^{i+1})$$

$$
\begin{array}{lcl}
N_2^i & \Longleftrightarrow & (N_{r+1}^{i+1}, N_{r+2}^{i+1}, ..., N_s^{i+1}) \\
... & ... & ... \\
N_m^i & \Longleftrightarrow & (N_{p+1}^{i+1}, N_{p+2}^{i+1}, ..., N_n^{i+1}).
\end{array}
$$

Eq. (1.9) gives two ways, "$\Longrightarrow$" from left to right corresponding to "from top to bottom" in the tree structure (Figure 1.4), and "$\Longleftarrow$" from right to left corresponding to "from bottom to top" in the same structure. In the top-down approach, the former way is used, and a document is divided into several regions each of which can be further divided into smaller sub-regions. Let $\Im$ be the set of objects which can be split into v disjoint subsets Θ^1, Θ^2, ..., Θ^p, ..., Θ^v,

$$\Theta^p \subset \Im, \qquad p = 1, 2, ..., v$$

such that

$$\bigcup_{p=1}^{v} \Theta^p = \Im, \qquad \forall_{p \neq q}(\Theta^p \cap \Theta^q = \phi).$$

A *C-function* [Watanabe, 1985] has been defined as

$$C(\Theta^p) \geq 0, \qquad p = 1, 2, ..., v$$

such that

$$C(\Theta^p \cup \Theta^q) \geq C(\Theta^p) + C(\Theta^q).$$

From Eq. (1.4), the strength of structure S_s will be

$$S_s(\Theta^p, \Theta^q) \equiv C(\Theta^p \cup \Theta^q) - C(\Theta^p) - C(\Theta^q) \geq 0. \tag{1.10}$$

The criterion of top-down splitting is that we should divide $\Theta^\tau = \Theta^p \cup \Theta^q$ into two subsets Θ^p and Θ^q such that the strength of structure S_s becomes minimum. This policy will maximize the intra-subset cohesion and minimize the inter-subset cohesion.

For multiple splitting, the strength of structure $S_s(\Theta^p, \Theta^q, ..., \Theta^y)$ can be derived by repeating Eq. (1.10):

$$
\begin{array}{l}
\Theta^\tau = \Theta^p \cup \Theta^q \cup ... \cup \Theta^y \\
\forall_{p \neq q}(\Theta^p \cap \Theta^q = \phi) \\
C(\Theta^\tau) \geq C(\Theta^p) + C(\Theta^q) + ... + C(\Theta^y) \\
S_s(\Theta^p, \Theta^q, ..., \Theta^y) = C(\Theta^\tau) - C(\Theta^p) - C(\Theta^q) - ... - C(\Theta^y) \geq 0.
\end{array} \tag{1.11}
$$

To achieve a good splitting, $S_s(\Theta^p, \Theta^q, ..., \Theta^y)$ should be minimized.

Many methods have been employed in the top-down approach, e.g. smearing [Johnston, 1974; Kanai et al., 1986; Wong et al., 1982], projection profile cut [Akiyama and Hagita, 1990; Iwaki et al., 1987; Krishnamoorthy et al., 1993; Lau and Leung, 1994; Nagy et al., 1993], Fourier transform detection [Hose and Hoshino, 1985], template [Dengel and Barth, 1988], and form definition language (FDL) [Fujisawa and Nakano, 1990; Higashino et al., 1986].

1.4.1.2 *Bottom-Up Approach*

Bottom-up (data-driven) approach progressively refines the data by layered grouping operations. The bottom-up approach is time consuming. But it is possible to develop algorithms which can be applied to a variety of documents [Akiyama and Hagita, 1990; Ciardiello et al., 1988; Doster, 1984; Fletcher and Kasturi, 1988; Masuda et al., 1985; Inagaki et al., 1984; Iwaki et al., 1987; O'Gorman, 1993; Wong et al., 1982].

Bottom-up approach corresponds to the direction of "$\Longleftarrow$" in Eq. (1.9). In this way, basic geometric components are extracted and connected into different groups in terms of their characteristics, then the groups are combined into larger groups, etc.

An analysis of this approach based on the entropy theory is given in terms of the *dynamic coalescence model* [Watanabe, 1985]. In this model, we start with $N(0)$ objects of equal "mass", suppose a region is formed by m original objects, such that this region has a mass m. $N(t)$ stands for the number of regions at time t. $X^{(\alpha)}$, $R^{(\alpha)}$ and $M^{(\alpha)}$ represent the position, size and mass of the α-th region respectively. We have

$$N(0) > N(t) > N(2t) > ...N(nt)$$

$$R^{(\alpha)} = \sqrt[n]{(M^{(\alpha)})}R_0$$

where R_0 indicates a constant called *coalescence parameter.*

The dynamic equation can be represented in the form of

$$\frac{dX^{\alpha}}{dt} = F^{(a)}$$

where

$$F^{(a)} = A \sum_{\beta} \frac{X^{(\beta)} - X^{(\alpha)}}{|X^{(\beta)} - X^{(\alpha)}|} f_{\alpha\beta}^{(1)} \tag{1.12}$$

$$f_{\alpha\beta}^{(1)} = (M^{(\beta)} M^{(\alpha)})^{\rho} g(|X^{(\beta)} - X^{(\alpha)}|).$$

If we want to include the second order effect in the equation in order to enhance the chain effect, then Eq. (1.12) can be replaced by the following formula

$$F^{(a)} = A \sum_{\beta} \frac{X^{(\beta)} - X^{(\alpha)}}{|X^{(\beta)} - X^{(\alpha)}|} [f_{\alpha\beta}^{(1)} + \varepsilon f_{\alpha\beta}^{(2)}] \tag{1.13}$$

where ε is a constant to be adjusted.

Two blocks α and β coalesce into a new block γ when they satisfy the following condition:

$$\begin{aligned} |X^{(\beta)} - X^{(\alpha)}| &= R^{(\beta)} - R^{(\alpha)} \\ X^{(\gamma)} &= \frac{X^{(\alpha)} M^{(\alpha)} + X^{(\beta)} M^{(\beta)}}{M^{(\beta)} + M^{(\alpha)}} \\ M^{(\gamma)} &= M^{(\beta)} + M^{(\alpha)}. \end{aligned} \tag{1.14}$$

There are two practical bottom-up methods: (1) neighborhood line density (NLD) indicating the complexity of characters and graphics [Iwaki et al., 1985; Iwaki et al., 1987; Kubota et al., 1984]; and (2) connected components analysis indicating the component properties of the document blocks [Bixler, 1988; Fletcher and Kasturi, 1988; Makino, 1983; Srihari et al., 1987].

1.4.2 *No-Hierarchical Methods*

Traditionally, two approaches have been used in document analysis, namely, top-down and bottom-up approaches [Tang et al., 1994]. Both approaches have their weaknesses:

- They are not effective for processing documents with high geometrical complexity. Specifically, the top-down approach can process only the simple documents which have specific format or contain some a priori information. It fails to process the documents which have complicated geometric structures as shown in Figures 1.5 and 1.6.

- To extract the geometric (layout) structure of a document, the top-down approach needs iterative operations to break the document into several blocks while the bottom-up approach needs to merge small components into large ones iteratively. Consequently, both approaches are time consuming.

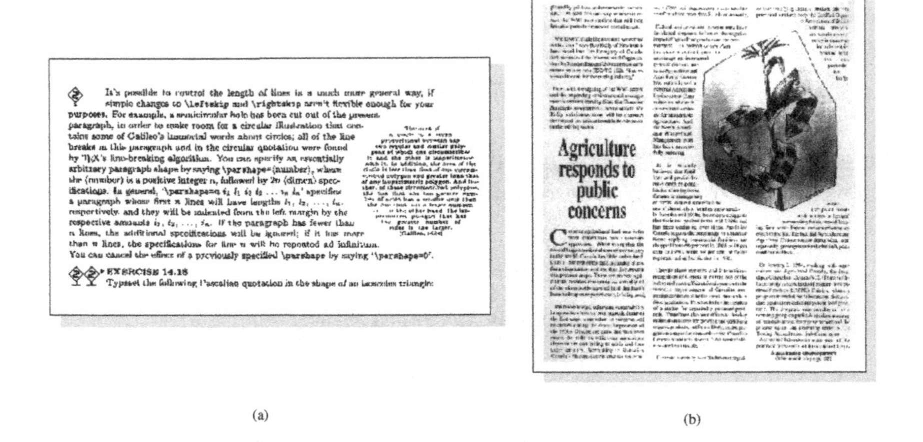

It's possible to control the length of lines in a much more general way, if simple changes to \leftskip and \rightskip aren't flexible enough for your purposes. For example, a semicircular hole has been cut out of the present paragraph, in order to make room for a circular illustration that contains some of Galileo's immortal words about circles; all of the line breaks in this paragraph and in the circular quotation were found by TeX's line-breaking algorithm. You can specify an essentially arbitrary paragraph shape by saying \parshape=⟨number⟩, where the ⟨number⟩ is a positive integer n, followed by $2n$ ⟨dimen⟩ specifications. In general, '\parshape= n i_1 l_1 i_2 l_2 ... i_n l_n' specifies a paragraph whose first n lines will have lengths $l_1, l_2, \ldots, l_n$, respectively, and they will be indented from the left margin by the respective amounts $i_1, i_2, \ldots, i_n$. If the paragraph has fewer than n lines, the additional specifications will be ignored; if it has more than n lines, the specifications for line n will be repeated ad infinitum. You can cancel the effect of a previously specified \parshape by saying '\parshape=0'.

▸ EXERCISE 14.18
Typeset the following Pascalian quotation in the shape of an isosceles triangle:

Agriculture responds to public concerns

(a) (b)

Fig. 1.5 Examples of document with complicated geometric structure.

1.4.2.1 *Modified Fractal Signature*

Tang et al. [Tang et al., 1995a; Tang et al., 1997b] presented a new approach based on modified fractal signature for document analysis. It does not need iterative breaking or merging, and can divide a document into blocks in only one step. This approach can be widely used to process various types of documents including even some with high geometrical complexity. An algorithm has been developed in [Tang et al., 1995a; Tang et al., 1997b], and briefly present as follows:

Algorithm 1.1 (fractal signature)

Input: a page of document image;
Output: the geometric structure of the document;

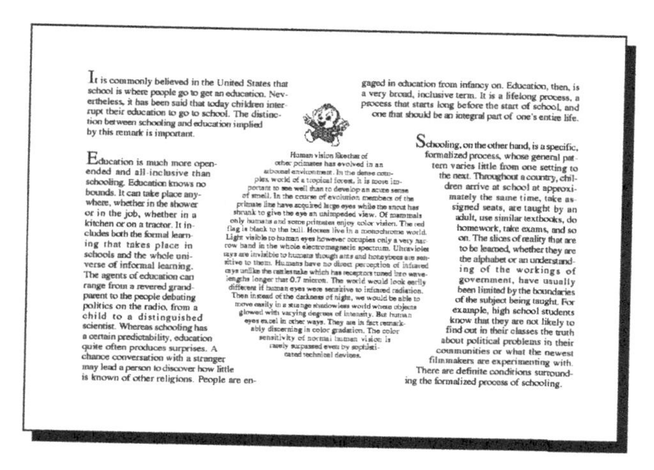

It is commonly believed in the United States that school is where people go to get an education. Nevertheless, it has been said that today children interrupt their education to go to school. The distinction between schooling and education implied by this remark is important.

Education is much more open-ended and all-inclusive than schooling. Education knows no bounds. It can take place anywhere, whether in the shower or in the job, whether in a kitchen or on a tractor. It includes both the formal learning that takes place in schools and the whole universe of informal learning. The agents of education can range from a revered grandparent to the people debating politics on the radio, from a child to a distinguished scientist. Whereas schooling has a certain predictability, education quite often produces surprises. A chance conversation with a stranger may lead a person to discover how little is known of other religions. People are engaged in education from infancy on. Education, then, is a very broad, inclusive term. It is a lifelong process, a process that starts long before the start of school, and one that should be an integral part of one's entire life.

Schooling, on the other hand, is a specific, formalized process, whose general pattern varies little from one setting to the next. Throughout a country, children arrive at school at approximately the same time, take assigned seats, are taught by an adult, use similar textbooks, do homework, take exams, and so on. The slices of reality that are to be learned, whether they are the alphabet or an understanding of the workings of government, have usually been limited by the boundaries of the subject being taught. For example, high school students know that they are not likely to find out in their classes the truth about political problems in their communities or what the newest filmmakers are experimenting with. There are definite conditions surrounding the formalized process of schooling.

Human vision likethat of other primates has evolved in an arboreal enviconment. In the dense complex world of a tropical forest, it is more important to see well than to develop an acute sense of smell. In the course of evolution members of the primate line have acquired large eyes while the snout has shrunk to give the eye an unimpeded view. Of mammals only humans and some primates enjoy color vision. The red flag is black to the bull. Horses live in a monochrome world. Light visible to human eyes however occupies only a very narrow band in the whole electromagnetic spectrum. Uhraviolet rays are invisible to humans though ants and honeybees are sensitive to them. Humans have no direct perception of infrared rays unlike the rattlesnake which has receptors tuned into wavelengths longer that 0.7 micron. The world would look eerily different if human eyes were sensitive to infrared radiation. Then instead of the darkness of night, we would be able to move easily in a strange shadowless world where objects glowed with varying degrees of intensity. But human eyes excel in other ways. They are in fact remarkably discerning in color gradation. The color sensitivity of normal human vision is rarely surpassed even by sophisticated technical devices.

Fig. 1.6 Example of document with complicated geometric structure.

Step-1 For x = 1 to X_{max} do

For y = 1 to Y_{max} do

F is mapped onto a gray-level function $g_k(x, y)$;

Step-2 For x = 1 to X_{max} do

For y = 1 to Y_{max} do

Substep-1 Initially, taking $\delta = 0$, the upper layer $u_0^k(x, y)$ and lower layer $b_0^k(x, y)$ of the blanket are chosen as the same as the gray-level function $g_k(x, y)$, namely:

$$u_0^k(x, y) = b_0^k(x, y) = g_k(x, y);$$

Substep-2 Taking $\delta = \delta_1$,

(a) $u_{\delta_1}(x, y)$ is computed according to the formula:

$$u_{\delta_1}(x, y) = max\left\{u_0(x, y) + 1, \quad \max_{|(i,j)-(x,y)| \leq 1} u_0(i, j)\right\};$$

(b) $b_{\delta_1}(x, y)$ is computed according to the formula:

$$b_{\delta_1}(x, y) = min\left\{b_0(x, y) - 1, \quad \min_{|(i,j)-(x,y)| \leq 1} b_0(i, j)\right\};$$

(c) The volume Vol_{δ_1} of the blanket is computed by the formula:

$$Vol_{\delta_1} = \sum_{x,y}(u_{\delta_1}(x,y) - b_{\delta_1}(x,y));$$

Substep-3 Taking $\delta = \delta_2$,

(a) $u_{\delta_2}(x,y)$ is computed according to

$$u_{\delta_2}(x,y) = max\left\{u_{\delta_1}(x,y)+1, \quad \max_{|(i,j)-(x,y)|\leq 1} u_{\delta_1}(i,j)\right\};$$

(b) $b_{\delta_2}(x,y)$ is computed according to

$$b_{\delta_2}(x,y) = min\left\{b_{\delta_1}(x,y)-1, \quad \min_{|(i,j)-(x,y)|\leq 1} b_{\delta_1}(i,j)\right\};$$

(c) The volume Vol_{δ_1} of the blanket is computed by

$$Vol_{\delta_2} = \sum_{x,y}(u_{\delta_2}(x,y) - b_{\delta_2}(x,y));$$

Step-3 The sub fractal signature A_δ^k is computed by the formula:

$$A_\delta^k = \frac{Vol_{\delta_2} - Vol_{\delta_1}}{2}.$$

Step-4 Combining sub fractal signatures A_δ^k, $k = 1, 2, ..., n$ into the whole fractal signature:

$$A_\delta = \bigcup_{k=1}^{n} A_\delta^k.$$

1.4.2.2 *Order Stochastic Filtering*

In this section, we present a new approach called Order Stochastic Filtering (OSF) [Ma et al., 2002]. It consists of two parts: (1) Median Order Statistic Filter (MedOSF), and (2) Maximum Order Statistic Filter (MaxOSF). This approach is more direct and much simpler. We use the MedOSF to remove the salt-pepper noise of the document and use the MaxOSF to do the page segmentation. In practice, they not only can adaptively process the documents with high geometrical complexity, but also save a lot of computing time.

Definition 1.9 The order statistic of a sample $(X_1, \cdots, X_n)$ are the sample values placed in ascending order. They are denoted by $(X_{(1)}, \cdots, X_{(n)})$ which satisfy

$$X_{(1)} \leq X_{(2)} \leq \cdots \leq X_{(n)}.$$

$X_{(n)}$ is the sample maximum of $(X_{(1)}, \cdots, X_{(n)})$ and we denote it as

$$X_{(n)} = \max(X_1, \cdots, X_n).$$

The sample median, which we denote by X_{med}, is a number such that approximately one-half of the observations are less than X_{med} and one-half are greater. In terms of the order statistics, X_{med} is defined by

$$X_{\text{med}} = \begin{cases} X_{(\frac{n+1}{2})} & if\ n\ is\ odd, \\ (X_{(\frac{n}{2})} + X_{(\frac{n}{2}+1)})/2 & if\ n\ is\ even. \end{cases}$$

(1) We use the MaxOSF, which is based on the maximum order statistic filter, to process the page segmentation. The details of the algorithm are presented in the following:

Algorithm 1.2 (MaxOSF)

Input: a page of document image.
Output: the geometric structure of the document.

Step 1 Scan a document image D_0 and get the binary image file (BMP file).

Step 2 Read the file bit by bit, then we can get a binary matrix M_0. The position of each matrix component stands for the location of the pixel in the image. 0's represent the white pixels and 1's represent the black pixels.

Step 3 Use the MaxOSF to process the binary matrix M_0, we can get another matrix M_1 which includes the information of the geometric structure of the document:

Substep 3.1 Let $x_{i,j}$ be a component of M_0 and $x_{i-1,j}$, $x_{i,j-1}$, $x_{i,j+1}$, $x_{i+1,j}$ be the four-connected neighborhood of $x_{i,j}$. Sort $x_{i-1,j}$, $x_{i,j-1}$, $x_{i,j}$, $x_{i,j+1}$, $x_{i+1,j}$ in ascending order.

Substep 3.2 Let $x'_{i,j}$ be a component of M_1, then

$$x'_{i,j} = \max\{x_{i-1,j},\ x_{i,j-1},\ x_{i,j},\ x_{i,j+1},\ x_{i+1,j}\} = x^{ij}_{(5)}.$$

Step 4 From matrix M_1, we can get the image D_1 which represents the geometric structure of the document.

(2) Sometimes the document image is corrupted by salt-pepper noise, we need to remove the noise first before we do the page segmentation. Here we use the MedOSF to remove the noise.

Algorithm 1.3 (MedOSF)

Input: a page of document image with salt-pepper-noise.
Output: a page of document without the noise.

Step 1 Scan a document image with salt–pepper–noise D_1 and get the binary image file (BMP file).

Step 2 Read the file bit by bit, then we can get a binary matrix M_1. The position of each matrix component stands for the location of the pixel in the image. 0's represent white pixels and 1's represent black pixels.

Step 3 Use MedOSF to process the binary matrix M_1, we can get another matrix M_2 which includes the information of the page for the clean document image.

Substep 3.1 Let $x_{i,j}$ be a component of M_0 and $x_{i-1,j}$, $x_{i,j-1}$, $x_{i,j+1}$, $x_{i+1,j}$ be the four-connected neighborhood of $x_{i,j}$. Sort $x_{i-1,j}$, $x_{i,j-1}$, $x_{i,j}$, $x_{i,j+1}$, $x_{i+1,j}$ in ascending order,

Substep 3.2 Let $x'_{i,j}$ be a component of M_1, then

$$x'_{i,j} = \text{median}\{x_{i-1,j},\ x_{i,j-1},\ x_{i,j},\ x_{i,j+1},\ x_{i+1,j}\} = x^{ij}_{(3)},$$

Step 4 From matrix M_2, we can get an image D_2 which represents the document without the noise.

1.4.3 *Web Document Analysis*

Web images play a crucial role in bringing visual impact to an otherwise plain text medium. Wed page designers regularly create page headers and titles as well as other semantically important textual entities in image form. However, using current technology it is not possible to analyse the text embedded in images on Web pages. This is a significant problem for a number of reasons [Karatzas and Antonacopoulos, 2003].

There is a number of approaches solving the problem of text segmentation and extraction. [Karatzas and Antonacopoulos, 2003] proposed two approaches: (1) Split-and-Merge approach, and (2) Fuzzy approach. They enable the extraction of text in complex situations such as in the presence of varying colour and texture.

XML (eXtensible Markup Language) is a standard language for defining structured documents of text-bades data. [Ishitani, 2003] proposed a technique for document transformation using OCR to generate various XML documents from printed documents. [Mukherjee et al., 2003] presented a method for automatic discovery of semantic structures in HTML document. [Ramachandran and Kashi, 2003] proposed an architecture for ink annotation on Web documents. [Hu and Bagga, 2003] presented an algorithm for automatic identification of story and preview images in news Web pages. [Jain and Namboodiri, 2003] developed a string matching-based method for indexing and retrieval of on-line handwritten documents on Web pages.

1.5 Document Understanding

As document analysis extracts geometric structures from a document image by using the knowledge about the general document and/or the specific document format, document understanding maps the geometric structures into logical structures considering the logical relationship between the objects in specific documents. There are several kinds of mapping methods in document understanding: [Tsujimoto and Asada, 1990] proposed a tree transformation method for understanding multi-article documents. [Toyoda et al., 1982] discussed the extraction of Japanese newspaper articles using a domain specific knowledge. [Inagaki et al., 1984] constructed a special purpose machine for understanding Japanese documents. [Higashino et al., 1986] proposed a flexible format understanding method, using a form definition language. Our research [Tang et al., 1991a; Tang et al., 1993b; Yan et al., 1991] has led to the development of a form description language for understanding financial documents. These mapping methods are based on specific rules applied to different documents with different formats. A series of document formatting rules are explicitly or implicitly used in all these understanding techniques.

In this section, document understanding based on *tree transformation*, document *formatting knowledge* and document *description language* will be discussed.

1.5.1 *Document Understanding Based on Tree Transformation*

This method defines document understanding as the transformation of a geometric structure tree into a logical structure tree [Tsujimoto and Asada, 1990].

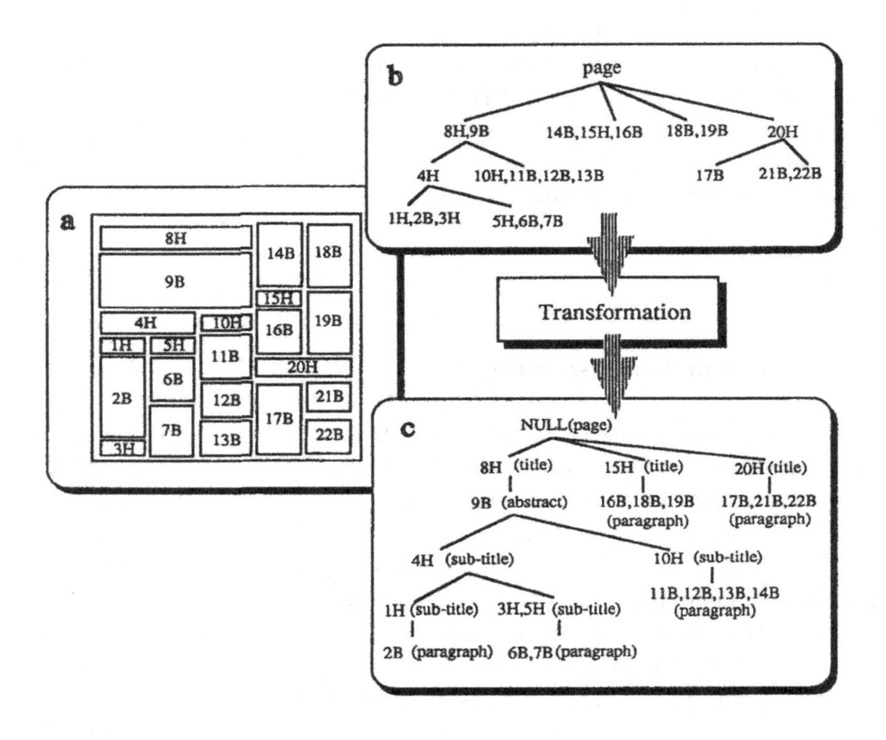

Fig. 1.7 Document understanding based on tree transformation.

A document has an obvious hierarchical geometric structure, represented by a tree as shown in Figure 1.7 And the logical structure of a document is also represented by a tree which is illustrated in Figure 1.7(c). In this example, three kinds of blocks are defined: H (head), B (body) and S (either body or head). During the transformation, a label is attached to each node. Labels include title, abstract, sub-title, paragraph, header, foot note, page number, and caption.

The transformation, which moves the nodes in the tree, is based on four transformation rules. These rules are created according to a layout designed according to the manner in which humans read. Rules 1 and 2 are based on the observation that a title should have a single set of paragraphs as a child in the logical structure. The paragraph body in another node is moved to the node under the body title by these rules. Rule 3 is mainly for the extraction of characters or sections headed by a sub-title. By rule 4, a unique class is attached to each node.

This method was implemented on a SUN-3 workstation. Pilot experiments were carried out using 106 documents taken from magazines, journals, newspapers, books, manuals, letters, scientific papers, and so on. The results show that only 12 out of 106 tested documents were not interpreted correctly.

1.5.2 *Document Understanding Based on Formatting Knowledge*

Since a logical structure can correspond to a variety of geometric structures, the generation of logical structure from the geometric structure is difficult. One of the promising solutions to this problem is use of *formatting knowledge.* The formatting rules may differ from each other because of the type of document and language to be used in it. However, for a specific kind of documents, once the formatting knowledge is acquired, its logical structure can be deduced. An example can be found in [Toyoda et al., 1982] where a method of extracting articles from Japanese newspapers has been proposed. In this method, six formatting rules of Japanese newspaper layout are summarized. An algorithm for extracting articles from Japanese newspaper has been designed based on the formatting knowledge.

Another example can be found in [Dengel, 1990] where a business letter processing approach has been developed. Because business letters are normally established in a single-column representation, letter understanding is mainly the identification of the logical objects, like sender, receiver, date, etc. In this approach, the logical objects of the letter are identified according to a *Statistical Database (SDB).* As the author reported, the SDB consists of about 71 rule packages derived from the statistical evaluation of a few hundred business letters.

Other knowledge, like the shape, size and pixel density etc. of the image block can also be used for document understanding. References [Downton and Leedham, 1990; Yeh et al., 1987] use statistical features of connected components to identify the address blocks on envelopes.

1.5.3 *Document Understanding Based on Description Language*

One of the most effective ways to describe the structures of a document is the use of a description language. [Higashino et al., 1986] detects the logical structure of a document and makes use of the knowledge rules represented by a *form definition language (FDL).* The basic concept of the form definition language is that both the geometric and logical structures of a document can be described in terms of a set of rectangular regions. For example, a part of a program in form definition language coded for the United Nations' (UN) documents is listed below:

```
(defform UN-DOC#
        (width 210) (height 297)
        (if (box (? ? ? ?)
                (mode IN Y LESS)
                (area (0 210 60 100))
                (include (160 210 1 5)))
            (form UN-DOC-A
                (0 210 0 297))
            (form UN-DOC-B
                (0 210 0 297))))
(defform UN-DOC-A ...)
(defform UN-DOC-B ...)
```

It means that the UN documents have a width of 210 mm and a height of 297 mm. The *if* predicate is one of the control structures. If the box predicate succeeds, the document named UN-DOC# is compared with UN-DOC-A and UN-DOC-B, and analyzed as UN-DOC-A. Otherwise, it is analyzed as UN-DOC-B. The box states that a rule line should exist inside the region (0 210 60 100) and satisfy the conditions that the width of the ruled line is between 160 mm and 210 mm and the height is between 1 mm and 5 mm. (defform UN-DOC-A ...) and (defform UN-DOC-B ...) will give the definition of the UN documents with and without a ruled line with these properties stated above.

According to the definition, a form dividing engine will analyze the document and produce the images of some logical objects, such as the organization which issued the document, document number, and section, etc. More details about this method can be found in [Higashino et al., 1986].

1.6 Form Document Processing

Form document is a type of special-purpose documents commonly used in our daily life. For example, millions of financial transactions take place every day. Associated with them are form documents such as bank cheques, payment slips and bills. For this specific type of documents, according to their specific characteristics, it is possible to use specific method to acquire knowledge from it.

1.6.1 *Characteristics of Form Documents*

Specific characteristics of form documents have been identified and analyzed in our early work [Tang et al., 1991a; Tang et al., 1993b; Yan et al., 1991] which are listed below:

- In general, form document may consist of straight lines which are oriented mostly in horizontal and vertical directions.
- The information that should be acquired from a form is usually the filled data. The filling positions can be determined by the above lines as references.
- Texts in form documents often contain a small set of known machine-printed, hand-printed and handwritten characters, such as legal and numeric amounts. They can be recognized with current character recognition techniques.

1.6.2 *Wavelet Transform Approach*

According to the above analysis, extraction of horizontal and vertical lines, which construct a form, plays a very important role. [Tang et al., 1997a] has applied wavelet transform to do so. A document image can be transformed into four sub-images, by the wavelet transform, namely:

- LL sub-image: both horizontal and vertical directions have low-frequencies.

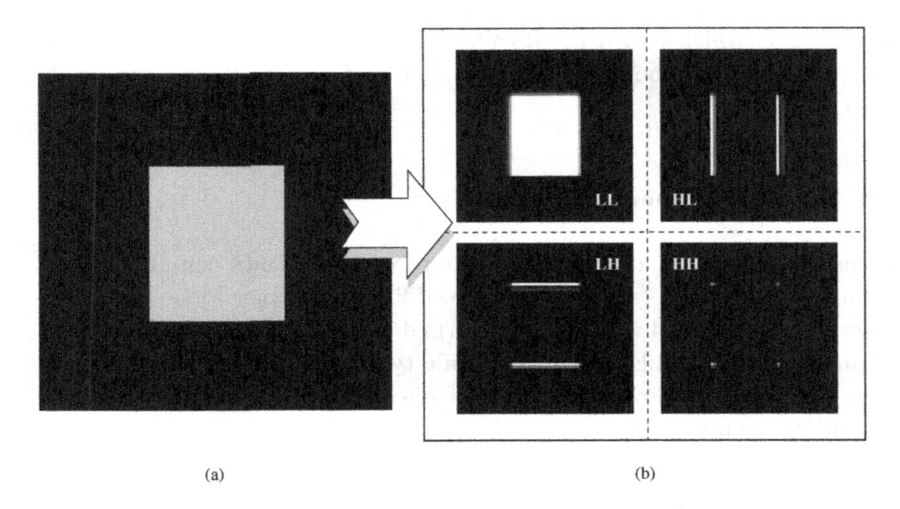

Fig. 1.8 A square image and its sub-images produced by wavelet transform.

- LH sub-image: the horizontal direction has low-frequencies, and the vertical one has high-frequencies.
- HL sub-image: the horizontal direction has high-frequencies, and the vertical one has low-frequencies.
- HH sub-image: both horizontal and vertical directions have high-frequencies.

We are interested in the LH and HL sub-images (Figure 1.8) The LH sub-image results from a filter which allows lower frequency components to pass through along the horizontal direction, and higher frequencies along the vertical direction. That is an "enhancing" effect on the vertical, and "smoothing" effect on the horizontal. The result of the HL sub-image is opposite to that of the LH one. The higher frequencies can pass through along the horizontal direction, and lower frequency components along the vertical direction. That is an "enhancing" effect on the horizontal, and "smoothing" effect on the vertical. A practical example can be found in Figure 1.9.

1.6.3 *Approach Based on Form Description Language*

A form document processing method based on form description language (FDL) has been proposed in [Tang et al., 1991a; Tang et al., 1993b; Tang

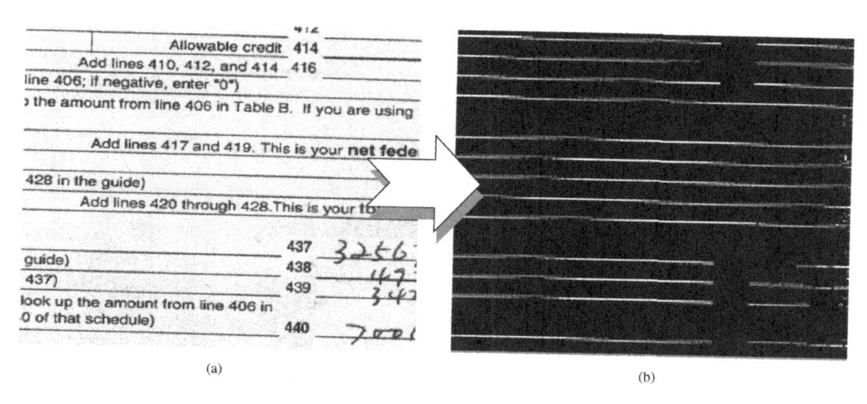
Allowable credit 414
Add lines 410, 412, and 414 416
line 406; if negative, enter "0")
the amount from line 406 in Table B. If you are using
Add lines 417 and 419. This is your net fede
428 in the guide)
Add lines 420 through 428. This is your to
437
guide) 438
437) 439
look up the amount from line 406 in
0 of that schedule) 440

(a) (b)

Fig. 1.9 Portion of USA federal tax return form and its wavelet LH sub-image.

et al., 1995b; Yan et al., 1991]. The FDL consists of two parts: (1) FSD which describes the structure of a document and (2) IDP which describes the items in a document. A block diagram of the FSD is illustrated in Figure 1.10, and a block diagram of the IDP is presented in Figure 1.11. The goal of this method is to extract the information of a form called *items* from the form document.

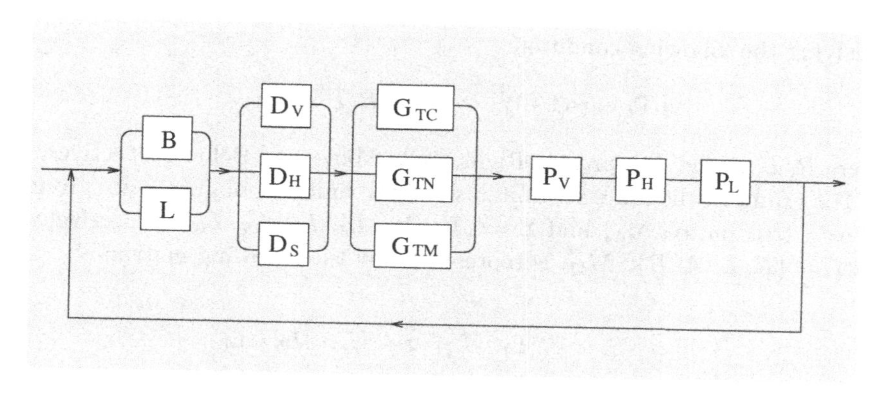

Fig. 1.10 Syntax diagram of the form structure description (FSD).

Suppose there exists a finite set of relations $\Gamma = \{\Gamma_1, \Gamma_2, ..., \Gamma_k\}$ between the finite set of items $\alpha = \{\alpha_1, \alpha_2, ..., \alpha_m\}$ and the finite set of graphs $\Sigma = \{\Sigma_1, \Sigma_2, ..., \Sigma_n\}$, and it can be represented by 0-Γ_i matrix. We call it

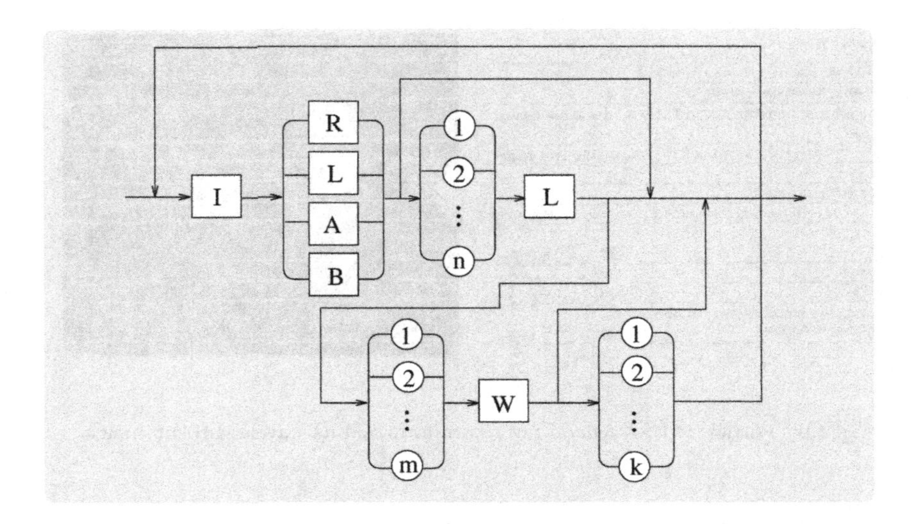

Fig. 1.11 Syntax diagram of the item description (IDP).

an Item Description Matrix: M_{ID}, such that

$$M_{ID} = \begin{cases} \Gamma_l & \text{if } (\alpha_i, \Sigma_j) \in \Gamma \\ 0 & \text{if } (\alpha_i, \Sigma_j) \notin \Gamma \end{cases}$$

satisfying the following condition:

$$\forall_l (\Gamma_l = (\alpha \Re \Sigma)), \qquad \Re = \{R, L, A, B\},$$

where R, L, A and B represent Right, Left, Above and Below respectively.

For example, the finite set of items and the finite set of graphs are given by $\alpha = \{\alpha_1,\ \alpha_2,\ \alpha_3,\ \alpha_4\}$ and $\Sigma = \{L_1,\ L_2,\ L_3,\ L_4,\ L_5,\ L_6\}$ respectively. Let $\Gamma = \{R,\ L,\ A,\ B\}$. M_{ID} is represented by the following matrix:

$$M_{ID} = \begin{array}{c} \\ \alpha_1 \\ \alpha_2 \\ \alpha_3 \\ \alpha_4 \end{array} \begin{array}{c} \begin{array}{cccccc} L_1 & L_2 & L_3 & L_4 & L_5 & L_6 \end{array} \\ \begin{bmatrix} A & 0 & 0 & L & 0 & 0 \\ B & A & 0 & R & L & 0 \\ 0 & 0 & B & 0 & R & L \\ 0 & 0 & B & 0 & 0 & R \end{bmatrix} \end{array} \tag{1.15}$$

Eq. (1.15) means that

1) α_1 is located above line L_1 and also on the left of line L_4;
2) α_2 is located below line L_1 and above line L_2 and also on the right of line L_4 and left of line L_5;
3) α_3 is located below line L_3 and also on the right of line L_5 and left of line L_6;
4) α_4 is located below line L_3 and also on the right of line L_6.

1.6.4 *Form Document Processing Based on Form Registration*

A form document processing system based on the pre-registered empty forms has been developed in [Nakano et al., 1986]. The process includes two steps: (1) empty form registration, and (2) data filled form recognition.

During the registration step, a form sample without any data is first scanned and registered with the computer. Through line enhancement, contour extraction and square detection, both the label and data fields are extracted. The relationships among these fields are then determined. Man-machine conversation is required during this registration process. The result of registration is stored as the format data of the form sample. During the recognition step, only the data fields are extracted according to the locations indicated by the format data.

1.6.5 *Form Document Processing System*

An automatic form-document processing system (FPS) has been described in [Casey et al., 1992]. It provides capabilities for automatically indexing form-document for storage/retrieval to/from a document library, and for capturing information from scanned form images using OCR. The FPS also provides capabilities for efficiently storing form images. The overall organization of FPS is shown in Figure 1.12, which contains two parallel paths, one for image applications such as retrieval, display and printing of a form-document, the other for data processing applications that deal with information contained on a form.

FPS consists of 6 major processing components:

- Defining form model;

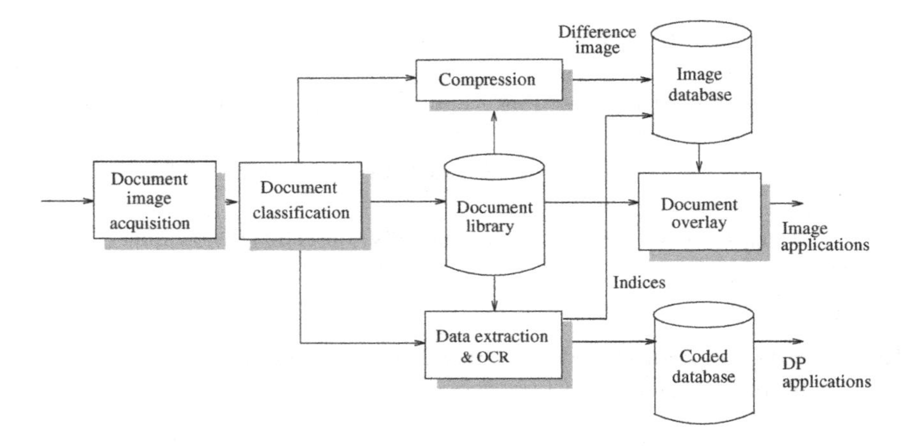

Fig. 1.12 An automatic form-document processing system.

- Storing the form model in a form library;
- Matching input form against the model stored in the form library;
- Registering the selected model to the input form;
- Converting the extracted image data to symbol code for input to data base;
- Removing the fixed part of a form, and retaining only the data filled in for storage.

1.7 Character Recognition and Document Image Processing

1.7.1 *Handwritten and Printed Character Recognition*

Character recognition including the identification of handwritten and printed characters is a major branch in the field of pattern recognition. A quite a number of articles, which deal with this branch, have been published. However, only a fewer have used wavelets [Lee et al., 1996; Tang et al., 1996; Tang et al., 1998b; Tang et al., 1998a; Wunsch and Laine, 1995]. In this sub-section, two publications [Lee et al., 1996; Wunsch and Laine, 1995] are briefly introduced. In addition, a specific chapter in this book is available to provide a detailed description of the character recognition with the wavelet theory.

1.7.1.1 *Extracting Multiresolution Features in Recognition of Handwritten Numerals with 2-D Haar Wavelet*

The well-known Haar wavelet is adequate for local detection of line segments and global detection of line structures with fast computation. [Lee et al., 1996] develops a method based on the Haar wavelet. It enables us to have an invariant interpretation of the character image at different resolutions and presents a multiresolution analysis in the form of coefficient matrices. Since the details of character image at different resolutions generally characterize different physical structures of the character, and the coefficients obtained from wavelet transform are very useful in recognizing unconstrained handwritten numerals. Therefore, in [Lee et al., 1996], wavelet transform with a set of Haar wavelets is used for multiresolution feature extraction in handwriting recognition.

In this way, we take

$$1 = \cos^2 \frac{\omega}{2} + \sin^2 \frac{\omega}{2}.$$

Let us write

$$\begin{aligned} |H(\omega)|^2 &= \cos^2 \frac{\omega}{2} = \frac{1+\cos\omega}{2} \\ &= \frac{1}{4}\left[1 + 2\cos\omega + \cos^2\omega + \sin^2\omega\right] \\ &= \frac{1}{4}\left[(1+\cos\omega)^2 + \sin^2\omega\right] \\ &= \left|\frac{1+\cos\omega - i\sin\omega}{2}\right|^2 = \left|\frac{1+e^{-i\omega}}{2}\right|^2, \end{aligned}$$

so that

$$|H(\omega+\pi)|^2 = \sin^2\frac{\omega}{2} = \frac{1-\cos\omega}{2} = \left|\frac{1-e^{-i\omega}}{2}\right|^2.$$

Thus, we can take

$$H(\omega) = \frac{1+e^{-i\omega}}{2} = \sum_{k=0}^{1} \frac{1}{\sqrt{2}} h_k e^{-i\omega k},$$

where

$$h_0 = \frac{1}{\sqrt{2}}, \qquad h_1 = \frac{1}{\sqrt{2}}.$$

The scaling function $\varphi(x)$ and wavelet function $\psi(x)$ can be represented by

$$\varphi(x) = \begin{cases} 1 & \text{if } 0 < x < 1 \\ 0 & \text{otherwise} \end{cases} \tag{1.16}$$

and

$$\begin{aligned} \psi(x) &= c_1\varphi(2x) - c_0\varphi(2x-1) \\ &= \varphi(2x) - \varphi(2x-1) \\ &= \begin{cases} 1 & \text{if } 0 < x < \frac{1}{2} \\ -1 & \text{if } \frac{1}{2} \leq x < 1 \\ 0 & \text{otherwise} \end{cases} \end{aligned} \tag{1.17}$$

An image of the handwritten character can be decomposed into its wavelet coefficients by using Mallat's pyramid algorithm. By using Haar wavelets, an image F is decomposed as follows:

$$F = \begin{bmatrix} & & & \\ & & & \\ & & w & x \\ & & y & z \end{bmatrix} \Rightarrow \begin{bmatrix} p & q & & \\ r & a & & b \\ & & & \\ & c & & d \end{bmatrix} \Rightarrow \begin{bmatrix} s & t & & \\ u & v & & b \\ & & & \\ & c & & d \end{bmatrix}$$

$$\begin{aligned} a &= 1/4(w+x+y+z), & b &= 1/4(w-x+y-z) \\ c &= 1/4(w+x-y-z), & d &= 1/4(w-x-y+z) \\ s &= 1/4(p+q+r+a), & t &= 1/4(p-q+r-a) \\ u &= 1/4(p+q-r-a), & v &= 1/4(p-q-r+a) \end{aligned} \tag{1.18}$$

In Eq. (1.18), the image $\{w, x, y, z\}$ is decomposed into image $\{a\}$, $\{b\}$, $\{c\}$, and $\{d\}$ at resolution 2^{-1}. The image $\{a\}$ corresponds to the lowest frequencies (D_1), $\{b\}$ gives the vertical high frequencies (D_2), $\{c\}$ the horizontal high frequencies (D_3), and $\{d\}$ the high frequencies in horizontal and vertical directions (D_4). Likewise, the image $\{p, q, r, a\}$ at resolution 2^{-1} is decomposed into image $\{s\}$, $\{t\}$, $\{u\}$, and $\{v\}$ at resolution 2^{-2}. This decomposition can be archived by convolving the 2×2 image array with the Haar masks as follows:

$$D_k^j = \frac{1}{4} I^{j+1} \otimes H_k \qquad k = 1, 2, 3, 4 \tag{1.19}$$

where, $\otimes$ is the convolution operator, D_k^j are decomposed images at resolution 2^j, I^{j+1} is 2×2 image at resolution 2^{j+1}, and H_k are Haar masks defined in Figure 1.13.

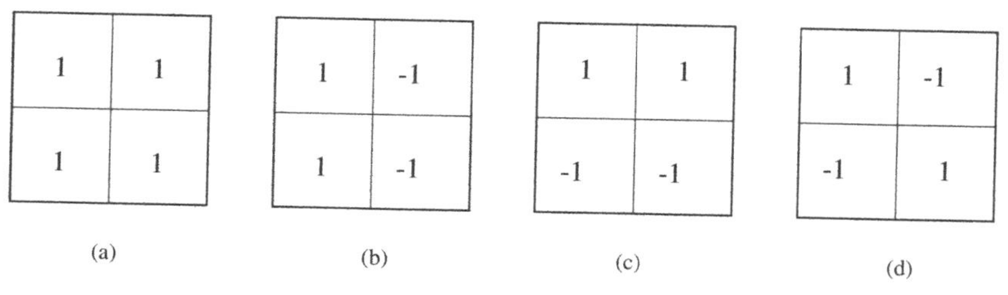

Fig. 1.13 Haar masks (a) Lowest frequencies, (b) Vertical high frequencies, (c) Horizontal high frequencies, (d) High frequencies in horizontal and vertical directions.

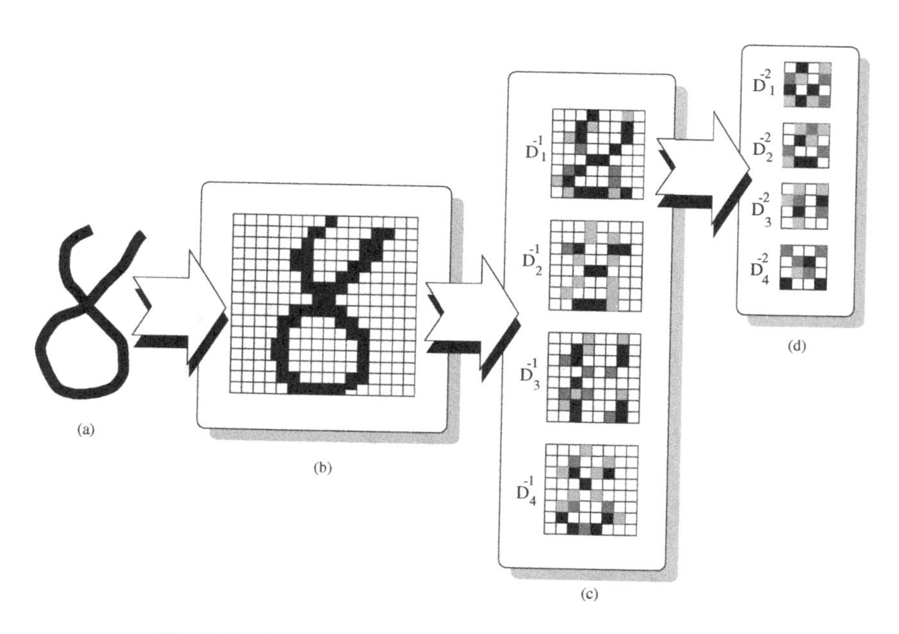

Fig. 1.14 Overview of multiresolution feature extraction.

In this application, the decomposed results at resolution 2^{-1} and 2^{-2} are used as multiresolution features . Figure 1.14 shows the process of multiresolution feature extraction. Figure 1.14(a) gives an input image of the handwritten number "8", its digitized image is shown in Figure 1.14(b). The decomposed feature vector at resolution 2^{-1} is illustrated in Figure 1.14(c), while the decomposed feature vector at resolution 2^{-2} is in Figure 1.14(d).

These features are used to recognize the unconstrained handwritten nu-

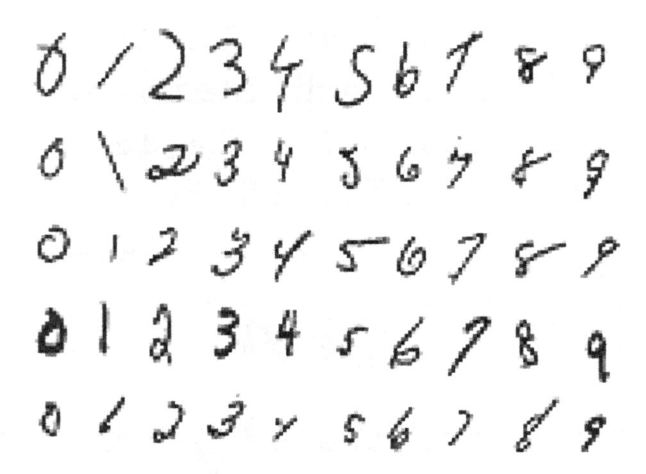

Fig. 1.15 Handwritten numeral database of Concordia University of Canada.

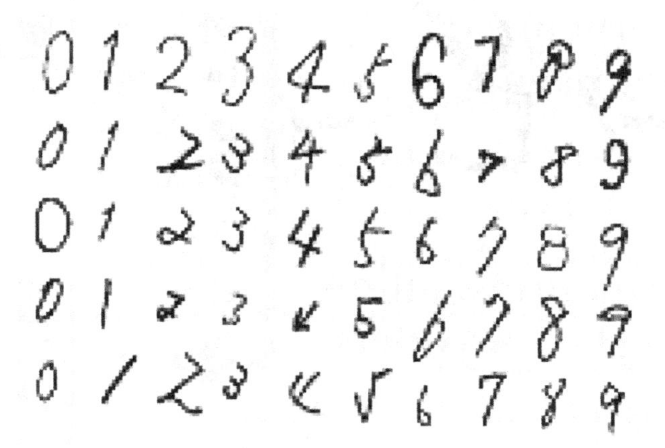

Fig. 1.16 Handwritten numeral database of Electro-Technical Laboratory of Japan.

merals from the database of Concordia University of Canada (Figure 1.15), Electro-Technical Laboratory of Japan (Figure 1.16), and Electronics and Telecommunications Research Institute of Korea (Figure 1.17). The error rates are 3.20%, 0.83%, and 0.75%, respectively. These results are shown that the proposed scheme is very robust in terms of various writing styles

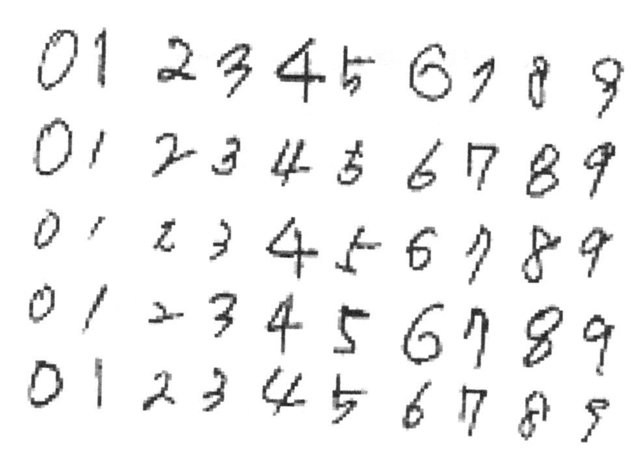

Fig. 1.17 Handwritten numeral database of Electronics and Telecommunications Research Institute of Korea.

and sizes. The detailed description of this application can be referred in [Lee et al., 1996].

1.7.1.2 *Recognition of Printed Kannada Text in Indian Languages*

An OCR system for Indian languages, especially for Kannada, a popular South Indian language, is developed in [Kunte and Samuel, 2007]. Some examples of Kannada language is presented in Figure 1.18.

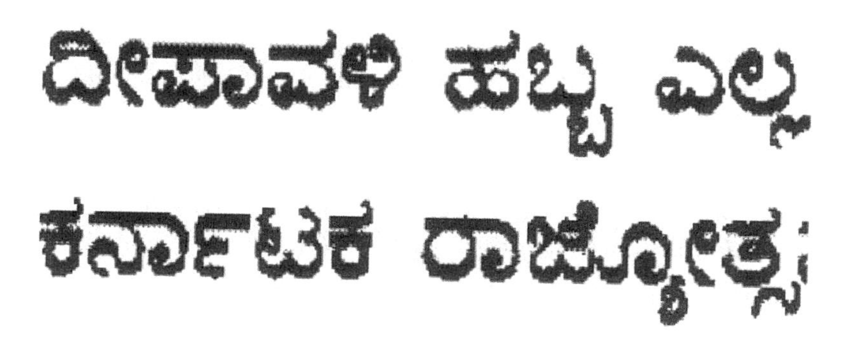

Fig. 1.18 Some examples of Kannada text in Indian languages [Kunte and Samuel, 2007].

1D discrete wavelet transform (DWT) is used for feature extraction.

Daubechies wavelet from the family of orthonormal wavelets is considered. A DWT when applied to a sequence of coordinates from the character contour returns a set of approximation coefficients and a set of detailed coefficients. The approximation coefficients correspond to the basic shape of the contour (low frequency components) and the detailed coefficients correspond to the details of the contour (high frequency components), which reflect the contour direction, curvature, etc. Neural classifiers are effectively used for the classification of characters based on wavelet features. The system methodology can be extended for the recognition of other south Indian languages, especially for Telugu.

1.7.1.3 *Wavelet Descriptors of Handprinted Characters*

Paper [Wunsch and Laine, 1995] presents a novel set of shape descriptors that represents a digitized pattern in concise way and that is particularly well-suited for the recognition of handprinted characters. The descriptor set is derived from the wavelet transform of a pattern's contour. The approach is closely related to feature extraction methods by Fourier Series expansion. The motivation to use an orthonormal wavelet basis rather than the Fourier basis is that wavelet coefficients provide localized frequency information, and that wavelets allow us to decompose a function into a multiresolution hierarchy of localized frequency bands. This paper describes a character recognition system that relies upon wavelet descriptors to simultaneously analyze character shape at multiple levels of resolution. The system was trained and tested on a large database of more than 6000 samples of handprinted alphanumeric characters. The results show that wavelet descriptors are an efficient representation that can provide for reliable recognition in problems with large input variability.

1.7.2 *Document Image Analysis Based on Multiresolution Hadamard Representation (MHR)*

A novel class of wavelet transform referred to as the multiresolution Hadamard representation (MHR) is proposed by [Liang et al., 1999] for document image analysis.

The multiresolution Hadamard representation (MHR) is a 2-D dyadic wavelet representation which employs two Hadamard coefficients $[1, 1, 1, 1]$ and $[1, -1, -1, 1]$ as shown in Figure 1.19. These coefficients are further normalized with respect of l^1 norm [Mallat, 1989a].

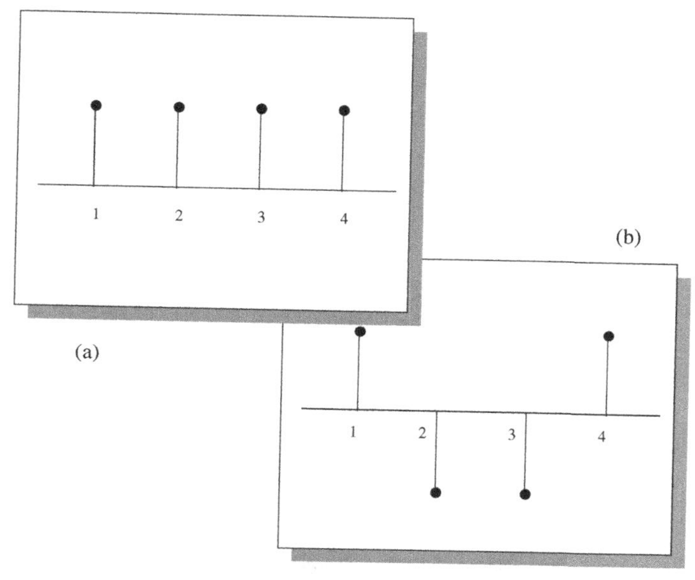

Fig. 1.19 The basic functions of the multiresolution Hadamard representation (MHR): (a) The low-pass filter $h(\cdot)$; (b) The high-pass filter $g(\cdot)$ [Liang et al., 1999].

The 2-D dyadic wavelet representation is produced by applying the 1-D filters $h(\cdot)$ and $g(\cdot)$to the 2-D image in both horizontal and vertical directions. This representation comprises four channels, namely, low-passed L, horizontal H, vertical V and diagonal D, at each level of transform. These channels can be defined by the following iterative formulas:

$$
\begin{aligned}
L_{u,v,j+1} &= \int_x \int_y L_{u,v,j} h(x-2u)h(y-2v), \\
H_{u,v,j+1} &= \int_x \int_y L_{u,v,j} h(x-2u)g(y-2v), \\
V_{u,v,j+1} &= \int_x \int_y L_{u,v,j} g(x-2u)h(y-2v), \\
D_{u,v,j+1} &= \int_x \int_y L_{u,v,j} g(x-2u)g(y-2v),
\end{aligned}
$$

where x and u are horizontal coordinates, y and v are vertical ones. Note that when $j=0$, $L_{u,v,j}$ denotes the original image.

The $L_{(j+1)}$ channel is obtained by the convolution of L_j with the 2-D filter $h(x)h(y)$ shown in Figure 1.20.

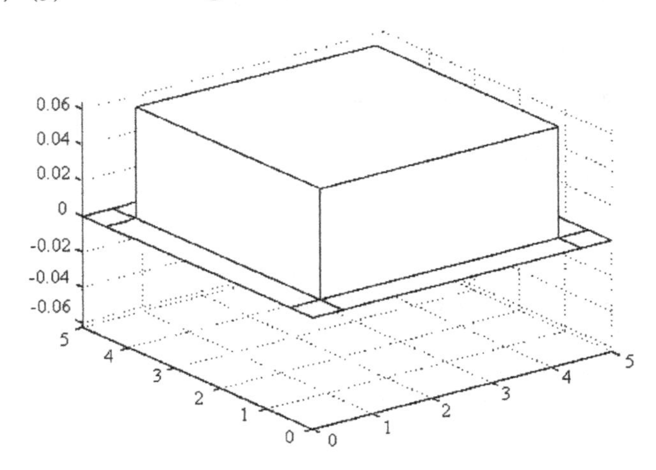

Fig. 1.20 The L channels of the 2-D filters $h(\cdot)g(\cdot)$ [Liang et al., 1999].

The $H_{(j+1)}(V_{(j+1)})$ channel is produced by the convolution of L_j with the 2-D filter $h(x)g(y)(h(y)g(x))$. Since the shape of $h(x)g(y)(h(y)g(x))$ is similar to a horizontal (vertical) bar, this 2-D filter serves as a detector for horizontal (vertical) bars on L_j, which can be graphically illustrated in Figure 1.21.

The $D_{(j+1)}$ channel is obtained by the convolution of L_j with the 2-D filter $g(x)g(y)$. It is displayed graphically in Figure 1.22.

In [Liang et al., 1999], the multiresolution Hadamard representation (MHR) is applied to document image analysis including the following processes:

- the exaction of half-tone picture,
- segmentation of document image into text blocks, and
- determination of character scales for each text block.

The transformed values of the vertical strokes of characters are very positive in some of the V channels, while, the horizontal strokes of characters react strongly in the H channels. The diagonal strokes of characters, on the other hand, react in both H and V channels. The pictures in newspapers are generally produced by half tones, where the tone of the pictures is produced by small block dots with varying densities. While characters

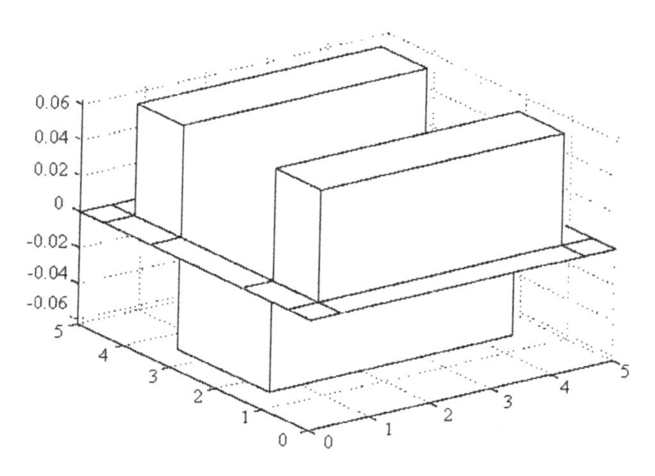

Fig. 1.21 The H and V channels of the 2-D filters $h(\cdot)g(\cdot)$ [Liang et al., 1999].

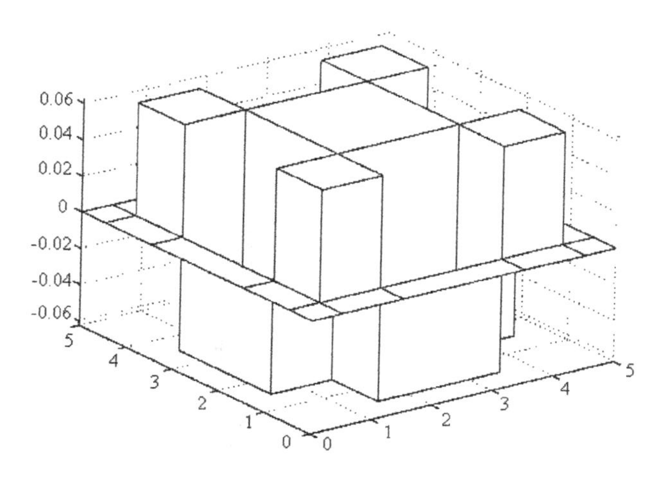

Fig. 1.22 The D channels of the 2-D filters $h(\cdot)g(\cdot)$ [Liang et al., 1999].

show up their strengths in the H and V channels, a half-tone picture reacts as a significant regular pattern in the D channel. This pattern is produced when the half-tone pictures are filtered by $(g(x)g(y))$.

In [Liang et al., 1999], the multiresolution Hadamard representation (MHR) is used to treat a portion of Chinese newspaper as shown in Figure 1.23. It contains Chinese characters of three different sizes. The MHR

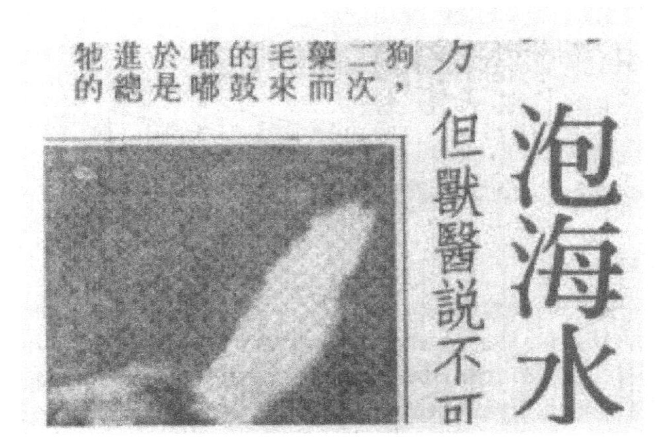

Fig. 1.23 A portion of Chinese newspaper [Liang et al., 1999].

Fig. 1.24 Three text blocks with different scales [Liang et al., 1999].

developed in [Liang et al., 1999] picks up three text blocks with different scales from the original image, which are presented in Figure 1.24. The multiresolution Hadamard representation of Figure 1.23 is illustrated in Figure 1.25, where the scale is $j = 1$.

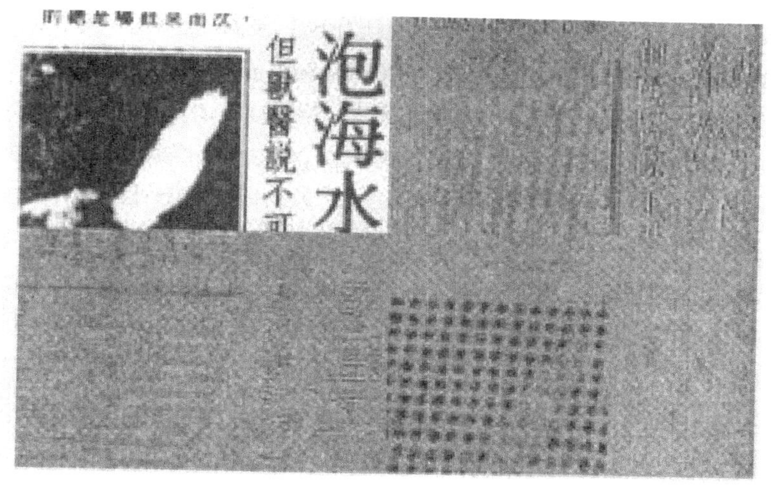

Fig. 1.25 The multiresolution Hadamard representation of Figure 1.23 at scale $j = 1$ [Liang et al., 1999].

The detailed description can be referred to [Liang et al., 1999].

1.8 Major Techniques

To implement the above approaches, a lot of practical techniques have been developed. In this section, the major techniques will be presented.

- Hough Transform,
- Skew Detection,
- Projection Profile Cuts,
- Run-Length Smoothing Algorithm (RLSA),
- Neighborhood Line Density (NLD),
- Connected Components Algorithm,
- Crossing Counting,
- Form Description Language (FDL),
- Texture Analysis,
- Wavelet Transform,
- Other Segmentation Techniques.

1.8.1 *Hough Transform*

The Hough transform maps points of Cartesian space (x, y) into sinusoidal curves in a (ρ, θ) space via the transformation:

$$\rho = xcos(\theta) + ysin(\theta).$$

Each time a sinusoidal curve intersects another at particular values of ρ and θ, the likelihood that a line corresponding to these $\rho\theta$ coordinate values is present in the original image also increases. An accumulator array (consisting of R rows and T columns) is used to count the number of intersections at various ρ and θ values. Those cells in the accumulator array with the highest number of counts will correspond to lines in the original image. Because text lines are actually thick lines of sparse density, Hough transform can be used to detect them and their orientation.

Three major applications of the Hough transform in document analysis are listed below:

- **Skew Detection**: An important application of Hough transform is skew detection. A typical method can be found in [Hinds et al., 1990]. It detects the document skew by applying Hough transform to a "burst image". At first, the resolution of the document image is reduced from 300 dpi (dots per inch) to 75 dpi. Next, a vertical and a horizontal burst image will be produced based on the reduced document image. Hough transform is then applied to either the vertical or the horizontal burst image according to the orientation of the document. Compared to the original image, the number of black pixels in the burst image is significantly reduced compared to the original image. It speeds up the skew detection procedure. In order to eliminate the negative effects of the large run-length contributed by the figures and black margins, only small run-lengths between 1 and 25 pixels are mapped to the (ρ, θ) space. The skew angle can then be calculated according to the accumulator array. In [Hinds et al., 1990], all skews have been detected correctly for the thirteen test images of five different types of documents.
- **Text block identification**: The accumulator array produced by the transform has different properties corresponding to the different contents of the document images. The high peaks in the array correspond to graphics in the document, while the cells with reg-

ular value and uniform width in the array correspond to texts in the document [Rastogi and Srihari, 1986; Srihari and Govindaraju, 1989]. Thus, the different document contents can be identified according to these properties.

- **Grouping the characters in a line for text/graph separation**: Hough transform can also be used to detect the text lines by means of grouping the characters together and separating them from the graphics [Fletcher and Kasturi, 1988].

1.8.2 *Techniques for Skew Detection*

Many techniques have been applied to skew detection, and many algorithms have been developed [O'Gorman and Kasturi, 1995; Zhu and Yin, 2002]. For example, Akiyama and Hagita [Akiyama and Hagita, 1990] developed an automated entry system for skewed documents. But it failed in a document which is mixed of text blocks, photographs, figures, charts, and tables. Hinds, Fisher and D'amato [Hinds et al., 1990] developed a document skew detection method using run-length encoding and the Hough transform. Nakano, Shima, Fujisawa, Higashino and Fujiwara [Nakano et al., 1990] proposed an algorithm for the skew normalization of a document image based on the Hough transform. These methods can handle documents in which the non-text regions are limited in size. Ishitani [Ishitani, 1993] proposed a method to detect skew for document images containing a mixture of text areas, photographs, figures, charts, and tables. To handle multi-skew problem, [Yu et al., 1995] developed a method using *Least Squares*, and the basic idea is presented below:

Given a set of N data points, i.e. the reference points of the text line, a linear function is assumed to exist between the dependent variable $f(x)$ and the independent variable x,

$$\begin{aligned} G(x) &= a_1 g_1(x) + a_2 g_2(x) + \ldots + a_m g_m(x) \\ &\approx a_1 + a_2 x \end{aligned} \tag{1.20}$$

They can therefore be solved to give

$$a_1 = \frac{\sum_{i=1}^{N} f(x_i) \sum_{i=1}^{N} x_i^2 - \sum_{i=1}^{N} x_i \sum_{i=1}^{N} f(x_i) x_i}{N \sum_{i=1}^{N} x_i^2 - (\sum_{i=1}^{N} x_i)^2} \tag{1.21}$$

$$a_2 = \frac{N \sum_{i=1}^{N} f(x_i)x_i - \sum_{i=1}^{N} x_i \sum_{i=1}^{N} f(x_i)}{N \sum_{i=1}^{N} x_i^2 - (\sum_{i=1}^{N} x_i)^2} \qquad (1.22)$$

Consequently, from a_1 and a_2, the slope of the text block as well as the skewed angle θ_b^i can be calculated. Furthermore, the skewed angle can be rotated to the correct position.

1.8.3 *Projection Profile Cuts*

Projection refers to the mapping of a two-dimensional region of an image into a waveform whose values are the sums of the values of the image points along some specified directions. A projection profile is obtained by determining the number of black pixels that fall onto a projection axis. Projection profiles represent a global feature of a document. They play a very important role in document element extraction, character segmentation and skew normalization.

Let $f(x, y)$ be a document image, and R stand for an area of the document image. Assume that $f(x, y) = 0$ lies outside the image. $\delta[...]$ denotes a delta function. $t = xsin\phi - ycos\phi$ gives the Euclidean distance of a line from the origin [Pavlidis, 1982]. If the projection angle from the x-axis is ϕ, the projection can be defined as follows:

$$p(\phi, t) = \int_R f(x, y)\delta[xsin\phi - ycos\phi - t]dxdy. \qquad (1.23)$$

Three directional projection profiles: (a) horizontal, (b) vertical and (c) diagonal, are commonly used:

$$\begin{aligned} p(0°, t) &= \int_R f(x, t)dx \\ p(90°, t) &= \int_R f(t, y)dy \\ p(45°, t) &= \int_R f(t, x - \sqrt{2t})dx \\ p(135°, t) &= \int_R f(t, \sqrt{2t} - x)dx. \end{aligned}$$

For a digitized image, the symbol $\int_R$ should be replaced by $\sum_R$.

All objects in a document are contained in rectangular blocks. Blanks are placed between these rectangles. Thus, the document projection profile is a waveform whose deep valleys correspond to the blank areas of the documents. A deep valley with a width greater than an established threshold, can be cut as the position corresponding to the edge of an object or a block. Because a document generally consists of several blocks, the process of projection should be done recursively until all of the blocks have been located. More details about various applications of this technique in the document analysis can be found in references [Akiyama and Hagita, 1990; Iwaki et al., 1987; Nagy et al., 1993; Tsuji, 1988; Viswanathan, 1990; Wang and Srihari, 1989].

1.8.4 *Run-Length Smoothing Algorithm (RLSA)*

The basic RLSA is applied to a binary sequence in which white pixels are represented by 0's and black pixels by 1's. It transforms a binary sequence x into an output sequence y according to the following rules:

1) 0's in x are changed to 1's in y, if the number of adjacent 0's is less than or equal to a predefined limit C.
2) 1's in x are unchanged in y.

For example, with $C = 4$ the sequence x is mapped into y as follows:

$$x \; : \; 00010000010100001000000011000$$
$$y \; : \; 11110000011111111000000011111$$

When applied to pattern arrays, the RLSA has the effect of linking together neighboring black areas that are separated by less than C pixels. With an appropriate choice of C, the linked areas will be regions of a common data type. The degree of linkage depends on the following factors: (a) the threshold value C, (b) the distribution of white and black pixels in the document, and (c) the scanning resolution. On the other hand, the RLSA may also be applied to the background. It has the effect of eliminating black pixels that are less than C in length [Johnston, 1974].

The choice of the smoothing threshold C is very important. Very small horizontal C values simply *close* individual characters. Slightly larger values of C merge together individual characters in a word, but are not large enough to bridge the space between two words. Too large values of C often cause sentences to join to non-text regions, or to connect to adjacent columns. In general, threshold C is set according to the character height, gap between words and interline spacing [Fisher et al., 1990; Johnston, 1974].

The RLSA was first proposed by Johnston [Johnston, 1974] to separate text blocks from graphics. It has also been used to detect long vertical and horizontal white lines [Abele et al., 1981; Wahl et al., 1981]. This algorithm was extended to obtain a bit-map of white and black areas representing blocks which contain various types of data [Wong et al., 1982]. Run-length smoothed document images can also be used as basic features for document analysis [Esposito et al., 1990; Fisher et al., 1990; Fujisawa and Nakano, 1990].

1.8.5 *Neighborhood Line Density (NLD)*

For every pixel on the document, its NLD is the sum complexity of its four directions.

$$NLD = \sum_{i \in \aleph} C_i$$

$$C_i = \sum_{i \in \aleph} (1/L_{ij})$$

$$\aleph = \{L, R, U, D\}$$

where, L, R, U and D stand for the four directions, i.e. left, right, up and down respectively. C_i indicates the complexity of a pixel for the direction i. L_{ij} represents the distance from the given pixel to its surrounding stroke j in the direction i.

Based to the following features, the NLD can be used to separate characters from graphics including the situation that some characters are touching the graphics: (1) NLD is higher for character fields than graphic fields, and (2) there are high peaks of NLD in the character fields and their height is affected by the character's size and pitch [Kubota et al., 1983].

The NLD consists of three processing steps. First, the NLD for all the black pixels of the input document is calculated using the method stated above. Second, an NLD emphasis processing is carried out in order to enlarge the NLD difference between the graphic fields and character fields.

The third step is thresholding, the pixels which have an NLD value greater than a threshold θ are classified as character fields, otherwise they are classified as graphic fields.

1.8.6 *Connected Components Analysis (CCA)*

A connected component is a set of connected black or white pixels such that an 8-connected path exists between any two pixels. Different contents of the document tend to have connected components with different properties. Generally, graphics consist of large connected components. Texts consist of connected components with regular and relatively smaller size. By analyzing these connected components, graphics and texts in the document can be identified, grouped together to different blocks and separated from each other.

The size and location of the connected component can be represented by a four-tuple [Toyoda et al., 1982]. The analysis of a document can be regarded as the process of merging these four-tuples. Taking newspaper as an example, its content is classified into several regions like index, abstract, article body, picture and figure, etc. During image analysis, the four-tuples are merged and classified into these regions using the features found in the regions. In [Toyoda et al., 1982], 13 features about the six regions of Japanese newspaper are summarized. According to these features, a table is created summarizing the properties of the four-tuples in each region. All the four-tuples can be classified and merged following the rules described in this table. Since the four-tuples contain information about the location of the components, all the regions can be classified and located at the end of the four-tuple merging process.

Two typical applications of the CCA in document processing can be illustrated below.

- **Envelope Processing**: An important application is automatic envelope processing [Bartneck, 1988; Bergman et al., 1988; Demjanenko et al., 1990; Downton and Leedham, 1990; Lee and Kim, 1994; Yeh et al., 1987]. By placing the connected components into several groups and further analyzing the components in them, CCA has been used to locate address blocks on envelopes [Yeh et al., 1987].
- **Mixed Text/Graphics Document Processing**: [Fletcher and

Kasturi, 1988] describes the development and implementation of a *Robust algorithm* where the CCA is successfully used to separate a text string from a mixed text/graphics document image. This algorithm consists of five steps: (a) Connected component generation, (b) Area/ratio filter, (c) Collinear component grouping, (d) Logical grouping of strings into words and phrases, and (e) Text string separation.

1.8.7 *Crossing Counts*

A crossing count is the number of times the pixel value turns from 0 (white pixel) to 1 (black pixel) along horizontal or vertical raster scan lines. It can be expressed as a vector whose components are defined as follows.

1) Horizontal crossing counts:

$$CC_h(i) = \sum_j \overline{f(i,j)} f(i,j+1).$$

2) Vertical crossing counts:

$$CC_v(j) = \sum_i \overline{f(i,j)} f(i+1,j).$$

Crossing counts can be used to measure document complexity. In [Akiyama and Hagita, 1990], crossing counts have been used as one of the basic features to separate and identify the document blocks.

1.8.8 *Form Definition Language (FDL)*

A top-down knowledge representation called *Form Definition Language (FDL)* was proposed in [Higashino et al., 1986] to describe the generic layout structure of document. The structure can be represented in terms of rectangular regions, each of which can be recursively defined in terms of smaller regions. An example is given in Figure 1.26. These generic descriptions are then matched to the preprocessed input document images. This method is powerful, but is complicated to implement. [Fujisawa and Nakano, 1990] developed a simplified version of FDL so that it may be implemented more easily.

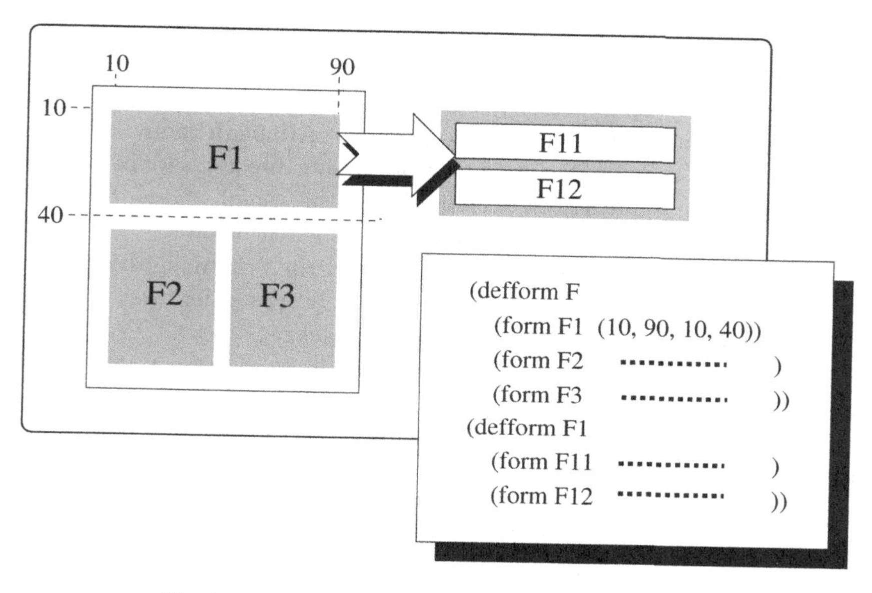

Fig. 1.26 Representation of structure using the FDL.

1.8.9 *Texture Analysis — Gabor Filters*

A text segmentation algorithm using Gabor filters for document processing has been proposed by [Jain and Bhattacharjee, 1992b]. The main steps of this algorithm are described below:

Step-1 Filter the input image through a bank of n even-symmetric Gabor filters [Malik and Perona, 1990], to obtain n *filtered images.*

Step-2 Compute the *feature image* consisting of the "local energy" estimates over windows of appropriate size around every pixel in each of the filtered images.

Step-3 Cluster the feature vectors corresponding to each pixel using a squared-error clustering algorithm to obtain a segmentation of the original input image into K clusters or segments.

This algorithm has been used for locating candidate regions of the destination address block (DAB) on images of letters and envelopes in [Jain and Bhattacharjee, 1992a]. It treats the document image as a multi-textured region in which the text on the envelope defines a specific texture and other non-text contents including blank regions produce different textures. Thus, the problem of locating the text in the envelope image is posed as a texture segmentation problem.

A great variety of methods in texture analysis have been developed for image processing and pattern recognition. Although many of these methods have not been used in document processing directly, a lot of texture classification and segmentation techniques can be useful. Some of those published since 1990 have been selected and are listed below:

A new set of textural measures derived from the texture spectrum has been presented in [He and Wang, 1991]. The proposed features contain more complete texture characteristics of the image.

Based on the texture spectrum, a texture edge detection method has been developed in [He and Wang, 1992]. The basic concept of the method is to use the texture spectrum as the texture measure of the image and combine it with the conventional edge detectors.

In [Pitas and Kotropoulos, 1992], two new methods have been described which use geometric proximity to reference points in region growing. The first one is based on Voronoi tessellation and mathematical morphology while the second one is based on the "radiation model" for region growing and image segmentation.

In [Mao and Jain, 1992], a multi resolution simultaneous autoregressive (MR-SAR) model has been presented for texture classification and segmentation.

1.8.10 *Wavelet Transform*

Wavelet theory is a versatile tool with very rich mathematical content and great applications including pattern recognition [Haley and Manjunath, 1999; Liang et al., 1999; Murtagh and Starck, 1998; Tang et al., 2001; Tang et al., 1998b; Tieng and Boles, 1997]; [Combettes and Pesquet, 2004; Daugman, 2003; El-Khamy et al., 2006; Jain and Merchant, 2004]; [Kouzani and Ong, 2003; Ksantini et al., 2006; Kubo et al., 2003; Kumar and Kumar, 2005]; [Kumar et al., 2003; Kunte and Samuel, 2007; Li, 2006; Li, 2008; Liang and Tjahjadi, 2006]; [Liao and Tang, 2005; Li, 2008; Muneeswaran et al., 2005; Moghaddam et al., 2005]; [Sharnia et al., 2004; Shankar et al., 2007; Tang and You, 2003; Yang et al., 2003]; [You et al., 2006; You and Tang, 2007]; [Dejey and Rajesh, 2010; Jacob et al., 2010; Zhang et al., 2010; Bhatnagar and Raman, 2010]; [Shi et al., 2010; Ehler and Koch, 2010; Yin and Liu, 2010]. What is a wavelet? The simplest answer is a "short" wave (wave + let ⇒ wavelet). The suffix "let" initially means "a small kind of" in a general English dictionary. However, in the mathematical term "wavelet", the meaning of the suffix "let" comes "short", which indicates that the duration of the function is very limited. In other words, it is said that wavelets are localized.

Definition 1.10 A function $\psi \in L^2(R)$ is called an admissible wavelet or a basic wavelet if it satisfies the following "admissibility " condition:

$$C_\psi := \int_R \frac{|\hat{\psi}(\xi)|^2}{|\xi|} d\xi < \infty. \tag{1.24}$$

The continuous (or integrable) wavelet transform with kernel ψ is defined by

$$(W_\psi f)(a,b) := |a|^{-\frac{1}{2}} \int_R f(t)\overline{\psi(\frac{t-b}{a})}dt, \quad f \in L^2(R), \tag{1.25}$$

where $a, b \in R$ and $a \neq 0$ are the dilation parameter and the translation parameter respectively [Tang et al., 2000; Tang, 2009].

1.8.11 *Other Segmentation Techniques*

Segmentation techniques can be roughly categorized as [Ahmed and Suen, 1987; Ahmed and Suen, 1982]; [Cesar and Shinghal, 1990; Fenrich, 1991]; [Kimura and Shridhar, 1991; Mitchell and Gillies, 1989]; [Shridhar and Badreldin, 1986]:

- projection-based,
- pitch-based,
- recognition-based,
- region-based.

The first two techniques are suitable for typewritten texts where characters are equally spaced and there is a significant gap between adjacent characters. In the "recognition-based" methods, segmentation is performed by recognizing a character in a sequential scan. For handwritten or hand-printed texts where variations in handwriting are unpredictable, the performance of these methods is dubious. The "region-based" method is the only alternative for the segmentation of totally unconstrained handwritten characters. This category of techniques consists of finding and analyzing the input image components as well as how these components are related in order to detect suitable regions for segmentation. Also, segmentation techniques can be categorized as the following types by the sorts of pixels to be worked on:

- Methods working on foreground pixels (black pixels) [Ahmed and Suen, 1982; Cesar and Shinghal, 1990; Fenrich, 1991; Kimura and Shridhar, 1991; Mitchell and Gillies, 1989; Shridhar and Badreldin, 1986],
- Methods working on background pixels (white pixels).

In the end of this chapter, a summary can be conducted below:

Every day, millions of documents including technical reports, government files, newspapers, books, magazines, letters, bank cheques, etc. have to be processed. A great deal of time, effort and money will be saved if it can be executed automatically. However, in spite of major advances in computer technology, the degree of automation in acquiring data from such documents is very limited and a great deal of manual labour is still needed in this area. Thus, any method which can speed up this process will make a significant contribution.

This chapter deals with the essential concepts of document analysis and understanding. It begins with a key concept, document structure. The importance of this concept can be seen in the whole chapter: constructing a geometric structure model and a logical structure model; considering document analysis as a technique of extracting the geometric structure; regarding document understanding as a mapping from the geometric structure into logical structure; etc. This chapter attempts to theoretically analyze document structure and top-down, bottom-up approaches which are commonly used in document analysis in terms of entropy function.

Some open questions and problems still exist, especially in document understanding. Any practical document can be viewed differently depending on its geometric structure space and logical structure space. Because there is no one-to-one mapping between these two spaces, it is difficult to find a correct mapping to transform a geometric structure into a logical one. For example, rules based on knowledge may vary in different documents, how to find the correct rules is a profound subject for future research.

Chapter 2

Basic Concepts of Fractal Dimension

The purpose of this chapter is to review some basic aspects of the fractal theory.

Fractal concepts have been becoming more important in pattern recognition, image processing and signal processing. Fractal capability to characterize complicated structures is suitable for processing complex patterns. Fractals can be used not only in image compression but also in image feature extraction and segmentation. In this section, the basic concepts of *fractals* are introduced, and three types of *fractal dimensions*, namely, Hausdorff dimension, box computing dimension and Minkowski dimension, which are related to our work are also discussed.

2.1 Definitions of Fractals

Before discussing the definitions of fractals, we will first introduce a mathematical terminology named *topological dimension*:

Definition 2.1 A collection $\mathbb{R}$ of subsets of the space X is said to have order $m+1$ if some point of X lies in $m+1$ elements of $\mathbb{R}$, and no point of X lies in more than $m+1$ elements of $\mathbb{R}$. Given a collection $\mathbb{R}$ of subsets of X, a collection $\Im$ is said to *refine* $\mathbb{R}$, or to be a *refinement* of $\mathbb{R}$, if each element B of $\Im$ is contained in at least one element of $\mathbb{R}$.

Now we define what we mean by the *topological dimension* of a space X.

Definition 2.2 A space X is said to be *finite-dimensional* if there is some integer m such that for every open covering $I\!R$ of X, there is an open covering $\Im$ of X that refines $I\!R$ and has order at most $m+1$. The *topological dimension* of X is defined to be the smallest value of m for which this statement holds. We use the notation $\dim_T X$ to represent the Topological dimension of X.

The topological dimension is a complicated and advanced mathematical topic, more details about it can be found in [Munkres, 1975].

What are fractals? There are many definitions, because it is very difficult to define fractal strictly. B. Mandelbrot gave two definitions in 1982 and 1986 respectively.

(1) The first definition from his original essay (1982) says:

Definition 2.3 A set F is called fractal set if its *Hausdorff dimension* ($\dim_H F$) is greater than the *Topological dimension* ($dim_T F$), namely:

$$\dim_H F > \dim_T F.$$

(2) In 1986, B. Mandelbrot defined the fractal as:

Definition 2.4 Fractal is a compound object, which contains several sub-objects. The global characteristic of this object is similar to the local characteristics of each sub-object.

(3) A more precise definition of the fractal set F can be provided below:

Definition 2.5 A set F is called fractal set if the following conditions are satisfied:

- The global characteristic of the set F is self-similar to the local characteristics of each sub-set, namely:

$$\Im(F) \sim \Im(f_i), \quad f_i \supset F,$$

where $\Im(.)$ stands for the characteristic of (.).
- The set F is infinitely separable, i.e.

$$F = \{f_1^1, f_2^1, ..., f_i^1, ..., f_n^1\},$$

$$f_i^1 = \{f_1^2, f_2^2, ..., f_k^2, ..., f_n^2\},$$

$$\cdots\cdots\cdots$$

$$f_k^m = \{f_1^{m+1}, f_2^{m+1}, ..., f_k^{m+1}, ..., f_n^{m+1}\}, \quad m+1 \to \infty.$$

- Usually, the fractal dimension of the set F is a fraction, and greater than the *Topological dimension* $\dim_T F$, namely:

$$\dim_H F > \dim_T F.$$

- In many cases the definition of F is recursive.

2.2 Hausdorff Dimension

There exists a variety of fractal dimensions, the most important one is *Hausdorff dimension.* Because it is based on a mathematical tool - *measure theory*, it is suitable for any sets and makes analyzing them easy.

2.2.1 *Hausdorff Measure*

Let U be a non-empty subset of n-dimensional Euclid space $\mathbb{R}^n$, and the diameter of U is defined as

$$|U| = sup\{|x-y| : x, y \in U\},$$

where $sup\{.\}$ stands for the supremum of $\{.\}$. Thus, the diameter of U is the greatest distance apart of any pair of points in U. If $\{U_i\}$ is a countable collection of sets of diameter at most δ that cover F, such that

$$\mathbb{R}(\delta) = \{U_i\} = \{U_i : i = 1, 2, ...\},$$

and

$$F \subset \bigcup_{i=1}^{\infty} U_i, \qquad 0 < |U_i| \leq \delta,$$

we say that $\{U_i\}$ is a δ-cover of F.

Suppose that $F \subset \mathbb{R}^n$ and s is a real number and $s \geq 0$. For any $\delta > 0$, we define

$$H_\delta^s(F) = \inf_{\mathbb{R}(\delta)} \left\{ \sum_{i=1}^{\infty} |U_i|^s : \{U_i\} \text{ is a } \delta\text{-cover of } F \right\}, \tag{2.1}$$

where the symbol $\inf\{.\}$ indicates the infimum of $\{.\}$. As δ decreases, the class of permissible covers of F in Eq. (2.1) is reduced. Consequently, the infimum $H^s_\delta(F)$ increases, and so approaches a limit as $\delta \to 0$. We have the following definition:

Definition 2.6 When $\delta \to 0$, the limit of $H^s_\delta(F)$ exists for any subset F of $\mathbb{R}^n$, and the limiting value can be (and usually) 0 or ∞. The *s-dimensional Hausdorff measure* of F can be defined by:

$$\begin{aligned} H^s(F) &= \lim_{\delta\to 0} H^s_\delta(F) \\ &= \lim_{\delta\to 0}\left[\inf_{\mathbb{R}(\delta)}\left\{\sum_{i=1}^{\infty}|U_i|^s : \{U_i\} \text{ is a } \delta\text{-cover of } F\right\}\right]. \end{aligned} \quad (2.2)$$

We can clearly prove that H^s is a measure. Specifically, $H^s(\phi) = 0$, and if $E \subset F$ then $H^s(E) \leq H^s(F)$. If $\{F_i\}$ is any countable collection of *Borel set*, such that

$$\bigcap_{i=1}^{\infty} F_i = \phi,$$

we have

$$H^s\left(\bigcup_{i=1}^{\infty} F_i\right) = \sum_{i=1}^{\infty} H^s(F_i).$$

Furthermore, if F is a Borel subset of $\mathbb{R}^n$, then the n-dimensional *Hausdorff measure* of F can be deduced as:

$$H^n(F) = \frac{\pi^{\frac{n}{2}}}{2^n(\frac{1}{2}n)!}Vol^n(F),$$

where

- $H^n(F)$ stands for the n-dimensional *Hausdorff measure* of F.
- $vol^n(F)$ represents the n-dimensional volume of F which is called *Lebesgue measure* of F.
  - vol^1 is length,
  - vol^2 is area,
  - vol^3 is the usual 3-dimensional volume.

Consequently, Hausdorff measures generalize the familiar ideas of length, area and volume. The scaling properties of length, area and volume are well known, which are fundamental to the theory of fractals.

Theorem 2.1 *Let F be a non-empty and bounded set in $\mathbb{R}^n$, and $\mu > 0$ then the set F scaled by a factor μ, i.e.*

$$H^s(\mu F) = \mu^s H^s(F) \tag{2.3}$$

where $\mu F = \{\mu g : g \in F\}$.

Proof If $\{U_i\}$ is a δ-cover of F, then $\{\mu U_i\}$ is a $\pi\delta$-cover of μF. Therefore, the following holds for any δ-cover $\{U_i\}$:

$$\begin{aligned} H^s_{\mu\delta} \leq \sum |\mu U_i|^s &= \mu^s \sum |U_i|^s \\ &\leq \mu^s H^s_\delta(F) \end{aligned}$$

Letting $\delta \to 0$ gives that

$$H^s(\mu F) \leq \mu^s H^s(F).$$

We replace μ by $1/\mu$, and F by μF, the opposite inequality required can be found.

The proof of the theorem is thus complete. ■

Theorem 2.2 *Let F be a non-empty and bounded set in $\mathbb{R}^n$, and $g : F \to \mathbb{R}^m$ be a mapping such that*

$$|g(x) - g(y)| \leq k|x - y|^\alpha, \qquad (x, y \in F) \tag{2.4}$$

where constants $k > 0$ and $\alpha > 0$. Hence, for each s, the following holds:

$$H^{s/\alpha}(g(F)) \leq k^{s/\alpha} H^s(F). \tag{2.5}$$

Proof If $\{U_i\}$ is a δ-cover of F, then we have

$$\{g(F \cap U_i)\} \text{ is an } \varepsilon - \text{cover of } g(F)$$

where, $\varepsilon = k\delta^\alpha$, because

$$|g(F \cap U_i)| \leq k|U_i|^\alpha.$$

Therefore, it follows that

$$\sum_i |g(F \cap U_i)|^{s/\alpha} \leq k^{s/\alpha} \sum_i |U_i|^s$$

such that

$$H_{\varepsilon}^{s/\alpha}(g(F)) \le k^{s/\alpha} H_{\delta}^{s}(F).$$

As $\delta \to 0$, so $\varepsilon \to 0$, producing Eq. (2.5).

The proof of the theorem is thus complete. ■

Definition 2.7 Recall Eq. 2.4

$$|g(x) - g(y)| \le k|x-y|^{\alpha}, \qquad (x, y \in F)$$

it is called as a Hlőder condition of exponent α, and implies that $g(.)$ is continuous. An particularly important case is $\alpha = 1$, so that Eq. (2.4) becomes

$$|g(x) - g(y)| \le k|x-y|, \qquad (x, y \in F)$$

g is called a Lipschitz mapping, and

$$H^{s}(g(F)) \le k^{s} H^{s}(F). \tag{2.6}$$

2.2.2 *Hausdorff Dimension*

Let us review Eq. (2.1). For any set F and $\delta < 1$, $H_{\delta}^{s}(F)$ is a non-increasing function of s. According to Eq. (2.2), it can be shown that $H^{s}(F)$ is also a non-increasing function of s. In fact, the stronger conclusion is that if $t > 0$ and $\{U_i\}$ is a δ-cover of F, we have

$$H_{\delta}^{t}(F) \le \sum_{i} |U_i|^{t} \le \delta^{t-s} \sum_{i} |U_i|^{s}. \tag{2.7}$$

We take the infimum, that is

$$H_{\delta}^{t}(F) \le \delta^{t-s} H_{\delta}^{s}(F).$$

Definition 2.8 Let $\delta \to 0$, if $H^{s}(F) < \infty$, then $H^{t}(F) = 0$ for $s < t$. Therefore, there exists a critical value of s, such that $H^{s}(F)$ jumps from ∞ to 0 at this point. This critical value is called the *Hausdorff Dimension* of F, and it is symbolized by $\dim_H F$.

Formally, we have

$$\dim_H F = \inf\{s : H^s(F) = 0\} = \sup\{s : H^s(F) = \infty\}, \qquad (2.8)$$

and

$$H^s(F) = \begin{cases} \infty & \text{if } s < \dim_H F \\ 0 & \text{if } s > \dim_H F. \end{cases}$$

If $s = \dim_H F$, probably $H^s(F)$ is 0 or ∞, or may satisfy

$$0 < H^s(F) < \infty.$$

A Borel set is called an *s-set* if the latter condition as shown in the above

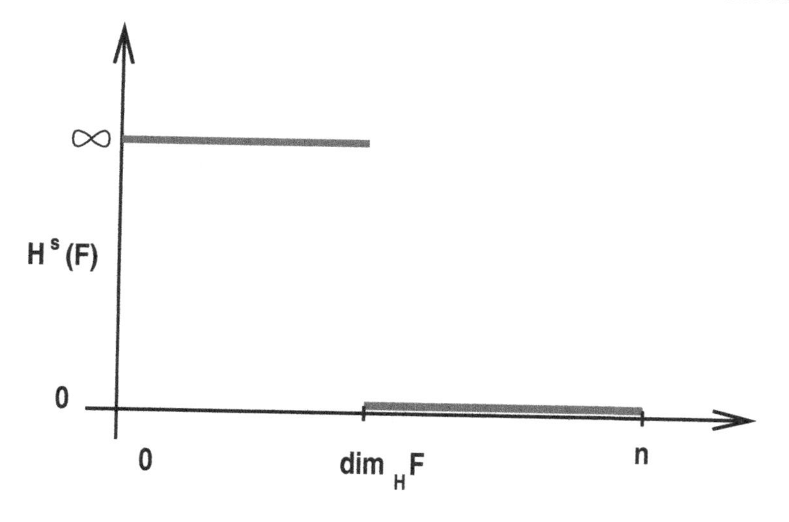

Fig. 2.1 Graph of $H^s(F)$ against s for a set F.

is satisfied. From Figure 2.1 we can see that the Hausdorff dimension is the value of s at which the jump from ∞ to 0 occurs. More details about Hausdorff dimension can be found in [Falconer, 1985; Falconer, 1990].

Hausdorff dimensions have many important properties. Here we are mainly interested in the two most important and useful properties of Hausdroff dimensions, such as monotonicity and countable stability.

- *Monotonicity*
  If $A \subset B$ then $dim_H A \leq dim_H B$. This is immediate from the measure property that $H^s(A) \leq H^s(B)$ for each s.

- *Countable stability*
If $F_1, F_2, \dots$, is a (countable) sequence of sets then $dim_H \bigcup_{i=1}^{\infty} F_i = sup_{1\leq i\leq\infty}\{dim_H F_i\}$. Certainly, $dim_H \bigcup_{i=1}^{\infty} F_i \geq dim_H F_j$ for all j from the monotonicity property. On the other hand, if $s > dim_H F_i$ for all i, then $H^s(F_i) = 0$, so that $H^s(\bigcup_{i=1}^{\infty} F_i) = 0$, giving the opposite inequality.

According to Theorem 2.2, the transformation properties of Hausdorff dimension can be found in the following theorems.

Theorem 2.3 *Let F be a non-empty and bounded set in $\mathbb{R}^n$, and suppose that $g : F \to \mathbb{R}^m$ satisfies a Hlőder condition*

$$|g(x) - g(y)| \leq k|x - y|^{\alpha}, \qquad (x, y \in F)$$

$$dim_H g(F) \leq (\frac{1}{\alpha}) dim_H F.$$

Proof According to Theorem 2.2, if $s > dim_H F$, the following holds:

$$H^{s/\alpha}(g(F)) \leq k^{s/\alpha} H^s(F) = 0.$$

Hence, we can obtain that

$$dim_H g(F) \leq (\frac{s}{\alpha}), \forall s > dim_H F.$$

The proof of the theorem is thus complete. ■

Corollary 1 *If $g : F \to \mathbb{R}^m$ satisfies*

$$|g(x) - g(y)| \leq k|x - y|, \qquad (x, y \in F),$$

which indicates that $g : F \to \mathbb{R}^m$ is a Lipschitz transformation, then we have

$$dim_H g(F) \leq dim_H F.$$

Proof According to Theorem 2.3, if we take $\alpha = 1$, the above is held. The proof of the corollary is therefore complete. ■

Corollary 2 *If $g : F \to I\!R^m$ satisfies*

$$k_1|x - y| \leq |g(x) - g(y)| \leq k_2|x - y|, \qquad 0 < k_1 \leq k_2 < \infty,$$

which indicates that $g : F \to I\!R^m$ is a bi-Lipschitz transformation, then we have

$$dim_H f(F) \leq dim_H F.$$

Proof Applying Corollary 1 to $g^{-1} : g(F) \to F$ gives the other inequality required for this corollary.

The proof of the corollary is therefore complete. ■

The fundamental property of Hausdorff dimension can be found from the above corollaries reveal, i.e. Hausdorff dimension is invariant under the bi-Lipschitz transformations. Therefore, if two sets have different dimensions, we cannot find a bi-Lipschitz mapping which maps from one set onto the other set. This implies that the situation in topology where various invariants, (such as homotopy or homology groups) are set up to distinguish between sets that are not homeomorphic: if the topological invariants of two sets differ then there cannot be a homeomorphism (continuous one-to-one mapping with continuous inverse) between the two sets. If there is a homeomorphism between these two sets, then these two sets are regarded as 'the same' in topology.

Specifically, for fractal geometry, if we have a Lipschitz mapping between two sets, we would say these two sets are the same. Just as topological invariants are applied to distinguish between non-homeomorphic sets, we can use some parameters, such as dimension, to identify between sets that are not bi-Lipschitz equivalent. Since bi-Lipschitz transformations are necessarily continuous, topological parameters provide the first step in this direction, whole Hausdorff dimension provide further distinguishing characteristics between fractals.

Theorem 2.4 *Let F be a non-empty and bounded set in $I\!R^n$, and $dim_H F < 1$ then the set F is totally disconnected.*

Proof Let α and β be distinct points of F. According to $g(z) = |z - \alpha|$ we can define a mapping:

$$g : I\!R^n \to [0, \infty)$$

Because g does not increase distances,

$$|f(z) - f(\psi)| \leq |z - \psi|$$

we have from Corollary 1 that

$$dim_H g(F) \leq dim_H F < 1.$$

Thus $g(F)$ is a subset of $I\!R^n$ of H^1-measure or length zero, and so has a dense complement. Choosing ϱ with $0 < \varrho < f(y)$ it follows that

$$F = \{z \in F : |z - \alpha| < \varrho\} \cup \{z \in F : |z - \alpha| > \varrho\}.$$

Therefore, F is contained in two disjoint open sets with α in one set and β in the other, so that α and β lie in different connected components of F.

The proof of the theorem is therefore complete. ∎

2.2.3 *Examples of Computing Hausdorff Dimension*

This sub-section is used to present how to compute the Hausdorff dimension of simple example of fractals.

Example 1

The middle third Cantor set is the most easily constructed fractals, which issues many typical characteristics of fractal [Falconer, 1990]. It is constructed from a unit interval by a sequence of deletion operations. Recall that $[a, b]$ denotes the set of real numbers x such that $a \leq x \leq b$. Let E_0 be the interval $[0, 1]$ as shown in Figure 2.2. Let E_1 be the set obtained by deleting the middle third of E_0, so that E_1 consists of the two intervals $[0, 1/3]$ and $[2/3, 1]$. Deleting the middle thirds of these intervals gives E_2; thus E_2 comprises the four intervals

$$[0, 1/9], [1/9, 1/3], [2/3, 7/9], [8/9, 1].$$

We continue in this way, with E_k, obtained by deleting the middle third of each interval in E_{k-1}. Thus E_k consists of 2^k intervals each of length 3^{-k}. The middle third Cantor set F consists of the numbers that are in E_k for all k; mathematically, F is the intersection

$$F = \bigcap_{k=0}^{\infty} E_k.$$

The Cantor set F may be thought of as the limit of the sequence of sets E_k as k tends to infinity. It is obviously impossible to draw the set F itself, with its infinitesimal detail, so 'pictures of F tend to be pictures of one of the E_k, which are a good approximation to F when k is reasonably large [Falconer, 1990].

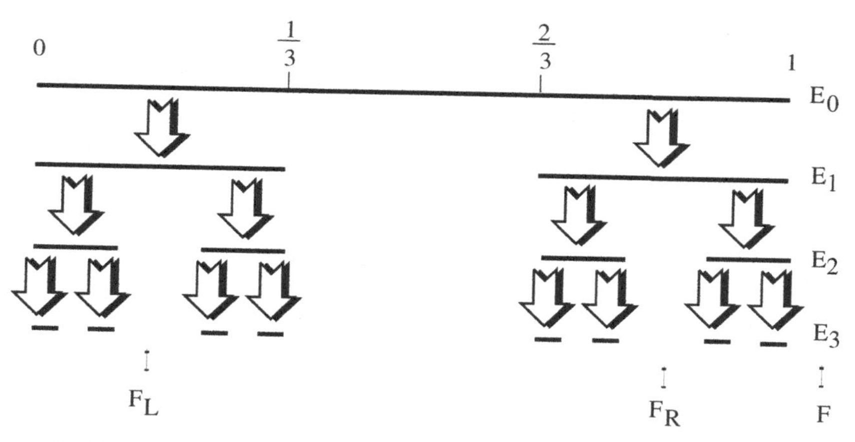

Fig. 2.2 Construction of the middle third Cantor set F, to do so we repeated remove the middle third of intervals. Here, F_L and F_R indicate the left and right parts of F, which are copies of F scaled by a factor 1/3.

At first glance it might appear that we have removed so much of the interval $[0, 1]$ during the construction of F, that nothing remains. In fact, F is an infinite and uncountable set, which contains infinitely many numbers in any neighbourhood of each of its points. The middle third Cantor set F consists precisely of those numbers in $[0, 1]$ whose base-3 expansion does not contain the digit 1, i.e. all numbers:

$$a_1 3^{-1} + a_2 3^{-2} + a_3 3^{-3} + ..., \text{with } \; a_i = 0 \; \text{ or 2 for each } i$$

To see this, note that to get E_1 from E_0 we remove those numbers with $a_1 = 1$, to get E_2 from E_1 we remove those numbers with $a_2 = 1$, and so on.

We look at the following case: Let F be the middle third Cantor set as mentioned above. If $s = 1og2/log3 = 0.6309...$, then

$$dim_H F = s\text{and}\frac{1}{2} \leq H^s(F) \leq 1.$$

Heuristic calculation:

The Cantor set F cracks into two parts, i.e. (1) a left part $F_L = F \cup [0, 1/3]$ and (2) a right part $F_R = F \cap [2/3, 1]$. We can find that the both parts are geometrically similar to F but scaled by a ratio $1/3$, and $F = F_L \cup F_R$ with this union disjoint. Therefore, for any s we have

$$H^s(F) = H^s(F_L) + H^s(F_R) = (\frac{1}{3})^s H^s(F) + (\frac{1}{3})^s H^s(F)$$

According to the scaling property of Hausdorff measures as shown in Theorem 6.1, Assuming that at the critical value $s = dim_H F$ we have $0 < H^s(F) < \infty$, we can split by $H^s(F)$ to get $1 = 2(\frac{1}{3})^s$ or $s = log2/log3$ [Falconer, 1990].

Rigorous calculation:

The intervals are installed to length $3^{-k}(k = 0, 1, 2, ...)$ such that the sets E_k can be constructed into F basic intervals. In this way, the cover $\{U_i\}$ of F consists of the 2^k intervals of E_k of length 3^{-k}. Thus, we have

$$H^s_{3^{-k}}(F) \leq \Sigma |U_i|^s = 2^k 3^{-ks} = 1, \text{ if } s = \log 2/\log 3.$$

Let us take $k \to \infty$, it results in $H^s(F) \leq 1$.

Next, we should prove that $H^s(F) \geq 1/2$, thus, the following will be proved

$$\sum |U_i|^s \geq \frac{1}{2} = 3^{-s} \text{ for any cover } \{U_i\} \text{ of } F. \tag{2.9}$$

Suppose the $\{U_i\}$ are intervals, we expand the intervals slightly and use the compactness of F, hence, we need only prove Eq. (2.9) if $\{U_i\}$ is a finite collection of closed subintervals of $[0, 1]$. For each U_i, let k be the integer such that

$$3^{-(K=1)} \leq |U_i| < 3^{-k}. \tag{2.10}$$

Because the basic intervals are separated into at least 3^{-k}, U_i can intersect at most one basic interval of E_k. If we have $j \geq k$, it should be that, U_i intersects at most

$$2^{j-k} = 2^j 3^{-sk} \leq 2^j 3^s |U_i|^s$$

basic intervals of E_j, by Eq. (2.10).

If we choose j large enough such that

$$3^{-(j+1)} \leq |U_i| \text{ for all } U_i,$$

then, since the $\{U_i\}$ intersect all 2^j basic intervals of length 3^{-j}', we can calculate the intervals, and obtain that

$$2^j \leq \sum_i 2^j 3^s |U_i|^s$$

which reduces to Eq. (2.9). The details of the example can be found in [Falconer, 1990].

Example 2

A plane analogue of the Cantor set, a 'Cantor dust' is shown in Figure 2.3. At each stage of the construction the squares are divided into 16 squares with a quarter of the side length, of which the same pattern of four squares is retained. Let F be the Cantor dust constructed from the unit square [Falconer, 1990].

Calculation:

Taking the obvious covering of F by the 4^k squares of side 4^{-k} (i.e. of diameter $\delta = 4^{-k}\sqrt{2}$) in E_k, we can reckon the kth stage of construction as the following estimate

$$H_\delta^1(F) \leq 4^k 4^{-k} \sqrt{2}$$

for the infimum in Eq. (2.1), which recalls

$$H_\delta^s(F) = \inf_{\mathbb{R}(\delta)} \left\{ \sum_{i=1}^{\infty} |U_i|^s : \{U_i\} \text{ is a } \delta\text{-cover of } F \right\}.$$

As $k \to \infty$ so $\delta \to 0$ giving $H_\delta^1(F) \leq \sqrt{2}$.

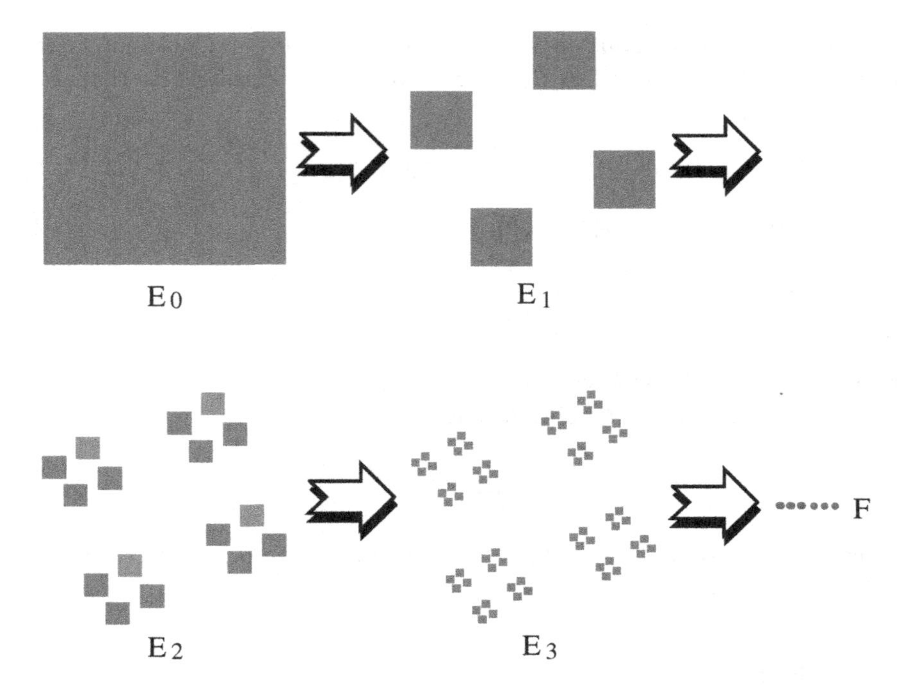

Fig. 2.3 Construction of the Cantor dust set F.

Let Φ denote orthogonal projection onto the x-axis. For the lower estimate, the orthogonal projection does not increase distances, i.e.

$$|\Phi x - \Phi y| \leq |x - y| \quad \text{if } x, y \in \mathbb{R}^2.$$

Therefor, Φ is a Lipschitz mapping. The projection of F on the x-axis, ΦF, is the unit interval $[0, 1]$, accordance with the construction of F, using Eq. (2.6):

$$H^s(g(F)) \leq k^s H^s(F),$$

we can produce

$$1 = length[0, 1] = H^1[0, 1] = H^1(\Phi F) \leq H^1(F).$$

The details of the example can be found in [Falconer, 1990].

Of the wide variety of fractal dimensions in use, Hausdorff dimension is the oldest and probably the most important. Hausdorff dimension has the advantage of being defined for any set, and is mathematically convenient, as it is based on measures, which are relatively easy to manipulate. A major disadvantage is that in many cases it is difficult to calculate or to estimate by computational methods. In practice, *box computing dimension* is convenient to apply. Therefore, our study will focus on the box computing dimension.

2.3 Box Computing Dimension

In this section, we will introduce an important concept - *dimension*, followed by box computing dimension (BCD).

2.3.1 *Dimensions*

Fundamental to most definitions of dimension is the idea of measurement at scale δ. For each δ, a set can be measured in a way that ignores irregularities of size less than δ, and we see how these measurements behave as $\delta \to 0$.

Suppose F is a plane curve, the measurement $M_\delta(F)$ denotes the number of sets (with length δ) which divide the set F. A dimension of F is determined by the power law obeyed by $M_\delta(F)$ as $\delta \to 0$. If

$$M_\delta(F) \sim \mathcal{K}\delta^{-s}, \tag{2.11}$$

for constants $\mathcal{K}$ and s, we might say that F has dimension s, and $\mathcal{K}$ can be considered as "s-dimensional length" of F.

Taking the logarithm of both sides in Eq. (2.11) yields the formula:

$$\log_2 M_\delta(F) \simeq \log_2 \mathcal{K} - s \log_2 \delta,$$

in the sense that the difference of the two sides tends to 0 with δ, we have

$$s = \lim_{\delta \to 0} \frac{\log_2 M_\delta(F)}{-\log_2 \delta}. \tag{2.12}$$

From the above equation, s can be regarded as a slope on a log-log scale [Falconer, 1990].

2.3.2 *Box Computing Dimension*

Box computing dimension or box dimension is one of the most widely used dimensions. Its popularity is largely due to its relative ease of mathematical calculation and empirical estimation.

Let F be a non-empty and bounded subset of $\mathbb{R}^n$, $\mho = \{\omega_i : i = 1, 2, 3, ...\}$ be covers of the set F. $N_\delta(F)$ denotes the number of covers, such that

$$N_\delta(F) = |\mho : d_i \leq \delta|,$$

where d_i stands for the diameter of the i-th cover. This equation means that $N_\delta(F)$ is the smallest number of subsets which cover the set F, and their diameters d_i's are not greater than δ (Figure 2.4).

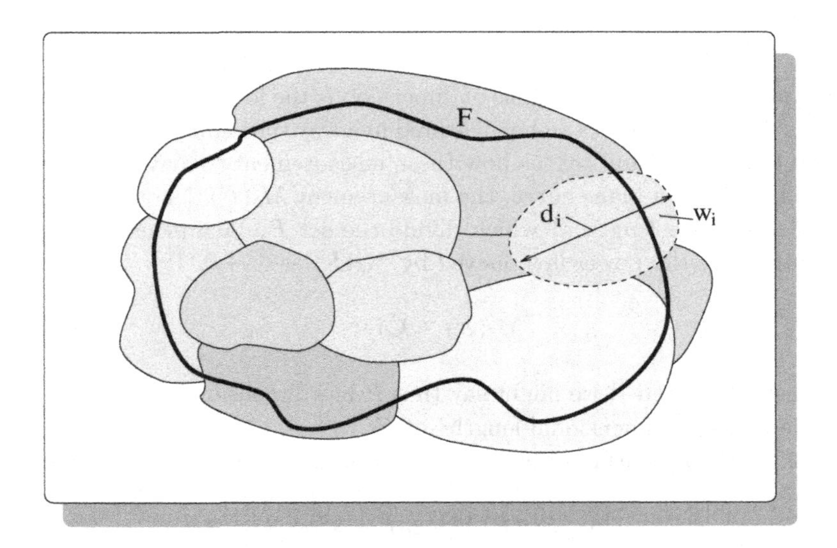

Fig. 2.4 Opening covers with diameters d_i's covering F.

The upper and lower bounds of the box computing dimension of F can be defined by the following formulas:

$$\underline{\dim}_B F = \liminf_{\delta \to 0} \frac{\log_2 N_\delta(F)}{-\log_2 \delta}, \tag{2.13}$$

$$\overline{\dim_B}F = \overline{\lim_{\delta\to 0}}\frac{\log_2 N_\delta(F)}{-\log_2 \delta}, \tag{2.14}$$

where the over line stands for the upper bound of dimension while the under line for lower bound. An example of the upper bound and lower bound is shown in Figure 2.5

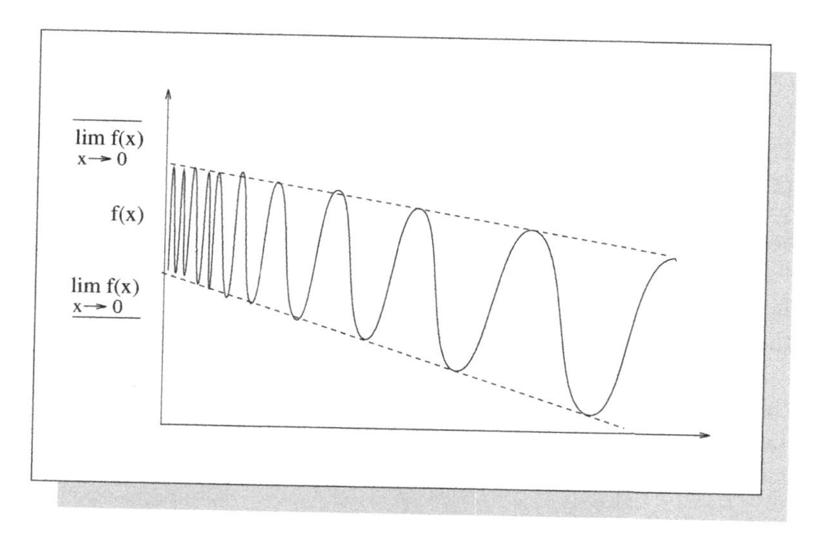

Fig. 2.5 Upper and lower bounds of a function.

Definition 2.9 If both the upper bound $\overline{\dim_B}F$ and the lower bound $\underline{\dim_B}F$ are equal, i.e.

$$\liminf_{\delta\to 0}\frac{\log_2 N_\delta(F)}{-\log_2 \delta} = \overline{\lim_{\delta\to 0}}\frac{\log_2 N_\delta(F)}{-\log_2 \delta},$$

the common value is called *box computing dimension* or *box dimension* of F, namely:

$$\dim_B F = \lim_{\delta\to 0}\frac{\log_2 N_\delta(F)}{-\log_2 \delta}. \tag{2.15}$$

There are five equivalent definitions of the box computing dimension, that can be found in the following theorem:

Theorem 2.5 *Let F be a non-empty and bounded set in $\mathbb{R}^n$, and the upper box dimension $\overline{\dim}_B F$, lower box dimension $\underline{\dim}_B F$ and box dimension $\dim_B F$ be represented by:*

$$\begin{aligned}
\underline{\dim}_B F &= \liminf_{\delta \to 0} \frac{\log_2 N_\delta(F)}{-\log_2 \delta}, \\
\overline{\dim}_B F &= \overline{\lim_{\delta \to 0}} \frac{\log_2 N_\delta(F)}{-\log_2 \delta}, \\
\dim_B F &= \lim_{\delta \to 0} \frac{\log_2 N_\delta(F)}{-\log_2 \delta}.
\end{aligned}$$

In the above definition, $N_\delta(F)$ can be considered as one of the following cases (Figure 2.6):

(i) the minimum number of closed balls of radius δ that cover F (Figure 2.6b);
(ii) the minimum number of cubes with side δ that cover F (Figure 2.6c);
(iii) the minimum number of sets with diameter D that cover F such that $D \leq \delta$ (Figure 2.6d);
(iv) the number of δ-mesh cubes which intersect F (Figure 2.6e);
(v) the maximum number of disjoint balls of radius δ with centers in F (Figure 2.6f).

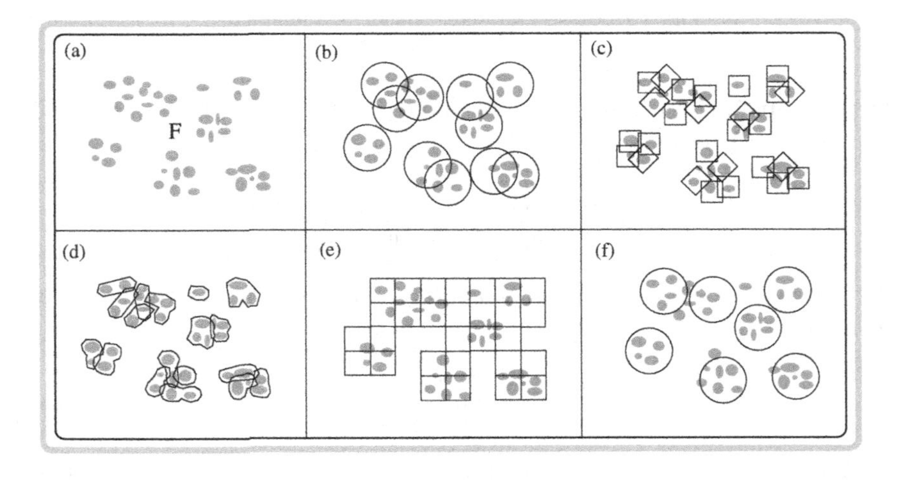

Fig. 2.6 Graphic illustration of the equivalent definitions of the BCD.

Proof This proof consists of four steps, namely:

1: (i) $\iff$ (ii), 2: (i) $\iff$ (iii), 3: (iii) $\iff$ (iv),
4: (iv) $\iff$ (v).

Step 1: (i) $\iff$ (ii)

A cube with side length δ must have only a ball of radius $\frac{\delta\sqrt{n}}{2}$ that covers this cube. On the other hand, any ball of radius δ must have a cube with side length 2δ that covers this ball. Therefore, the definitions (i) and (ii) are equivalent.

Step 2: (i) $\iff$ (iii)

Let $U_1, U_2, ..., U_{N_\delta(F)}$ be the sets with diameter at most δ that cover F, and $B_1, B_2, ..., B_{N'_\delta(F)}$ be the closed balls of radius δ that cover F. There exists a B_j for any U_i, such that $U_i \subset B_j$. Thus, it follows that

$$N'_\delta(F) \le N_\delta(F).$$

On the other hand, any closed ball B_j of radius δ can be regarded as a set of diameter at most 2δ. Thus, it can be found that

$$N_{2\delta}(F) \le N'_\delta(F).$$

Therefore, the definitions (i) and (iii) are equivalent.

Step 3: (iii) $\iff$ (iv)

Consider δ-mesh cubes in $I\!R^n$

$$[m_1\delta, \quad (m_1+1)\delta] \times [m_2\delta, \quad (m_2+1)\delta] \times \cdots\cdots \times [m_n\delta, \quad (m_n+1)\delta],$$

where $m_1, m_2, ..., m_n$ are positive integers. In the case of $n = 1$, a "cube" in $I\!R^1$ is a closed interval, while a "cube" in $I\!R^2$ is a square when $n = 2$. Suppose that

- $N'_\delta(F)$ is the smallest number of cubes in the δ-mesh that intersect F.
- $N_{\delta\sqrt{n}}(F)$ is the smallest number of sets that cover F, and the maximum diameter of them is $\delta\sqrt{n}$.

Because any cube in the δ-mesh can be regarded as a set, in which, the maximum diameter is $\delta\sqrt{n}$, we have

$$N_{\delta\sqrt{n}}(F) \le N'_\delta(F).$$

On the other hand, any set of diameter with a maximum value of δ must be covered in the cubes in the δ-mesh, and the largest number of cubes is 3^n. Thus, the following inequality is true

$$N'_\delta(F) \leq 3^n N_\delta(F).$$

According to the above two inequalities, we can conclude that the definitions (iii) and (iv) are equivalent.

Step 4: (iv) $\Longleftrightarrow$ (v)

Let B_1, B_2, ..., $B_{N'_\delta(F)}$ be the disjoint balls of radius δ with centers in F. If $x \in F$, then the distance between x and some B_i does not extend the value of δ, otherwise, the balls of radius δ with centers in x can be added into the set $\{ B_1, B_2, ..., B_{N'_\delta(F)} \}$ to grow the series of balls. It is clear that $N'_\delta(F)$ number of closed balls of radius 2δ with the same centers as that of B_i can cover F. Meanwhile, the closed balls of 2δ can be considered as the sets with a maximum diameter 4δ. Therefore, we have the following inequality:

$$N_{4\delta}(F) \leq N'_\delta(F).$$

On the other hand, suppose that U_1, U_2, ..., $U_{N_\delta(F)}$ are the sets of maximum diameter at δ that cover F. Because B_1, B_2, ..., $B_{N'_\delta(F)}$ are the disjoint balls of radius δ with centers in F. Obviously, the center of each B_i must belong to some U_j, and $U_j \subset B_i$. Furthermore, B_i's are disjoint each other, thus, U_j's which are covered in B_i are also disjoint each other. Therefore, we have the following inequality:

$$N'_\delta(F) \leq N_\delta(F).$$

According to these two inequalities, we can conclude that the definitions (iv) and (v) are equivalent.

The proof of the theorem is thus complete. ■

Theorem 2.6 *Let F be a non-empty and bounded set in $\mathbb{R}^n$, and it satisfies that $1 < H^s(F)$ when $s = \dim_H F$, we have*

$$\dim_H F \leq \underline{\dim}_B F \leq \overline{\dim}_B F$$

Proof If F can be covered in the sets B_1, B_2, ..., $B_{N_\delta(F)}$, then

$$\begin{aligned} H_\delta^s(F) &= \inf\left\{\sum_{i=1}^{\infty}|U_i|^s : \{U_i\} \text{ is a } \delta\text{-cover of } F\right\} \\ &\leq \sum_{i=1}^{N_\delta(F)} |B_i|^s \\ &= N_\delta(F)\delta^s. \end{aligned}$$

When $s = \dim_H F$, we get that

$$1 < H_\delta^s(F) = \lim_{\delta\to 0} H_\delta^s(F).$$

Thus, when $\delta \to 0$, we have

$$1 < H_\delta^s(F) \leq N_\delta(F)\delta^s.$$

Taking the logarithm of both sides in this inequality yields

$$0 \leq \log_2 N_\delta(F) + s\log_2 \delta.$$

From the inequality

$$s \leq \frac{\log_2 N_\delta(F)}{-\log_2 \delta},$$

it follows that

$$\dim_H F \leq \frac{\log_2 N_\delta(F)}{-\log_2 \delta}.$$

Furthermore, we have

$$\dim_H F \leq \underline{\dim}_B F \leq \overline{\dim}_B F.$$

The proof of the theorem is thus complete. ■

2.3.3 *Minkowski Dimension*

To facilitate the application of the box dimension to digital images, we will introduce *Minkowski dimension* which is suitable for processing digital images in computers.

Definition 2.10 Let F be a non-empty and bounded set in $\mathbb{R}^n$. For a constant s, if $\delta \to 0$, the limit of $Vol^n(F_\delta)/\delta^{n-s}$ is positive and bounded, we say that F has s dimension of *Minkowski dimension*, and is symbolized by $\dim_M F$. Here, $Vol^n(F_\delta)$ is called *Lebesgue Measure.*

The relationship between the box dimension and Minkowski dimension can be provided by the following theorems.

Theorem 2.7 *Let $\overline{F}$ be a non-empty and bounded set in $\mathbb{R}^n$. Then we have*

$$\underline{\dim}_B F = n - \overline{\lim_{\delta \to 0}} \frac{\log_2 Vol^n(F_\delta)}{\log_2 \delta},$$

$$\overline{\dim}_B F = n - \liminf_{\delta \to 0} \frac{\log_2 Vol^n(F_\delta)}{\log_2 \delta},$$

where F_δ stands for δ-parallel body of F, and $Vol^n(F_\delta)$ denotes n-dimensional volume of F_δ.

Proof If F can be covered by $N_\delta(F)$ number of closed balls of radius δ, then F_δ can be covered by balls of radius 2δ with the same centers. Thus

$$Vol^n(F_\delta) \leq N_\delta(F) C_n (2\delta)^n,$$

where, C_n denotes the volume of a unit ball. We take the logarithm of both sides in the above inequality, and then divide them by $\log_2 \delta$, giving

$$\frac{\log_2 Vol^n(F_\delta)}{\log_2 \delta} \geq \frac{\log_2 2^n C_n + n \log_2 \delta + \log_2 N_\delta(F)}{\log_2 \delta}.$$

Taking the limits of both sides yields

$$\begin{aligned}
\overline{\lim_{\delta \to 0}} \frac{\log_2 Vol^n(F_\delta)}{\log_2 \delta} &\geq \overline{\lim_{\delta \to 0}} \left(\frac{\log_2 2^n C_n}{\log_2 \delta} + n + \frac{\log_2 N_\delta(F)}{\log_2 \delta} \right) \\
&= \overline{\lim_{\delta \to 0}} \left(n - \frac{\log_2 N_\delta(F)}{-\log_2 \delta} \right) \\
&= n - \lim_{\delta \to 0} \frac{\log_2 N_\delta(F)}{-\log_2 \delta} \\
&= n - \underline{\dim}_B F.
\end{aligned}$$

Therefore, we obtain

$$\underline{\dim}_B F \geq n - \overline{\lim_{\delta \to 0}} \frac{\log_2 Vol^n(F_\delta)}{\log_2 \delta}. \tag{2.16}$$

On the other hand, suppose $N_\delta(F)$ is the number of disjoint balls of radius δ, and their centers are in F. It can be shown that

$$N_\delta(F) C_n \delta^n \leq Vol^n(F_\delta).$$

We take the logarithm of both sides in the above inequality, and then divide them by $\log_2 \delta$, getting

$$\frac{\log_2 2^n C_n + n \log_2 \delta + \log_2 N_\delta(F)}{\log_2 \delta} \geq \frac{\log_2 Vol^n(F_\delta)}{\log_2 \delta}.$$

Taking the limits of both sides yields

$$\begin{aligned} \overline{\lim_{\delta \to 0}} \frac{\log_2 Vol^n(F_\delta)}{\log_2 \delta} &\leq \overline{\lim_{\delta \to 0}} \left(\frac{\log_2 2^n C_n}{\log_2 \delta} + n + \frac{\log_2 N_\delta(F)}{\log_2 \delta} \right) \\ &= \overline{\lim_{\delta \to 0}} \left(n - \frac{\log_2 N_\delta(F)}{-\log_2 \delta} \right) \\ &= n - \lim_{\delta \to 0} \frac{\log_2 N_\delta(F)}{-\log_2 \delta} \\ &= n - \underline{\dim}_B F. \end{aligned}$$

Therefore, we obtain

$$\underline{\dim}_B F \leq n - \overline{\lim_{\delta \to 0}} \frac{\log_2 Vol^n(F_\delta)}{\log_2 \delta}. \tag{2.17}$$

According to Eqs. (2.16) and (2.17), we therefore have

$$\underline{\dim}_B F = n - \overline{\lim_{\delta \to 0}} \frac{\log_2 Vol^n(F_\delta)}{\log_2 \delta}.$$

Similar to the above proof, we obtain

$$\overline{\dim}_B F = n - \liminf_{\delta \to 0} \frac{\log_2 Vol^n(F_\delta)}{\log_2 \delta}.$$

This establishes our proof. ■

From Theorem 2.7, we can derive Theorem 2.8, which is a very important theorem. It shows that the box computing dimension is equal to the Minkowski dimension in a non-empty and bounded set in $\mathbb{R}^n$.

Theorem 2.8 *Let F be a non-empty and bounded set in $\mathbb{R}^n$. We have*

$$\dim_B F = \dim_M F.$$

Proof According to definition 2.10, the set F has s dimension of Minkowski dimension, which means that there exists a constant $\beta > 0$, so that

$$\lim_{\delta \to 0} \frac{Vol^n(F_\delta)}{\delta^{n-s}} = \beta.$$

Taking the limits of both sides yields

$$\begin{aligned} \log_2 \beta &= \log_2 \lim_{\delta \to 0} \frac{Vol^n(F_\delta)}{\delta^{n-s}} \\ &= \lim_{\delta \to 0} \log_2 \frac{Vol^n(F_\delta)}{\delta^{n-s}} \\ &= \lim_{\delta \to 0} \left[\log_2 Vol^n(F_\delta) - (n-s) \log_2 \delta\right]. \end{aligned}$$

Furthermore, we can derive

$$\begin{aligned} \frac{\log_2 \beta}{\lim\limits_{\delta \to 0} \log_2 \delta} &= \frac{\lim\limits_{\delta \to 0} \left[\log_2 Vol^n(F_\delta) - (n-s) \log_2 S\right]}{\lim\limits_{\delta \to 0} \log_2 \delta} \\ &= \lim_{\delta \to 0} \left[\frac{\log_2 Vol^n(F_\delta)}{\log_2 \delta}\right]. \end{aligned}$$

It is clear that

$$\frac{\log_2 \beta}{\lim\limits_{\delta \to 0} \log_2 \delta} = 0.$$

Hence, we get

$$s = n - \lim_{\delta \to 0} \frac{\log_2 Vol^n(F_\delta)}{\log_2 \delta},$$

and Theorem 2.7 shows that

$$\dim_B F = n - \lim_{\delta \to 0} \frac{\log_2 Vol^n(F_\delta)}{\log_2 \delta}.$$

Therefore, we can conclude that

$$\dim_B F = \dim_M F.$$

The proof of the theorem is thus complete. ■

In this book, the Minkowski dimension will be used for automatic analysis of documents.

It is important to understand the relationship between box-counting dimension and Hausdorff dimension. If F can be covered by $N_\delta(F)$ sets of diameter δ, then, from definition, i.e Eq. (2.1), and recall here

$$H_\delta^s(F) = \inf_{\mathbb{R}(\delta)} \left\{ \sum_{i=1}^{\infty} |U_i|^s : \{U_i\} \text{ is a } \delta\text{-cover of } F \right\},$$

we have

$$H_\delta^s(F) \leq N_\delta(F)\delta^s.$$

If

$$1 < H^s(F) = \lim_{\delta \to 0} H_\delta^s(F)$$

then $\log N_\delta(F) + s \log \delta > 0$ if δ is sufficiently small. Thus

$$s \leq \frac{\underline{\lim}_{\delta \to 0} \log N_\delta(F)}{-\log \delta},$$

therefore, we can obtain

$$\dim_H F \leq \underline{\dim}_B F \leq \overline{dim}_B F, \quad \text{for any } F \subset \mathbb{R}. \tag{2.18}$$

We do not in general get equality here. Although Hausdorff and box dimensions are equal for many 'reasonably regular' sets, there are plenty of examples where this inequality is strict.

Roughly

$$\dim_B F = \lim_{\delta \to 0} \frac{\log_2 N_\delta(F)}{-\log_2 \delta}$$

speaking says that $N_\delta(F) \cong \delta^{-s}$ for small δ, where $s = dim_B F$. More precisely, it says that

$$N_\delta(F)\delta^s \to \infty, \quad \text{if } s < \dim_B F,$$

$$N_\delta(F)\delta^s \to 0, \quad \text{if } s > \dim_B F,$$

However,

$$N_\delta(F)\delta^s = \inf\left\{\sum_i \delta^s : \{U_i\} \text{ is a (finite) } \delta\text{-cover of } F,\right\}$$

which should be compared with

$$H_\delta^s(F) = \inf\left\{\sum_i |U_i|^s : \{U_i\} \text{ is a } \delta\text{-cover of } F,\right\}$$

which occurs in the definition of Hausdorff measure and dimension. In calculating Hausdorff dimension, we assign different weights $|U_i|^s$ to the covering sets U_i, whereas for the box dimensions we use the same weight δ^s for each covering set. Box dimensions may be thought of as indicating the efficiency with which a set may be covered by small sets of equal size, whereas Hausdorff dimension involves coverings by sets of small but perhaps widely varying size.

There is a temptation to introduce the quantity

$$v(F) = \underline{\lim}_{\delta\to 0} N_\delta(F)\delta^s,$$

but this does not give a measure on subsets of $\mathbb{R}$. As we shall see, one consequence of this is that box dimensions have a number of unfortunate properties, and can be awkward to handle mathematically.

Since box dimensions are determined by coverings by sets of equal size they tend to be easier to calculate than Hausdorff dimensions.

Let F be the middle third Cantor set as shown in Figure 2.2. Then, we can obtain

$$\underline{\dim}_B F = \overline{\dim}_B F = \frac{log2}{log3}.$$

Calculation:

The obvious covering by the 2^k intervals of E_k of length 3^{-k} gives that

$$N_\delta(F) \le 2^k \quad \text{if } 3^{-k} < \delta \le 3^{-k+1}.$$

According to Theorem 2.5:

$$\overline{\dim}_B F = \overline{\lim_{\delta\to 0}} \frac{\log_2 N_\delta(F)}{-\log_2 \delta},$$

we know that

$$\overline{\dim}_B F = \overline{\lim_{\delta\to 0}} \frac{\log_2 N_\delta(F)}{-\log_2 \delta} \leq \overline{\lim_{k\to\infty}} \frac{\log 2^k}{\log 3^{k-1}} = \frac{\log 2}{\log 3}.$$

On the other hand, any interval of length δ with $3^{-k-1} \leq \delta < 3^{-k}$ intersects at most one of the basic intervals of length 3^{-k} used in the construction of F. There are 2^k such intervals so at least 2^k intervals of length δ are required to cover F. Hence $N_\delta(F) \geq 2^k$ leading to $\underline{\dim}_B F \geq log2/log3$.

Thus, at least for the Cantor set, $dim_H F = dim_B F$.

2.3.4 *Properties of Box Counting Dimension*

Box dimension has the following elementary properties, which mirror those of Hausdorff dimension. These properties can be verified in the similar method.

- A smooth m-dimensional submanifold of $\mathbb{R}^n$ has $\dim_B F = m$.
- $\underline{\dim}_B$ and $\overline{\dim}_B$ are monotonic.
- $\overline{\dim}_B$ is finitely stable, i.e.

$$\overline{\dim}_B(E \cup F) = \max\{\overline{\dim}_B E, \overline{\dim}_B F\}$$

  though $\underline{\dim}_B$ is not.
- $\underline{\dim}_B$ and $\overline{\dim}_B$ are Lipschitz invariant. This is so because, if $|f(x)-f(y)| \leq c|x-y|$ and F can be covered by $N_\delta(F)$ sets of diameter at most δ, then the $N_\delta(F)$ images of these sets under F form a cover by sets of diameter at most $c\delta$, thus $\dim_B f(F) \leq \dim_B F$. Similarly, box dimensions behave just like Hausdorff dimensions under bi-Lipschitz and Hölder transformations.

Theorem 2.9 *Let F be a non-empty and bounded set in $\mathbb{R}^n$, and $\bar{F}$ denote the closure of F that means the smallest closed subset of $\mathbb{R}^n$ containing F, we have*

$$\underline{\dim}_B \bar{F} = \underline{\dim}_B F,$$

$$\overline{\dim}_B \bar{F} = \overline{\dim}_B F,$$

Proof Let $B_1, ..., B_k$ be a finite collection of closed balls of radii δ. If the closed set

$$\bigcup_{i=1}^{k} B_i \supset F, \text{ and } \bigcup_{i=1}^{k} B_i \supset \bar{F}.$$

Hence the smallest number of closed balls of radius δ that cover F is enough to cover the larger set $\bar{F}$. The result follows

The proof of the theorem is thus complete. ■

An direct result of this theorem is that if F is a dense subset of an open region of $\mathbb{R}^n$, we can obtain that

$$\underline{\dim}_B F = \overline{\dim}_B F = n.$$

For instance, let F be the countable set of rational numbers between 0 and 1. Then $\bar{F}$ is the entire interval [0, 1], such that

$$\underline{\dim}_B F = \overline{\dim}_B F = 1.$$

Therefor, the countable sets can have non-zero box dimension. Moreover, the box-counting dimension of each rational number, which can be considered to be a one-point set, is zero, On the other hand, the countable union of these singleton sets has dimension 1. Consequently, it is not generally true that

$$\bigcup_{i=1}^{\infty} F_i = \sup_i \dim_B F_i.$$

This severely restricts the application of box dimension.

Example 2.1 $F = 0, 1, \frac{1}{2}, \frac{1}{3}, \cdots$ is a compact set with $\dim_B F = \frac{1}{2}$

Calculation. If I $|U| = \delta$ and k is the integer, which satisfies the condition

$$\frac{1}{(k-1)k} > \delta \geq \frac{1}{k(k+1)},$$

thus, U can cover at most one of the points $\{1, \frac{1}{2}, \cdots, \frac{1}{k}\}$. Consequently, at least k sets of diameter δ are required to cover F, then it is true that

$$\frac{\log N_\delta(F)}{-\log \delta} \geq \frac{\log k}{\log k(k+1)}$$

If we take different values of δ, we can obtain

$$\underline{\dim}_B F \geq \frac{1}{2}, \text{ if } \delta \to 0.$$

On the other hand, if $\frac{1}{2} > \delta > 0$, take k such that

$$\frac{1}{(k-1)k} > \delta \geq \frac{1}{k(k+1)}.$$

Then $(k+1)$ intervals of length δ cover $[0, \frac{1}{k}]$, leaving $k-1$ points of F which can be covered by another $k-1$ intervals. we can obtain

$$\frac{\log N_\delta(F)}{-\log \delta} \leq \frac{\log 2k}{\log k(k-1)}$$

and produces

$$\overline{\dim}_B F \leq \frac{1}{2}$$

The calculation is thus complete.

Nevertheless, as well as being convenient in practice, box dimensions are very useful in theory. If, as often happens, it can be proved that a set has equal box and Hausdorff dimensions, the interplay of these definitions can be used to powerful effect.

2.4 Basic Methods for Calculating Dimensions

In this section we bring together some of the basic methods that can be used for dimension calculations. For most fractals, 'obvious' upper estimates of dimension can be obtained by the natural coverings with small sets.

Theorem 2.10 *Suppose that F can be covered by n_k sets of diameter at most δ_k with $\delta_k \to 0$ as $k \to \infty$. We have*

$$\dim_H F \leq \underline{\dim}_B F \leq \varliminf_{k\to\infty} \frac{\log n_k}{-\log \delta_k}$$

and

$$\dim_B F \leq \overline{\lim_{k\to\infty}} \frac{\log n_k}{-\log \delta_k} \; \textit{for} \; \exists 0 < c < 1, \; \textit{if} \; \delta_{k+1} \geq c\delta_k$$

then, we can obtain

$$H^s(F) < \infty,$$

if $n_k\delta_k^s$ *remains bounded as* $k \to \infty$.

Proof According to the definitions, the inequalities for the box-counting dimension are proved. For the last part, $H^s_{\delta_k}(F) \leq n_k\delta_k^s$, therefore we can obtain that $H^s_{\delta_k}(F)$ tends to a finite limit $H^s(F)$ as $k \to \infty$.

The proof of the theorem is thus complete. ■

For the previous example, in the case of the middle third Cantor set the natural coverings by 2^k intervals of length 3^{-k} give $\dim_H F \leq \log 2/\log 3$.

The 'obvious' upper bound of the Hausdorff dimension of a set turns out to be the actual value. But, the demonstration of it may be difficult. To get an upper bound, we need to evaluate sums of the form $\sum |U_i|^s$ for specific coverings $\{U_i\}$ of F. we also need to demonstrate that $\sum |U_i|^s$ is greater than some positive constant for all δ-coverings of F for a lower bound. Especially, if we use Hausdorff dimension, not box dimension, we need consider that case, i.e. some of the U_i are very small and others have relatively large diameter. It prohibits sweeping estimates for $\sum |U_i|^s$ such as those available for upper bounds.

However, we can avoid these difficulties by proving that no individual set U can cover too much of F compared with its size measured as $|U|^s$. We know that $\sum |U_i|^s$ cannot be too small if $\{U_i\}$ covers the whole of F. The usual way to do this is to concentrate a suitable mass distribution μ on F and compare the mass $\mu(U)$ covered by U with $|U|^s$ for each U.

Theorem 2.11 *Let* ξ *be a mass distribution on* F *and suppose that for some* s *there are numbers* $k > 0$ *and* $\alpha > 0$ *such that*

$$\xi(U) \leq k|U|^s \tag{2.19}$$

for all sets U *with* $|U| \leq \alpha$. *Then*

$$H^s(F) \geq \xi(F)/k$$

and

$$s \leq \dim_H F \leq \underline{\dim}_B \leq \overline{\dim}_B F.$$

Proof If $\{U_i\}$ is any cover of F then

$$0 < \xi(F) = \xi(\bigcup_i U_i) \leq \sum_i \xi(U_i) \leq k \sum_i |U_i|^s. \tag{2.20}$$

If we take infima,

$$H^s_\alpha(F) \geq \xi(F)/c \text{ if } \alpha \text{ is small enough,}$$

therefore we can obtain

$$H^s(F) \geq \xi(F)/c.$$

The proof of the theorem is thus complete. ∎

This theorem is called "mass distribution principle" which gives a quick lower estimate for the Hausdorff dimension of the middle third Cantor set F as shown in Figure 2.2.

Let ξ be the natural mass distribution on F, so that each of the 2^m basic intervals of length 3^{-m} in E_m in the construction of F, carry a mass 2^{-m}. Let U be a set with $|U| < 1$ and let m be the integer such that

$$3^{-m+1} \leq |U| < 3^{-m}.$$

Then U can intersect at most one of the intervals of E_m, thus, we have

$$\xi(U) \leq 2^{-m} = (3^{-m})^{\frac{\log 2}{\log 3}} \leq (3|U|)^{\frac{\log 2}{\log 3}}$$

We also have

$$H^{\frac{\log 2}{\log 3}}(F) > 0$$

according to the mass distribution principle giving

$$\dim_H F \geq \frac{\log 2}{\log 3}.$$

Example 2.2 Let F_1 be the product of the middle third Cantor set F and the unit interval, such that $F_1 = F \times [0,1] \subset I\!R^2$. We can obtain that

$$\dim_B F_1 = \dim_H F_1 = 1 + \frac{\log 2}{\log 3} = s,$$

with $0 < H^s(F_1) < \infty$.

Calculation:

For each m, there is a covering of F by 2^m intervals of length 3^{-m}. A column of 3^m squares of side 3^{-m} covers the part of F_1 above each such interval, when we take these all together, F_1 may be covered by $2^m 3^m$ squares of side 3^{-m}. Hence, we can get

$$H^s_{3^{-m}\sqrt{2}}(F_1) \leq 2^m 3^m (3^{-m}\sqrt{2})^s,$$

therefore,

$$H^s(F_1) \leq 2^{\frac{s}{2}}$$

and

$$\dim_H F_1 \leq \underline{\dim}_B F_1 \leq \overline{\dim}_B F_1 \leq s.$$

A mass distribution ξ on F_1 can be difined by taking the natural mass distribution on F described above and 'spreading it' uniformly along the intervals above F. Each basic interval of F of side 3^{-m} having mass 2^{-m}. Hence, if U is a rectangle, with sides parallel to the coordinate axes, of height h, above a basic interval of F of side 3^{-m}, then, we know that $\xi(U) = h2_{-m}$. Any set U is contained in a square of side $|U|$ with sides parallel to the coordinate axes. If

$$3^{-(m+1)} \leq |U| < 3^{-m}$$

then U lies above at most one basic interval of F of side 3^{-m}, finally we can obtain

$$\xi(U) \leq |U| 2^{-m} \leq |U| 3^{-\frac{m \log 2}{\log 3}} \leq |U| (3|U|)^{\frac{\log 2}{\log 3}} \leq 3^{\frac{\log 2}{\log 3}} |U|^s.$$

According to Theorem 2.11, i.e. the Mass distribution principle, we have $H^s(F_1) > 0$.

The calculation is thus complete.

The following general construction of a subset of $I\!R$ may be thought of as a generalization of the Cantor set construction. Let

$$[0,1] = E_0 \supset E_1 \supset E_2 \supset \cdots$$

be a decreasing sequence of sets, with each E_m a union of a finite number of disjoint closed intervals, which are called basic intervals, with each interval of E_m containing at least two intervals of E_{m+1}, and the maximum length of intervals in E_m tending to 0 as $k \to 0$. Then the set F is a totally disconnected subset of $[0,1]$, so that

$$F = \bigcap_{k=0}^{\infty} E_m \tag{2.21}$$

which is generally a fractal.

In the following examples, the upper estimates for $\dim_H F$ depend on the number and size of the basic intervals, whilst the lower estimates depend on their spacing. For these to be equal, the intervals of E_{m+1} must be 'nearly uniformly distributed' inside the intervals of E_m .

Example 2.3 Let s be a number strictly between 0 and 1. Suppose that the E_m in the general construction (2.21) have the following property: for each basic interval Γ of E_m, the intervals $\Gamma_1, \cdots, \Gamma_n (n \geq 2)$ of E_{m+1} contained in Γ are of equal length and equally spaced, the lengths can be represented by formula

$$|\Gamma_i|^s = \frac{1}{m}|\Gamma|^s \qquad (1 \leq i \leq m) \tag{2.22}$$

with the left-hand ends of Γ_1 and Γ coinciding, and the right-hand ends of Γ_m and Γ coinciding. Then

$$\dim_H F = s \text{ and } 0 < H^s(F) < \infty.$$

Notice that n may be different for different intervals Γ in the construction, hence, the intervals of E_m may have widely differing lengths.

Calculation:

According to Eq. (2.22) with Γ, Γ_i, we can obtain

$$|\Gamma|^s = \sum_{i=1}^{m} |\Gamma_i|^s \tag{2.23}$$

This inductively can be applied to the intervals of E_m for successive m, thus we can know that, for each m, it follows that

$$1 = \sum |\Gamma_i|^s,$$

where the sum is over all the intervals in E_m. The intervals of E_m cover F. Because that the maximum interval length tends to 0 as $m \to \infty$, we can conclude taht

$$H_\delta^s(F) \leq 1$$

for sufficiently small δ giving $H^s(F) \leq 1$.

We distribute a mass ξ on F in such a way that $\xi(\Gamma) = |\Gamma|^s$ whenever Γ is a basic interval. In this way, we start with unit mass on $[0, 1]$, we divide this equally between each interval of E_1, the mass on each of these intervals is divided equally between each subinterval of E_2, and so on. Eq. (2.23) ensures that we obtain a mass distribution on F with $\xi(\Gamma) = |\Gamma|^s$ for every basic interval. We estimate $\xi(U)$ for any interval U with end-points in F. Let Γ be the smallest basic interval that contains U; suppose that Γ is an interval of E_m, and let $\Gamma_i, \cdots, \Gamma_m$ be the intervals of E_{m+1} contained in Γ. In succession, U intersects a number $\eta \leq 2$ of the Γ_i, otherwise U is contained in a smaller basic interval. The space between consecutive Γ_i is

$$\begin{aligned}
\frac{|\Gamma| - n|\Gamma_i|}{n-1} &= \frac{|\Gamma|(1 - n|\Gamma_i|/|\Gamma|)}{n-1} \\
&= \frac{|\Gamma|(1 - n^{1-\frac{1}{s}})}{n-1} \\
&\geq \frac{c_s|\Gamma|}{n}.
\end{aligned}$$

We apply Eq. (2.22), and $c_s = (1 - 2^{1-1/s})$. Therefore, we can get

$$|U| \geq \frac{\eta - 1}{n} c_s |\Gamma| \geq \frac{\eta}{2n} c_s |\Gamma|.$$

We use Eq. (2.23)

$$\xi(U) \leq \eta \xi(\Gamma_i) = \eta |\Gamma_i|^s = \frac{\eta}{n} |\Gamma|^s$$

$$\leq 2^s c_s^{-s} (\frac{\eta}{n})^{1-s} |U|^s \leq 2^s c_s^{-s} |U|^s \tag{2.24}$$

For any interval U with end-points in F, this is true, and so for any set U. We can apply Eq. (2.24) to the smallest interval containing $U \cap F$, and the Mass distribution principle 2.20, $H^s(F)$. A more careful estimate of $\xi(U)$ in this example leads to $H^s(F) = 1$.
The calculation is thus complete.

We call the sets obtained when n is kept constant throughout the construction of this example uniform Cantor sets. These provide a natural generalisation of the middle third Cantor set.

Example 2.4 Suppose that $m \geq 2$ is an integer and $0 < \lambda < 1/m$. Let F be the constructed set, where each basic interval Γ is replaced by m equally spaced subintervals of lengths $\lambda|\Gamma|$, the ends of Γ coinciding with the ends of the extreme subintervals. Thus, we can obtain

$$\dim_H F = \dim_B F = \log m/(-\log \lambda),$$

and

$$0 < H^{\log m/(-\log \lambda)}(F) < \infty.$$

Calculation:

In the above example, we take m constant and $s = \log m/(-\log \lambda)$ to produce the set F. Eq. (2.22) becomes

$$(\lambda|\Gamma|)^s = (1/m)|\Gamma|^s,$$

which is satisfied identically, thus $\dim_H F = s$. For the box dimension, in the usual way, F is covered by the m^k basic intervals of length λ^{-k} in E_k for each k. Thus, we can obtain

$$\dim_B F \leq \log m/(-\log \lambda).$$

The calculation is thus complete.

Example 2.5 This example is another case of the general construction. In the general construction (2.21), suppose each interval of E_{k-1} contains at least m_k intervals of $E_k (k = 1, 2, \cdots)$ which are separated by gaps of at least ε_k, and suppose $0 < \varepsilon_{k+1} < \varepsilon_k$ for each k. Then, we can find that

$$\dim_H F \geq \underline{\lim}_{k\to\infty} \frac{\log(m_1 \cdots m_{k-1})}{-\log(m_k \varepsilon_k)} \tag{2.25}$$

Calculation:

Assume that each set E_{k-1} contains exactly m_k intervals of E_k; if not we may throw out excess intervals to get smaller sets E_k and F for which this is so. In this way, a mass distribution μ on F by assigning a mass of $(m_1 \cdots m_k)^{-1}$ to each of the $m_1 \cdots m_k$ basic intervals of E_k can be defined.

Suppose U is an interval with $0 < |U| < \varepsilon_1$; we estimate $\mu(U)$. We assume that k is the integer such that $\varepsilon_k \leq |U| < \varepsilon_{k-1}$. The number of intervals of E_k that intersect U can be computed as below:

Case 1: The number of intervals of E_k that intersect U is at most m_k, since U intersects at most one interval of E_{k-1}

Case 2: The number of intervals of E_k that intersect U is at most $|U|/\varepsilon_k + 1 \leq 2|U|/\varepsilon_k$, since the intervals of E_k have gaps of at least ε_k between them. Each interval of E_k supports mass $(m_1 \cdots m_k)^{-1}$ so that

$$\begin{aligned} \xi(U) &\leq (m_1, ..., m_k)^{-1} min\{2|U|/\epsilon_k, m_k\} \\ &\leq (m_1, ..., m_k)^{-1} (2|U|/\epsilon_k)^s m_k^{1-s}, \quad \text{for any } 0 \leq s \leq 1. \end{aligned}$$

Therefore, we can obtain

$$\frac{\xi(U)}{|U|^s} \leq 2^s/(m_1 \cdots m_{k-1}) m_k^s \varepsilon_k^s$$

which is bounded above by a constant provided that

$$s < \underline{\lim}_{k\to\infty} \log(m_1 \cdots m_{k-1})/ - \log(m_k \varepsilon_k))$$

The calculation is thus complete.

We suppose that the intervals of E_k are all of length δ_k, and that each interval of E_{k-1} contains exactly m_k intervals of E_k, which are 'roughly equally spaced' in the sense that $m_k\varepsilon_k \geq c\delta_{k-1}$, where $c > 0$ is a constant. In this case, Eq. (2.25) will be

$$\dim_H F \geq \varliminf_{k\to\infty} \frac{\log(m_1\cdots m_{k-1})}{-\log c - \log\delta_{k-1}} = \varliminf_{k\to\infty} \frac{\log(m_1\cdots m_{k-1})}{-\log\delta_{k-1}}$$

However, E_{k-1} comprises $m_1\cdots m_{k-1}$ intervals of length δ_{k-1}, hence, this expression equals the upper bound for $\dim_H F$. Thus, in case of that the intervals are well spaced, we can obtain the equality in Eq. (2.25).

Example 2.6 Let $\alpha_l, \alpha_2, \ldots$ be a rapidly increasing sequence of integers, and $0 < s < 1$, so that

$$m_{k+1} \geq \max\{m_k^k, 3\alpha_k^{1/s}\} \text{ for each } k.$$

Assume $H_k \subset R$ consist of equally spaced equal intervals of lengths $\alpha_k^{-l/s}$ with the midpoints of consecutive intervals distance α_k^{-1} apart for each k. Thus, we have

$$\dim_H F = \dim_B F = s, \text{ where } F = \cap_{k=1}^{\infty} H_k.$$

Calculation:

The set $F\cap[0,1]$ is contained in at most α_k+1 intervals of length $\alpha_k^{-1/s}$, because $F \subset H_k$ for each k. According to (2.10), we have

$$\overline{\dim}_B(F\cap[0,1]) \leq \overline{\lim} k \to \infty \log(\alpha_k+1)/-\log\alpha_k^{-1/s} = s.$$

Similarly,

$$\overline{dim_B}(F\cap[n,\alpha+1]) \leq s \text{ for any } \alpha \in Z,$$

As a countable union of such sets, F has $\overline{dim_B}F \leq s$.

We consider that (1) $E_0 = [0,1]$ for $k \geq 1$, and (2) E_k consist of the intervals of H_k that are completely contained in E_{k-1}. Thus, we know that each interval Γ of E_{k-1} contains at least

$$n_k|\Gamma| - 1 \geq \alpha_k\alpha_{k-1}^{-1/s} - 1 \text{ intervals of } E_k,$$

which are separated by gaps of at least

$$\alpha_k^{-1} - \alpha_k^{-1/s} \geq \frac{1}{2}n_k^{-1} \text{ if } k \text{ is large enough.}$$

Using Example 2.5, and noting that replacing

$$\alpha_k \alpha_{k-l}^{-1/s} - 1 \text{ by } \alpha_k n_{k-l}^{-1/s}$$

does not affect the limit,

$$\dim_H F \geq \dim_H \cap_{k=1}^{\infty} E_k = \varliminf_{k\to\infty} \frac{\log(\alpha_1 \cdots \alpha_{k-2})^{1-1/s}\alpha_{k-1}}{-\log(\alpha_k n_{k-1}^{-1/s} \frac{1}{2}\alpha_k^{-1})}$$

$$= \varliminf_{k\to\infty} \frac{\log(\alpha_1 \cdots \alpha_{k-2})^{1-1/s} + \log \alpha_{k-1}}{\log 2 + (\log \alpha_{k-1})/s}$$

Provided that α_k is sufficiently rapidly increasing, the terms in log α_{k-1} in the numerator and denominator of this expression are dominant, so that $\dim_H F \geq s$, as required.

The calculation is thus complete.

The Mass distribution is based on a simple idea, however, it can be applied to finding Hausdorff and box dimensions. Some important variations of the method can be found.

Theorem 2.12 *Let L be a family of balls contained in some bounded region of $\mathbb{R}^n$. Further, we have a finite or countable disjoint sub-collection $\Re_i$, which satisfies the following*

$$\bigcup_{\Re \in L} \Re \subset \bigcup_i \widetilde{\Re}_i \tag{2.26}$$

where $\widetilde{\Re}_i$ is the closed ball concentric with $\Re_i$ and of four times the radius.

Proof We consider a special case, i.e. assume L is a finite family. The basic idea of proving this case is the same in the general case. We select the $\{\Re_i\}$ inductively. Let $\Re_1$ be a ball in L of maximum radius. Suppose that $\Re_1, \cdots, \Re_{k-1}$ have been chosen. We also consider $\Re_k$ to be the largest ball in L that does not intersect $\Re_1, \cdots, \Re_{k-1}$. When no such ball remains, this process terminates. Obviously, the selected balls are disjoint, we must examine this theorem holds. If $\Re \in L$, we know that either $\Re = \Re_i$ for some i, or $\Re$ intersects one of the $\Re_i$ with $|\Re_i| \geq |\Re|$. If we can not find this case, then we can choose $\Re$ instead of the first ball $\Re_k$ with $|\Re_i| < |\Re|$. In either case, $\Re \subset \widetilde{\Re}_i$, we have Eq. (2.26).

The proof of the theorem is thus complete. ■

Theorem 2.13 *If we have the following assumptions:*

(a) ξ *is a mass distribution on* $\mathbb{R}^n$,
(b) $F \subset \mathbb{R}^n$ *is a Borel set,*
(c) $0 < \beta < \infty$ *is a constant.*

Then, we can obtain

(a) *If* $\overline{\lim}_{r\to 0}\xi(\Re_r(x))/r^s < \beta$ *for all* $x \in F$ *then* $\mathbb{H}^s(F) \geq \xi(F)/\beta$
(b) *If* $\overline{\lim}_{r\to 0}\xi(\Re_r(x))/r^s > \beta$ *for all* $x \in F$ *then* $\mathbb{H}^s(F) \leq 2^s\xi(\mathbb{R}^n)/\beta$.

Proof (a) For each $\delta > 0$, we assume that

$$F_\delta = \{x \in F : \xi(\Re_r(x)) < (\beta - s)r^s \text{ for all } 0 < r \leq \delta \text{ for some } \varepsilon > 0\}$$

Let $\{U_i\}$ be a δ-cover of F and thus of F_δ. For each U_i containing a point x of F_δ, the ball$B$ with centre x and radius $|U_i|$ certainly contains U_i. According to the definition of F_δ, we can get

$$\xi(U_i) \leq \xi(\Re) < \beta|U_i|^s$$

so that

$$\xi(F_\delta) \leq \sum_i \{\xi(U_i) : U_i \text{ intersects } F_\delta\} \leq \beta\sum_i |U_i|^s$$

Since $\{U_i\}$ is any δ-cover of F, it follows that

$$\xi(F_\delta) \leq \beta\mathbb{H}^s_\delta(F) \leq \beta\mathbb{H}^s(F).$$

However, F_δ increases to F as δ decreases to 0, thus, we can obtain

$$\xi(F) \leq c\mathbb{H}^s(F).$$

(b) We will prove a weaker version of (b) with 2^s replaced by 8^s, but the basic idea is similar. Suppose first that F is bounded. Fix $\delta > 0$ and let $\mathbb{C}$ be the collection of balls

$$\{B_r(x) : x \in F, 0 < r \leq \delta \text{ and } \xi(\Re_r(x)) > \beta r^s\}$$

According to the hypothesis of (b), i.e.

$$F \subset \cup_{\Re \in \mathbb{C}}.$$

Applying Covering lemma to the collection $\mathbb{C}$, there is a sequence of disjoint balls $\Re_i \in \mathbb{C}$ such that

$$\cup_{\Re \in \beta} \Re \subset_i \widetilde{\Re}_i,$$

where $\widetilde{\Re}_i$ is the ball concentric with $\Re_i$ but of four times the radius. Thus $\{\widetilde{\Re}_i\}$ is an 8δ-cover of F. Therefore, we can obtain

$$\mathbb{H}^s_{8\delta}(F) \leq \sum_i |\widetilde{\Re}_i|^s \leq 4^s \sum_i |\Re_i|^s \leq 8^s \beta^{-1} \sum_i \xi(\Re_i) \leq 8^s \beta^{-1} \xi(\mathbb{R}^n)$$

If we assume that $\delta \to 0$, we can get

$$\mathbb{H}^s(F) \leq 8^s \beta^{-1} \xi(\mathbb{R}^n) < \infty.$$

Finally, if F is unbounded and

$$\mathbb{H}^s(F) > 8^s \beta^{-1} \mu(\mathbb{R}^n),$$

the $\mathbb{H}^s$-measure of some bounded subset of F will also exceed this value, contrary to the above.

The proof of the theorem is thus complete. ■

If this limit exists, it is immediate from this that

$$\dim_H F = \lim_{r \to 0} \log \xi(\Re_r(x)) / \log r.$$

The densities $\lim_{r \to 0} \log \xi(\Re_r(x))/r^s$ that occur in this theorem can be applied to define the dimension of a set. Often a fractal F is naturally endowed with a mass distribution ξ. If the mass of small balls obeys a law

$$\log \mu(F \cap \Re_r(x)) / \log r \to s \text{ as } r \to 0 \text{ for all } x \in F,$$

then the Hausdorff dimension of F equals s.

More details of this chapter can be found in books [Falconer, 1985; Falconer, 1990].

Chapter 3

Basic Concepts of Wavelet Theory

The wavelet theory consists of two parts, namely, the wavelet transform and the wavelet basis. In this chapter, we will organize it into two sections: (1) continuous wavelet transforms, and (2) multiresolution analysis and wavelet bases.

3.1 Continuous Wavelet Transforms

In this section, we organize it into three subsections: (1) the general theory of wavelet transform, (2) the filtering properties of the wavelet transforms, and (3) the characterization of Lipschitz regularity of signals by wavelet transforms.

3.1.1 *General Theory of Continuous Wavelet Transforms*

In Fourier transform

$$\mathcal{F}f(\xi) = \int_{\mathbb{R}} f(\xi - x)\overline{e^{i\xi x}}dx,$$

if we replace $e^{i\xi x}$, the dilation of the basic wave function e^{ix}, with $\psi(\frac{t-b}{a})$, the translation and dilation of the basic wavelet $\psi(x)$, then, the transform is referred as a continuous wavelet transform (also called an integral wavelet transform), which is defined as follows:

Definition 3.1 A function $\psi \in L^2(\mathbb{R})$ is called an admissible wavelet or a basic wavelet if it satisfies the following "admissibility" condition:

$$C_\psi := \int_{\mathbb{R}} \frac{|\hat{\psi}(\xi)|^2}{|\xi|} d\xi < \infty. \tag{3.1}$$

The continuous (or integrable) wavelet transform with kernel ψ is defined by

$$(W_\psi f)(a,b) := |a|^{-\frac{1}{2}} \int_{\mathbb{R}} f(t)\overline{\psi(\frac{t-b}{a})}dt, \qquad f \in L^2(\mathbb{R}), \tag{3.2}$$

where $a, b \in \mathbb{R}$ and $a \neq 0$ are the dilation parameter and the translation parameter respectively.

Before showing the functions of the admissibility condition Eq. (3.1), we would like to discuss some basic properties of the continuous wavelet transform at first. According to the admissibility condition Eq. (3.1), if $\psi \in L^1(\mathbb{R})$, it can be mathematically inferred that

$$\hat{\psi}(0) = 0$$

and

$$\hat{\psi}(\xi) \to 0 \quad (|\xi| \to \infty).$$

This indicates that the function ψ is a bandpass filter.

The characteristics of the localized components of a signal $f(t)$ can be described by the continuous wavelet transform.

- Because of the damp of $\psi(x)$ at infinity, the localized characteristic of f near $x = b$ is described by Eq. (3.2). It will be clearer if we extremely assume that the $\psi(x)$ always be zero out of $[-1, 1]$. Then, for all $x \notin [b - |a|, b + |a|]$, we have

  $$\psi(\frac{t-b}{a}) = 0.$$

  Therefore,

  $$\begin{aligned} (W_\psi f)(a,b) &= |a|^{-\frac{1}{2}} \int_{\mathbb{R}} f(t)\overline{\psi(\frac{t-b}{a})}dt \\ &= |a|^{-\frac{1}{2}} \int_{b-|a|}^{b+|a|} f(t)\overline{\psi(\frac{t-b}{a})}dt. \end{aligned}$$

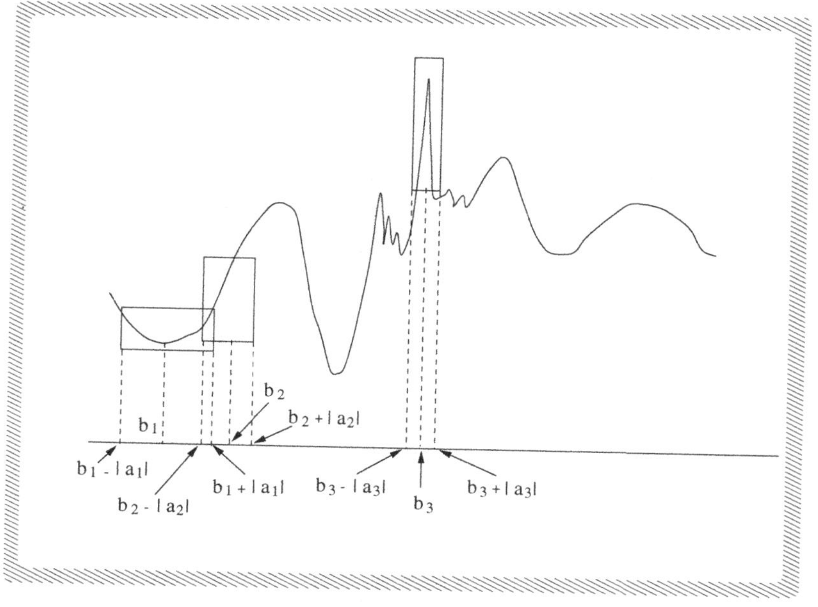

Fig. 3.1 Time-frequency localization.

This means that $(W_\psi f)(a,b)$ is completely determined by the behaviors of f in $[b-|a|, b+|a|]$ with the center b (Figure 3.1). It is said that $(W_\psi f)(a,b)$ describes only the localized characteristic of f in $[b-|a|, b+|a|]$. The smaller a, the better the localized characteristic of f, which can also be found in Figure 3.1.

- On the other hand, by the basic properties of Fourier transform, we have:

$$\begin{aligned}(W_\psi f)(a,b) &= \frac{|a|^{-\frac{1}{2}}}{2\pi}\int_{\mathbb{R}} \hat{f}(\xi)(\psi(\frac{\cdot - b}{a}))^{\wedge}(\xi)d\xi \\ &= \frac{|a|^{-\frac{1}{2}}}{2\pi}\int_{\mathbb{R}} \hat{f}(\xi)|a|e^{-ib\xi}\hat{\psi}(a\xi)d\xi \\ &= \frac{|a|^{\frac{1}{2}}}{2\pi}\int_{\mathbb{R}} \hat{f}(\xi)e^{-ib\xi}\hat{\psi}(a\xi)d\xi.\end{aligned}$$

There is only a phase difference between $e^{-ib\xi}\hat{\psi}(a\xi)$ and $\hat{\psi}(a\xi)$. The bandpass property of ψ ensures that the energy of $\hat{\psi}(a\xi)$ concentrates on two bands, while the bandwidth depends on the scale parameter a with positive ratio. Hence, the localization of $\hat{f}$ is illustrated by $(W_\psi f)(a,b)$, and $\hat{f}$ gets worse localization with a smaller a.

ψ can be viewed as a basic window function. Through this window, we can not observe the integrated f or $\hat{f}$, whereas the local performances of f and $\hat{f}$ are very clear. Moving the translation parameter b, each part of f and $\hat{f}$ can be traveled. Meanwhile, by adjusting a, we can observe f and $\hat{f}$ with different localization. The later is the so-called focus property of the wavelet.

For the convenience of our discussion, under the meaning of statistics, we define the center of basic window ψ as

$$x^*_\psi := \frac{1}{\|\psi\|_2^2} \int_{\mathbb{R}} t|\psi(t)|^2 dt. \tag{3.3}$$

Moreover, we define the radius of the window , which is the statistic width of the energy of ψ to its center, as

$$\Delta_\psi := \frac{1}{\|\psi\|_2} \left[\int_{\mathbb{R}} (t - x^*)^2 |\psi(t)|^2 dt \right]^{1/2}, \tag{3.4}$$

where $\|\psi\|_2$ is the norm of ψ, i.e.,

$$\|\psi\|_2 := \left(\int_{\mathbb{R}} |\psi(x)|^2 dx \right)^{1/2}.$$

The window of ψ (also called the time window of ψ in general) is

$$[x^*_\psi - \Delta_\psi, x^*_\psi + \Delta_\psi].$$

In the same way, the window of $\hat{\psi}$ (also called the frequency window of ψ) is

$$[x^*_{\hat{\psi}} - \Delta_{\hat{\psi}}, x^*_{\hat{\psi}} + \Delta_{\hat{\psi}}].$$

The following square region

$$[x^*_\psi - \Delta_\psi, x^*_\psi + \Delta_\psi] \times [x^*_{\hat{\psi}} - \Delta_{\hat{\psi}}, x^*_{\hat{\psi}} + \Delta_{\hat{\psi}}]$$

is referred to the time-frequency window of ψ. For $a, b \in \mathbb{R}, a \neq 0$, a set of wavelets can be generated by the basic wavelet ψ as follows:

$$\psi_{a,b}(t) := |a|^{-\frac{1}{2}} \psi(\frac{t-b}{a}). \tag{3.5}$$

Thus, the time-frequency window of $\psi_{a,b}$ is

$$[b+ax_\psi^* - |a|\Delta_\psi, b+ax_\psi^* + |a|\Delta_\psi] \times \left[\frac{x_{\hat{\psi}}^*}{a} - \frac{1}{|a|}\Delta_{\hat{\psi}}, \frac{x_{\hat{\psi}}^*}{a} + \frac{1}{|a|}\Delta_{\hat{\psi}}\right]. \quad (3.6)$$

The procedure of reasoning is below:

$$\begin{aligned}
x_{\psi_{a,b}}^* &= \frac{1}{\|\psi\|_2^2}\int_{\mathbb{R}} t|a|^{-1}|\psi(\frac{t-b}{a})|^2 dt \\
&= \frac{|a|^{-1}}{\|\psi\|_2^2}\int_{\mathbb{R}} (ax+b)|\psi(x)|^2|a|dx \\
&= \frac{1}{\|\psi\|_2^2}\left[a\int_{\mathbb{R}} x|\psi(x)|^2 dx + b\|\psi\|_2^2\right] \\
&= ax_\psi^* + b;
\end{aligned}$$

$$\begin{aligned}
\Delta_{\psi_{a,b}} &= \frac{1}{\|\psi\|_2}\left[\int_{\mathbb{R}} (t-ax_\psi^*-b)^2|a|^{-1}|\psi(\frac{t-b}{a})|^2 dt\right]^{1/2} \\
&= \frac{1}{\|\psi\|_2}\left[\int_{\mathbb{R}} (ax-ax_\psi^*)^2|\psi(x)|^2|a|^{-1}|a|dx\right]^{1/2} \\
&= \frac{|a|}{\|\psi\|_2}\left[\int_{\mathbb{R}} (x-x_\psi^*)^2|\psi(x)|^2 dx\right]^{1/2} \\
&= |a|\Delta_\psi;
\end{aligned}$$

$$\hat{\psi}_{a,b}(\xi) = \int_{\mathbb{R}} |a|^{-\frac{1}{2}}\psi(\frac{t-b}{a}e^{-i\xi t}dt = |a|^{\frac{1}{2}}e^{-ib\xi}\hat{\psi}(a\xi);$$

$$\|\hat{\psi}_{a,b}\|_2^2 = |a|\int_{\mathbb{R}} |\hat{\psi}(a\xi)|^2 d\xi = \|\hat{\psi}\|_2^2;$$

$$\begin{aligned}
x_{\hat{\psi}_{a,b}}^* &= \frac{1}{\|\hat{\psi}_{a,b}\|_2^2}\int_{\mathbb{R}} \xi|a||\hat{\psi}(a\xi)|^2 d\xi \\
&= \frac{1}{a}\frac{1}{\|\hat{\psi}_{a,b}\|_2^2}\int_{\mathbb{R}} \xi|\hat{\psi}(\xi)|^2 d\xi \\
&= \frac{1}{a}x_{\hat{\psi}}^*;
\end{aligned}$$

$$\Delta_{\hat{\psi}_{a,b}} = \frac{1}{\|\hat{\psi}\|_2}\left[\int_{\mathbb{R}} (\xi - \frac{1}{|a|}x_{\hat{\psi}}^*)^2|a||\hat{\psi}(a\xi)|^2 d\xi\right]^{1/2}$$

$$
\begin{aligned}
&= \frac{1}{|a|}\frac{1}{\|\hat{\psi}\|_2}\left[\int_{\mathbb{R}}(|a|\xi - x^*_{\hat{\psi}})^2|\hat{\psi}(a\xi)|^2|a|d\xi\right]^{1/2} \\
&= \frac{1}{|a|}\frac{1}{\|\hat{\psi}\|_2}\left[\int_{\mathbb{R}}(\xi - x^*_{\hat{\psi}})^2|\hat{\psi}(\xi)|^2d\xi\right]^{1/2} \\
&= \frac{1}{|a|}\Delta_{\hat{\psi}}.
\end{aligned}
$$

From Eq. (3.6), we can easily find out, when $|a|$ changes to be smaller, the time window of ψ becomes more narrow, whereas the frequency window becomes wider. The area of its time-frequency window is a constant, which is irrelevant to a and b.

$$(2\Delta_{\psi_{a,b}})(2\Delta_{\hat{\psi}_{a,b}}) = 4\Delta_\psi\Delta_{\hat{\psi}}.$$

From the above discussion, we find an intrinsic fact: for a basic wavelet ψ, it is impossible to achieve perfect localization in both the time domain and the frequency domain simultaneously. Is there any ψ could make the area of the time-frequency to be small enough? Unfortunately, the following theorem, the famous Heisenberg uncertainty principle, gave a negative answer to this question.

Theorem 3.1 *Let $\psi \in L^2(\mathbb{R})$ satisfy $x\psi(x)$ and $\xi\hat{\psi}(\xi) \in L^2(\mathbb{R})$. Then*

$$\Delta_\psi\Delta_{\hat{\psi}} \geq \frac{1}{2}.$$

Furthermore, the equality in the above equation holds if and only if

$$\psi(x) = ce^{iax}g_\alpha(x-b),$$

where $c \neq 0, \alpha > 0,\ a, b \in \mathbb{R}$ and $g_\alpha(x)$ is the Gaussian function defined by

$$g_\alpha(x) := \frac{1}{2\sqrt{\pi\alpha}}e^{-\frac{x^2}{4\alpha}}. \tag{3.7}$$

Proof We assume that the window centers of ψ and $\hat{\psi}$ are 0 without losing generality. Based on the basic knowledge of Fourier analysis, we have

$$(\Delta_\psi\Delta_{\hat{\psi}})^2 = \frac{\left(\int_{\mathbb{R}} t^2|\psi(t)|^2dt\right)\left(\int_{\mathbb{R}}\xi^2|\hat{\psi}(\xi)|^2d\xi\right)}{\|\psi\|_2^2\|\hat{\psi}\|_2^2}$$

$$
\begin{aligned}
&= \frac{\left(\int_{\mathbb{R}} t^2|\psi(t)|^2dt\right)\left(\int_{\mathbb{R}} |\hat{\psi}'(\xi)|^2 d\xi\right)}{\|\psi\|_2^2\|\hat{\psi}\|_2^2} \\
&= \frac{\|x\psi(x)\|_2^2\|\hat{\psi}'(x)\|_2^2}{\|\psi\|_2^2\|\hat{\psi}\|_2^2} \\
&= \frac{\|x\psi(x)\|_2^2\ 2\pi\|\psi'(x)\|_2^2}{\|\psi\|_2^2\ 2\pi\|\psi\|_2^2} \\
&\geq \frac{\|x\psi(x)\psi'(x)\|_2^2}{\|\psi\|_2^4} \\
&\geq \frac{1}{\|\psi\|_2^4}\left|Re\int_{\mathbb{R}} x\psi(x)\overline{\psi'(x)}dx\right|^2 \\
&= \frac{1}{\|\psi\|_2^4}\left|\frac{1}{2}\int_{\mathbb{R}} x\frac{d}{dx}|\psi(x)|^2dx\right|^2 \\
&= \frac{1}{\|\psi\|_2^4}\left(\frac{1}{2}\int_{\mathbb{R}} |\psi(x)|^2dx\right)^2 \\
&= \frac{1}{4}.
\end{aligned}
$$

Thus, we have

$$\Delta_\psi\Delta_{\hat{\psi}} \geq \frac{1}{2}.$$

Further, by the conditions, which preserve the equal-sign in the Holder inequality, we know that the equalities in above reasoning will be tenable, if and only if there is a constant α such that

$$
\begin{cases}
-Re(x\psi(x)\overline{\psi'(x)}) = |x\psi(x)\overline{\psi'(x)}|, \\
|x\psi(x)| = 2\alpha|\psi'(x)|, \\
\|\psi\|_2 \neq 0
\end{cases}.
$$

First, by the second equality, we have

$$x\psi(x) = 2\alpha\psi'(x)e^{i\theta(x)},$$

where $\theta(x)$ is a real-valued function. Second, by the first equality, we have

$$-x\psi(x)\overline{\psi'(x)} \geq 0.$$

Then,

$$-2\alpha|\psi'(x)|^2e^{i\theta(x)} \geq 0.$$

It infers $e^{i\theta(x)} = -1$. Thus, we have

$$x\psi(x) = -2\alpha\psi'(x).$$

By resolving this ordinary differential equation, we can obtain

$$\psi(x) = ce^{-\frac{x^2}{4\alpha}}.$$

Finally, by the third equality, we know that $c \neq 0$.

It should be mentioned, at the beginning of our proof, we have assumed that the window centers of ψ and $\hat{\psi}$ are 0. Otherwise, consider

$$\tilde{\psi}(x) := e^{-iax}\psi(x+b),$$

where $a := x^*_{\hat{\psi}}$, $b := x^*_{\psi}$. The centers of the time window and the frequency window of $\tilde{\psi}(x)$ are 0. It means that the above reasoning is available to $\tilde{\psi}(x)$. It is easy to know that

$$\Delta_{\psi} = \Delta_{\tilde{\psi}}, \quad \Delta_{\hat{\psi}} = \Delta_{\hat{\tilde{\psi}}},$$

Hence, $\Delta_{\psi}\Delta_{\hat{\psi}} \geq \frac{1}{2}$. The equality holds if and only if

$$\tilde{\psi}(x) = ce^{-\frac{x^2}{4\alpha}},$$

i.e.

$$\psi(x) = ce^{i(x-b)a}e^{-\frac{(x-b)^2}{4\alpha}},$$

where $c \neq 0$. Replacing $\frac{ce^{iab}}{2\sqrt{2\pi\alpha}}$ with c, we have

$$\psi(x) = ce^{iax}g_{\alpha}(x-b),$$

where $c \neq 0, \alpha > 0,\ a, b \in \mathbb{R}$ and $g_{\alpha}(x)$ is the Gaussian function defined by Eq. (3.7). This establishes the theorem. ■

An important fact is supported by this theorem: no matter what wavelet ψ we choose, it can not achieve perfect localization in both the time domain and the frequency domain simultaneously. The time window and the frequency window just like the length and width of a rectangle with a fixed area. When the time window is narrow, the frequency window must be wide. By the contrary, when the time window is wide, the frequency window must be narrow (see Figure 3.2). In fact, the narrow time window and the wide frequency window are good for the high frequency part of a signal,

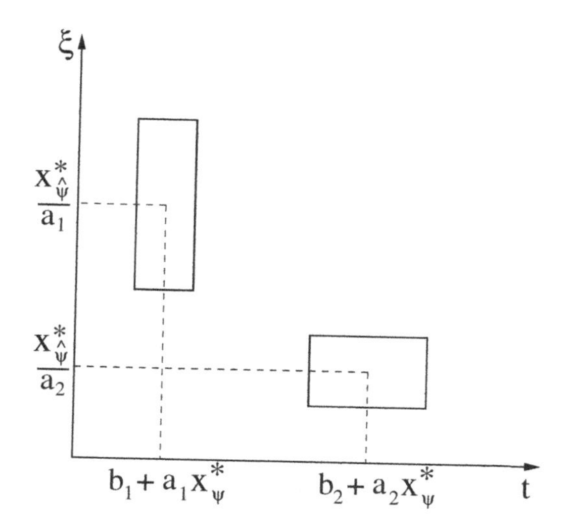

Fig. 3.2 Time-frequency windows, $a_1 < a_2$.

whereas the wide time window and the narrow frequency window are good for the low frequency part of a signal. Therefore, this property of wavelet just meets the demand of signal analysis. It can be applied for self-adaptive signal analysis.

The area of the time-frequency window of Gaussian is the smallest, which is $4\Delta_\psi\Delta_{\hat{\psi}} = 2$. Therefore, for the localized time-frequency analysis, Gaussian function is the best. It has been a traditional tool for signal analysis.

As the same as Fourier transform, wavelet transform is invertible. The inverse wavelet transform can be viewed as the reconstruction of the original signal. First of all, we need to prove the following theorem, which specifies that wavelet transform keeps the law of energy-conservation.

Theorem 3.2 *Let ψ be a basic wavelet. Then*

$$\int_{\mathbb{R}}\int_{\mathbb{R}} W_\psi f(a,b)\overline{W_\psi g(a,b)}\frac{dadb}{a^2} = C_\psi\langle f,\ g\rangle$$

holds for any $f,\ g \in L^2(\mathbb{R})$, where C_ψ is a constant defined by Eq. (3.1), $\langle f,\ g\rangle$ is the inner product of f and g. Particularly, for $g = f$ we have

$$\int_{\mathbb{R}}\int_{\mathbb{R}} |W_\psi f(a,b)|^2\frac{dadb}{a^2} = C_\psi\|f\|_2^2.$$

Proof Obviously, the second equality is a specific instance of the first equality at $g = f$. Thus, only the first equality needs to be proved. It is easy to know that

$$(\psi(\frac{t-b}{a}))\hat{}(\xi) = |a|e^{-ib\xi}\hat{\psi}(a\xi).$$

Denote that

$$\begin{cases} F(x) := \hat{f}(x)\overline{\hat{\psi}(ax)} \\ G(x) := \hat{g}(x)\overline{\hat{\psi}(ax)} \end{cases}$$

Then,

$$\begin{aligned} (W_\psi f)(a,b) &= \langle f(x), |a|^{-\frac{1}{2}}\psi(\frac{x-b}{a})\rangle \\ &= \frac{1}{2\pi}\langle \hat{f}(\xi), |a|^{\frac{1}{2}}e^{-b\xi}\hat{\psi}(a\xi)\rangle \\ &= \frac{1}{2\pi}|a|^{\frac{1}{2}}\int_{\mathbb{R}} \hat{f}(\xi)\overline{\hat{\psi}(a\xi)}e^{b\xi}d\xi \\ &= \frac{1}{2\pi}|a|^{\frac{1}{2}}\int_{\mathbb{R}} F(\xi)e^{b\xi}d\xi \\ &= \frac{1}{2\pi}|a|^{\frac{1}{2}}\hat{F}(-b). \end{aligned}$$

In the same way,

$$(W_\psi g)(a,b) = \frac{1}{2\pi}|a|^{\frac{1}{2}}\hat{G}(-b).$$

Therefore,

$$\begin{aligned} &\int_{\mathbb{R}}(W_\psi f)(a,b)\overline{(W_\psi g)(a,b)}db \\ &= (\frac{1}{2\pi})^2|a|\int_{\mathbb{R}}\hat{F}(-b)\overline{\hat{G}(-b)}db \\ &= (\frac{1}{2\pi})^2|a|\int_{\mathbb{R}}\hat{F}(b)\overline{\hat{G}(b)}db \\ &= \frac{1}{2\pi}|a|\int_{\mathbb{R}}F(x)\overline{G(x)}dx. \end{aligned}$$

Furthermore,

$$\int_{\mathbb{R}}\int_{\mathbb{R}}(W_\psi f(a,b)\overline{W_\psi g(a,b)}\frac{dadb}{a^2}$$

$$
\begin{aligned}
&= \int_{\mathbb{R}} \frac{1}{2\pi}|a| \int_{\mathbb{R}} F(x)\overline{G(x)}dx\frac{da}{a^2} \\
&= \frac{1}{2\pi}\int_{\mathbb{R}}\int_{\mathbb{R}} \frac{|\psi(ax)|^2}{|a|}\hat{f}(x)\overline{\hat{g}(x)}dxda \\
&= \frac{1}{2\pi}\int_{\mathbb{R}}\left(\int_{\mathbb{R}} \frac{|\psi(ax)|^2}{|ax|}d(ax)\right)\hat{f}(x)\overline{\hat{g}(x)}dx \\
&= \frac{1}{2\pi}\left(\int_{\mathbb{R}} \frac{|\psi(\xi)|^2}{|\xi|}d\xi\right)\int_{\mathbb{R}}\hat{f}(x)\overline{\hat{g}(x)}dx \\
&= \frac{1}{2\pi}C_\psi\langle\hat{f},\hat{g}\rangle \\
&= C_\psi\langle f,g\rangle.
\end{aligned}
$$

This completes our proof. ∎

According to the above theorem, the inverse wavelet transform, which is also called the reconstruction formula, can be formally inferred. We consider

$$\overline{W_\psi g(a,b)} = \langle |a|^{-\frac{1}{2}}\psi(\frac{x-b}{a}),\ g\rangle.$$

If we give up the exactness in mathematics temporarily, we have

$$
\begin{aligned}
C_\psi\langle f,g\rangle &= \int_{\mathbb{R}}\int_{\mathbb{R}} W_\psi f(a,b)\langle |a|^{-\frac{1}{2}}\psi(\frac{x-b}{a}),\ g\rangle\frac{dadb}{a^2} \\
&= \langle \int_{\mathbb{R}}\int_{\mathbb{R}} W_\psi f(a,b)|a|^{-\frac{1}{2}}\psi(\frac{x-b}{a})\frac{dadb}{a^2},\ g\rangle.
\end{aligned}
$$

Because that g can be any function in $L^2(\mathbb{R})$, we can write

$$f(x) = C_\psi^{-1}\int_{\mathbb{R}}\int_{\mathbb{R}} W_\psi f(a,b)|a|^{-\frac{1}{2}}\psi(\frac{x-b}{a})\frac{dadb}{a^2}. \tag{3.8}$$

This is the formal inverse wavelet transform, by which we can reconstruct the original signal from the wavelet transform $W_\psi f(a,b)$. The above reasoning is not exact in mathematics. The exact inverse formula is studied as follows.

Theorem 3.3 *Let ψ be a basic wavelet. Then, for any $f \in L^2(\mathbb{R})$, the inverse wavelet transform Eq. (3.8) holds in the sense of $L^2(\mathbb{R})$-norm, namely:*

$$\left\| f(x) - C_\psi^{-1}\int_{|a|\geq A} da \int_{|b|\leq B} (W_\psi f)(a,b)|a|^{-\frac{1}{2}}\psi(\frac{x-b}{a})\frac{dadb}{a^2}\right\|_2 \to 0\ ,$$

as $A \to 0$, $B \to \infty$, where C_ψ is a constant defined by Eq. (3.1).

Proof For any $A, B > 0$, we prove the following equality at first:

$$\begin{aligned}
&\langle \int\int_{|a|\geq A,|b|\leq B} (W_\psi f)(a,b)|a|^{-\frac{1}{2}}\psi(\frac{x-b}{a})\frac{dadb}{a^2},\ g(x)\rangle \\
= &\int\int_{|a|\geq A,|b|\leq B} (W_\psi f)(a,b)\overline{(W_\psi g)(a,b)}\frac{dadb}{a^2}.
\end{aligned}$$

In fact,

$$\begin{aligned}
&\int_{\mathbb{R}}\int\int_{|a|\geq A,|b|\leq B} \left|(W_\psi f)(a,b)|a|^{-\frac{1}{2}}\psi(\frac{x-b}{a})g(x)\right| \frac{dadb}{a^2}dx \\
&\leq \int\int_{|a|\geq A,|b|\leq B} |(W_\psi f)(a,b)||a|^{-\frac{1}{2}}(\int_{\mathbb{R}} |\psi(\frac{x-b}{a})|^2 dx)^{\frac{1}{2}}\|g\|_2\frac{dadb}{a^2}dx \\
&= \|\psi\|_2\|g\|_2 \int\int_{|a|\geq A,|b|\leq B} |(W_\psi f)(a,b)|\frac{dadb}{a^2} \\
&\leq \|\psi\|_2\|g\|_2 \left(\int\int_{|a|\geq A,|b|\leq B} |(W_\psi f)(a,b)|^2\frac{dadb}{a^2}\right)^{1/2} \\
&\quad \left(\int\int_{|a|\geq A,|b|\leq B} \frac{dadb}{a^2}\right)^{1/2} \\
&= \|\psi\|_2\|g\|_2 C_\psi^{1/2}\|f\|_2 \left(4B\int_A^\infty \frac{da}{a^2}\right)^{1/2} \\
&= \|\psi\|_2\|g\|_2\|f\|_2 \left(\frac{4B}{A}C_\psi\right)^{1/2} < \infty,
\end{aligned}$$

where C_ψ is the constant defined by Eq. (3.1). By the Fubini theorem in real-analysis [Rudin, 1974], we have

$$\begin{aligned}
&\langle \int\int_{|a|\geq A,|b|\leq B} (W_\psi f)(a,b)|a|^{-\frac{1}{2}}\psi(\frac{x-b}{a})\frac{dadb}{a^2},\ g(x)\rangle \\
&= \int_{\mathbb{R}}\int\int_{|a|\geq A,|b|\leq B} (W_\psi f)(a,b)|a|^{-\frac{1}{2}}\psi(\frac{x-b}{a})\overline{g(x)}\frac{dadb}{a^2}dx \\
&= \int\int_{|a|\geq A,|b|\leq B} (W_\psi f)(a,b)|a|^{-\frac{1}{2}}\int_{\mathbb{R}}\psi(\frac{x-b}{a})\overline{g(x)}dx\frac{dadb}{a^2} \\
&= \int\int_{|a|\geq A,|b|\leq B} (W_\psi f)(a,b)\overline{(W_\psi g)(a,b)}\frac{dadb}{a^2}.
\end{aligned}$$

This is the equality that we intend to prove. Therefore,

$$\begin{aligned}
&\left\| f(x) - C_\psi^{-1} \int_{|a|\geq A} da \int_{|b|\leq B} (W_\psi f)(a,b)|a|^{-\frac{1}{2}}\psi(\frac{x-b}{a})\frac{dadb}{a^2} \right\|_2 \\
&= C_\psi^{-1} \sup_{\|g\|_2=1} \Big| C_\psi \langle f,\ g\rangle - \\
&\quad \langle \int\int_{|a|\geq A, |b|\leq B} (W_\psi f)(a,b)|a|^{-\frac{1}{2}}\psi(\frac{x-b}{a})\frac{dadb}{a^2},\ g(x)\rangle \Big| \\
&= C_\psi^{-1} \sup_{\|g\|_2=1} \left| C_\psi \langle f,\ g\rangle - \int\int_{|a|\geq A, |b|\leq B} (W_\psi f)(a,b)\overline{(W_\psi g)(a,b)}\frac{dadb}{a^2} \right| \\
&= C_\psi^{-1} \sup_{\|g\|_2=1} \left| \int\int_{|a|<A \ or \ |b|>B} (W_\psi f)(a,b)\overline{(W_\psi g)(a,b)}\frac{dadb}{a^2} \right| \\
&\leq C_\psi^{-1} \sup_{\|g\|_2=1} \left(\int\int_{\mathbb{R}^2} |(W_\psi g)(a,b)|^2 \frac{dadb}{a^2} \right)^{1/2} \\
&\quad \left(\int\int_{|a|<A \ or \ |b|>B} |(W_\psi f)(a,b)|^2 \frac{dadb}{a^2} \right)^{1/2} \\
&= C_\psi^{-1} \sup_{\|g\|_2=1} C_\psi^{1/2} \|g\|_2 \left(\int\int_{|a|<A \ or \ |b|>B} |(W_\psi f)(a,b)|^2 \frac{dadb}{a^2} \right)^{1/2} \\
&= C_\psi^{-1/2} \left(\int\int_{|a|<A \ or \ |b|>B} |(W_\psi f)(a,b)|^2 \frac{dadb}{a^2} \right)^{1/2} \\
&\to 0\,, \quad (A \to 0,\ B \to \infty).
\end{aligned}$$

The proof of the theorem is complete. ■

Note: The result of Theorem 3.3 is not so clear as formula Eq. (3.8). It is important to determine whether formula Eq. (3.8) is tenable.

By the result of Theorem 3.3, the following conclusion can be mathematically inferred: as $A \to 0,\ B \to \infty$,

$$f_{A,B}(x) := C_\psi^{-1} \int_{|a|\geq A} da \int_{|b|\leq B} (W_\psi f)(a,b)|a|^{-\frac{1}{2}}\psi(\frac{x-b}{a})\frac{dadb}{a^2}$$

converges to $f(x)$ in measure ([Rudin, 1974]). According to Riesz theorem ([Rudin, 1974]), there are two sequences $A_k \to 0+$ and $B_k \to +\infty$, such that $f_{A_k,B_k}(x) \to f(x)$, $(k \to \infty)$, *a.e.* $x \in \mathbb{R}$. That is

$$f(x) = \lim_{k\to\infty} C_\psi^{-1} \int_{|a|\geq A_k} da \int_{|b|\leq B_k} (W_\psi f)(a,b)|a|^{-\frac{1}{2}}\psi(\frac{x-b}{a})\frac{dadb}{a^2}.$$

If the right integral of Eq. (3.8) exists, Eq. (3.8) holds *a.e.* $x \in \mathbb{R}$. Based on this conclusion, we have

Theorem 3.4 *Let ψ be a basic wavelet. Then for any $f \in L^2(\mathbb{R})$, Eq. (3.8) holds a.e. $x \in \mathbb{R}$ if*

$$\int_{\mathbb{R}}\int_{\mathbb{R}} \left|(W_\psi f)(a,b)|a|^{-\frac{1}{2}}\psi(\frac{x-b}{a})\right| \frac{dadb}{a^2} < \infty, \tag{3.9}$$

where C_ψ is a constant, which can be defined by Eq. (3.1).

The following theorem gives a sufficient condition such that Eq. (3.9) holds.

Theorem 3.5 *Let $\psi \in L^1(\mathbb{R})$ be a basic wavelet. For any $f \in L^2(\mathbb{R})$, if there exists a non-negative measurable function $F(a)$, such that*

$$|W_\psi f(a,b)| \leq F(a), \quad \text{and} \quad \int_{\mathbb{R}} \frac{1}{|a|^{3/2}} F(a)da < \infty,$$

then, Eq. (3.9) holds. Consequently, the inverse wavelet transform Eq. (3.8) holds a.e. $x \in \mathbb{R}$. Furthermore, if $\psi(x)$ is continuous, the right part of Eq. (3.8) is a continuous function on $\mathbb{R}$.

Proof It is known that

$$\begin{aligned}
&\int_{\mathbb{R}}\int_{\mathbb{R}} \left|(W_\psi f)(a,b)|a|^{-\frac{1}{2}}\psi(\frac{x-b}{a})\right| \frac{dadb}{a^2} \\
&\leq \int_{\mathbb{R}}\int_{\mathbb{R}} F(a)\left|\psi(\frac{x-b}{a})\right| \frac{1}{|a|^{5/2}} dadb \\
&= \int_{\mathbb{R}} F(a)\frac{1}{|a|^{5/2}}\left(\int_{\mathbb{R}}\left|\psi(\frac{x-b}{a})\right| db\right) da \\
&= \|\psi\|_1 \int_{\mathbb{R}} \frac{1}{|a|^{3/2}} F(a)da \\
&< \infty.
\end{aligned}$$

Thus, Eq. (3.9) holds. If $\psi(x)$ is continuous, we denote

$$w_x(a,b) := (W_\psi f)(a, x-b)|a|^{-\frac{1}{2}}\psi(\frac{b}{a})\frac{1}{a^2}.$$

Then

$$|w_x(a,b)| \leq F(a)\left|\psi(\frac{b}{a})\right|\frac{1}{|a|^{5/2}} \in L^1(\mathbb{R}^2).$$

$\forall x_n,\ x \in \mathbb{R},\ x_n \to x\ (n \to \infty)$, by the Lebesgue dominated convergence theorem (see [Rudin, 1974]), we have:

$$\lim_{n\to\infty}\int_{\mathbb{R}}\int_{\mathbb{R}} w_{x_n}(a,b)dadb = \int_{\mathbb{R}}\int_{\mathbb{R}} w_x(a,b)dadb.$$

Note that

$$\begin{aligned}
&\lim_{n\to\infty}\int_{\mathbb{R}}\int_{\mathbb{R}}(W_\psi f)(a,b)|a|^{-\frac{1}{2}}\psi(\frac{x_n-b}{a})\frac{dadb}{a^2}\\
=\ &\lim_{n\to\infty}\int_{\mathbb{R}}\int_{\mathbb{R}} w_{x_n}(a,b)dadb\\
=\ &\int_{\mathbb{R}}\int_{\mathbb{R}} w_x(a,b)dadb,
\end{aligned}$$

therefore, the right part of Eq. (3.8), $\int_{\mathbb{R}}\int_{\mathbb{R}} w_x(a,b)dadb$, is a continuous function on $\mathbb{R}$. Our proof is complete. ■

Note: Let ψ be a basic wavelet, $\alpha > 0$ and $f \in L^2(\mathbb{R})$. If there exists constants $C,\ \delta > 0$, such that

$$|W_\psi f(a,b)| \leq C|a|^{\alpha+\frac{1}{2}}, \qquad (\forall b \in \mathbb{R}),$$

for $|a| < \delta$, the conditions of the previous theorem are satisfied. In fact, it will be clear if we let

$$F(a) := \begin{cases} C|a|^{\alpha+\frac{1}{2}}, & |a| < \delta; \\ \|f\|_2\|\psi\|_2, & |a| \geq \delta \end{cases}$$

The necessity of the admissibility condition Eq. (3.1) is very important. It ensures the existence of the inverse wavelet transform. As a conclusion, it is feasible to apply the wavelet transform to signal analysis, image processing and pattern recognition. It is easy to see that this condition is rather weak and is almost equivalent to $\hat{\psi}(0) = 0$, or

$$\int_{\mathbb{R}} \psi(x)dx = 0. \tag{3.10}$$

Actually, if $\psi(x) \in L^2(\mathbb{R})$ and there exists $\alpha > 0$, such that $(1 + |x|)^\alpha\psi(x) \in L^1(\mathbb{R})$, then, Eq. (3.1) must be equivalent to Eq. (3.10). This conclusion is proved as below:

By $(1+|x|)^\alpha \psi(x) \in L^1(\mathbb{R})$, we know that $\psi(x) \in L^1(\mathbb{R})$. It means that $\hat{\psi}(\xi)$ is a continuous function in $\mathbb{R}$. If Eq. (3.1) is tenable, then $\hat{\psi}(0) = 0$. It indicates that Eq. (3.10) holds. Contrarily, if Eq. (3.10) holds, we assume $\alpha \leq 1$ without losing generality, then

$$\begin{aligned}
|\hat{\psi}(\xi)| &= \left| \int_{\mathbb{R}} \psi(x) e^{-i\xi x} dx \right| \\
&= \left| \int_{\mathbb{R}} \psi(x) [e^{-i\xi x} - 1] dx \right| \\
&\leq \int_{\mathbb{R}} \left| \psi(x) 2 \sin \frac{\xi x}{2} \right| dx \\
&\leq |\xi|^\alpha \int_{|\xi x| \leq 1} |x^\alpha \psi(x)| dx + |\xi|^\alpha \int_{|\xi x| > 1} |x^\alpha \psi(x)| dx \\
&= |\xi|^\alpha \int_{\mathbb{R}} |x^\alpha \psi(x)| dx \\
&= C|\xi|^\alpha,
\end{aligned}$$

where $C := \int_{\mathbb{R}} |x^\alpha \psi(x)| dx$. Therefore,

$$C_\psi := \int_{\mathbb{R}} \frac{|\hat{\psi}(\xi)|^2}{|\xi|} d\xi \leq C^2 \int_{|\xi \leq 1} |\xi|^{2\alpha - 1} d\xi + \int_{|\xi| > 1} |\hat{\psi}(\xi)|^2 d\xi < \infty.$$

It specifies that Eq. (3.1) holds. This finished our proof. ■

In general, the wavelets applied to practice have good damping and satisfy the condition $(1+|x|)^\alpha \psi(x) \in L^1(\mathbb{R})$ $(\alpha > 0)$. As a result, in engineering, the wavelet is often defined as the function in $L^2(\mathbb{R})$ which satisfies the condition Eq. (3.10). This definition is not exact in mathematics, however, it is harmless in the application of wavelet. The condition Eq. (3.10) is a objective specification of the vibration of ψ. The damping and the vibration are two basic characteristics of basic wavelets.

In signal analysis, only the positive frequency is need to be considered. That is $a > 0$. With this premise, we have another theorem.

Theorem 3.6 *Let ψ be a basic wavelet satisfying*

$$\int_0^\infty \frac{|\hat{\psi}(\xi)|^2}{\xi} d\xi = \int_0^\infty \frac{|\hat{\psi}(-\xi)|^2}{\xi} d\xi = \frac{1}{2} C_\psi < \infty.$$

Then, for any $f,\ g \in L^2(\mathbb{R})$, we have

$$\int_0^\infty \int_{\mathbb{R}} W_\psi f(a,b)\overline{W_\psi g(a,b)}\frac{dadb}{a^2} = \frac{1}{2}C_\psi\langle f,\ g\rangle.$$

In particular,

$$\int_0^\infty \int_{\mathbb{R}} |W_\psi f(a,b)|^2\frac{dadb}{a^2} = \frac{1}{2}C_\psi\|f\|_2^2.$$

The proof of this theorem is similar to those of Theorem 3.2 and Theorem 3.3.

3.1.2 *The Continuous Wavelet Transform as a Filter*

Let ψ be a basic wavelet and we denote

$$\tilde{\psi}(x) := \overline{\psi(-x)}. \tag{3.11}$$

We define scale wavelet transform as follows:

$$W_s f(x) := W_s^{\tilde{\psi}} f(x) := (f * \tilde{\psi})(x). \tag{3.12}$$

By the definition of the wavelet transform, for $a > 0$, we have

$$\begin{aligned}(W_\psi f)(a,b) \quad &:= \quad \int_R f(t)a^{-\frac{1}{2}}\overline{\psi(\frac{t-b}{a})}dt \\ &= \quad \int_R f(t)a^{-\frac{1}{2}}\tilde{\psi}(\frac{b-t}{a})dt \\ &= \quad a^{\frac{1}{2}}(f * \tilde{\psi}_a)(b),\end{aligned}$$

where $\tilde{\psi}_a(x) := \frac{1}{a}\tilde{\psi}(\frac{x}{a})$. It specifies that the wavelet transform is actually a convolution, which is also called a filter in engineering.

The design of filters is very important in the filter theory. There are two kinds of digital filters, namely, the finite impulse response (FIR) and the infinite impulse response (IIR). The former is easy to be realized and has good time localization. Therefore, in the above scale wavelet transforms, the basic wavelet ψ, which is the kernel of the transform, should be compactly supported. Although Heisenberg uncertainty principle has specified that the area of the time-frequency domain can not be arbitrarily small, the localization in the time-frequency domain still can be perfect, if ψ and its Fourier transform $\hat{\psi}$ have compact support simultaneously. Unfortunately, from the following theorem, we can find that this condition cannot be satisfied.

Theorem 3.7 *If $f \in L^2(\mathbb{R})$ is a non-zero function, f and its Fourier transform $\hat{f}$ cannot be compactly supported simultaneously.*

Proof If $\hat{f}$ is compactly supported, supp $\hat{f} \subset [-B, B]$, then

$$f(z) := \frac{1}{2\pi} \int_{-B}^{B} \hat{f}(\xi) e^{i\xi z} d\xi$$

is an analytic function on complex plane $\mathbb{C}$. Because of the zero-isolation of non-zero analytic functions, f cannot be compactly supported. This finishes our proof. ∎

We have known that the Fourier transform $\hat{\psi}$ of a compactly supported basic wavelet ψ cannot be compactly supported. Now, we wish its damping property would be good enough. It is essentially equivalent to that the smoothness of ψ would be good enough. The following theorem ensures this fact.

Theorem 3.8 *Let m be a non-negative integer and $H^m(\mathbb{R})$ be the Sobolev space of order m defined by*

$$H^m(\mathbb{R}) := \{f \mid f, f', \cdots, f^{(m)} \in L^2(\mathbb{R})\}.$$

Then, $f \in H^m(\mathbb{R})$ if and only if

$$\int_{\mathbb{R}} (1 + |x|^m) |\hat{f}(x)| dx < \infty. \tag{3.13}$$

Proof This is a fundamental fact in the theory of Sobolev spaces. The detailed proof can be found in [Gilbarg and Trudinger, 1977]. ∎

According to the definition of $H^m(\mathbb{R})$, $f \in H^m(\mathbb{R})$ indicates that f has certainly differentiableness and smoothness (or regularity). However, Eq. (3.13) indicates that $\hat{f}$ has damping of order m. Therefore, when we choose a basic wavelet ψ as a filter function, in order to ensure that ψ is good for localized analysis in the time-frequency, we must consider both its damping (or compact support) and its regularity.

Another very important property of filters is the linear phase. With the linear phase or the generalized linear phase, a filter can avoid distortion.

The linear phase is mathematically defined as follows.

Definition 3.2 $f \in L^2(\mathbb{R})$ is said to have liner phase if its Fourier transform satisfies

$$\hat{f}(\xi) = \epsilon|\hat{f}(\xi)|e^{-ia\xi},$$

where $\epsilon = 1$ or -1 and a is a real constant. f is said to have generalized linear phase if there exists a real function $F(\xi)$ and two real constants a and b, such that

$$\hat{f}(\xi) = F(\xi)e^{-i(a\xi+b)}.$$

Obviously, if f has linear phase, it also has generalized linear phase. The generalized linear phase will degenerate to the linear phase if and only if $e^{-ib} = \pm 1$ and the real function $F(\xi)$ keeps its sign, i.e., identically positive or identically negative.

According to the above mathematical definition, the linear phase or the generalized linear phase of ψ is an attribute in Fourier transform domain. In the following, we will demonstrate that the generalized linear phase is equivalent to a symmetry of ψ itself.

Theorem 3.9 *$f \in L^2(\mathbb{R})$ has generalized phase if and only if f is skew-symmetric at $a \in \mathbb{R}$, i.e., there exists a constant $b \in \mathbb{R}$ such that*

$$e^{ib}f(a+x) = \overline{e^{ib}f(a-x)}, \qquad x \in \mathbb{R}.$$

In particular, for a real function f, it has generalized linear phase if and only if it is symmetric or antisymmetric at $a \in \mathbb{R}$. More precisely,

$$f(a+x) = f(a-x), \qquad x \in \mathbb{R},$$

or

$$f(a+x) = -f(a-x), \qquad x \in \mathbb{R}.$$

Proof It is easy to see that f has generalized phase, if and only if two constants a and $b \in \mathbb{R}$ exist, such that

$$\hat{f}(\xi)e^{i(a\xi+b)} = F(\xi)$$

is a real function. That is

$$\hat{f}(\xi)e^{i(a\xi+b)} = \overline{\hat{f}(\xi)e^{i(a\xi+b)}}, \qquad (\xi \in \mathbb{R}),$$

which is equivalent to

$$\int_{\mathbb{R}} \hat{f}(\xi)e^{i(a\xi+b)}e^{i\xi x}d\xi = \overline{\int_{\mathbb{R}} \hat{f}(\xi)e^{i(a\xi+b)}e^{-i\xi x}d\xi}, \qquad (x \in \mathbb{R}).$$

i.e.,

$$e^{ib}f(a+x) = \overline{e^{ib}f(a-x)}, \qquad (x \in \mathbb{R}).$$

If f is a real function, e^{i2b} must be a real function. It means $e^{i2b} = 1$ or -1. Thus

$$f(a+x) = f(a-x), \qquad x \in \mathbb{R},$$

or

$$f(a+x) = -f(a-x), \qquad x \in \mathbb{R}.$$

It specifies that f is symmetric or antisymmetric on a. Our proof is complete. ■

Summarily, as a filter, in order to be good for the localized analysis in the time domain, the basic wavelet ψ should have good damping or have compact support. On the other hand, ψ should have good regularity or smoothness, to obtain a good property for the localized analysis in the frequency domain. At last, ψ should be skew-symmetric to avoid the distortion.

3.1.3 *Description of Regularity of Signal by Wavelet*

In mathematics, a signal is actually a function. A stationary signal always corresponds to a smooth function, while a transient one refers to a singularity. In fact, the concepts of the stability and singularity of the signal are ambiguous without using the tool of mathematics. In this section, we will introduce an accurate description of the regularity of the signal mathematically employing the exponent of Lipschitz.

Definition 3.3 Let α satisfy $0 \le \alpha \le 1$. A function f is called uniformly Lipschitz α over the interval (a, b), if there exists a positive constant K, such that

$$|f(x_1) - f(x_2)| \le K|x_1 - x_2|^{\alpha}, \quad \forall x_1, x_2 \in (a, b),$$

and we denote $f \in C^{\alpha}(a, b)$. The constant α is called Lipschitz exponent.

Function f to be uniformly Lipschitz α over (a, b) is also equivalent to the following definition:

Definition 3.4 Let α satisfy $0 \le \alpha \le 1$. A function f is called uniformly Lipschitz α over the interval (a, b) if

$$K := \sup_{x_1, x_2 \in (a,b), x_1 \neq x_2} \frac{|f(x_1) - f(x_2)|}{|x_1 - x_2|} < \infty. \tag{3.14}$$

It shows that the variation of f depends on $|x_1 - x_2|^\alpha$ over (a, b). Function f has neither the "fracture" points over (a, b), nor sharp transient points. The shape of the f depends on α, and the sharpness of the shape can be decided by the constant K in Eq. (3.14).

In the above definition, the Lipschitz α is confined within $\alpha \le 1$. Otherwise, when $\alpha > 1$, the above definition looses its meaning, because in this case, the function will be too smooth. Since the definition of the Lipschitz regularity with $\alpha \le 1$ does not deal with the differentiation, we can extend the definition of the Lipschitz exponent to $\alpha > 1$.

Definition 3.5 Let $\alpha > 1$, and n be the largest number which is less than α. A function f is uniformly Lipschitz α over (a, b), if f is nth differentiable and $f^{(n)} \in C^{\alpha-n}(a, b)$. It is symbolized by $f \in C^\alpha(a, b)$.

The concept of uniformly Lipschitz α can be extended to $\alpha < 0$, and the definition can be modified as:

Definition 3.6 Let α be $-1 \le \alpha < 0$. If the primitive function of f

$$F(x) := \int_a^x f(t)dt$$

is uniformly Lipschitz $(\alpha + 1)$ over (a, b), f is said to be uniformly Lipschitz α over(a, b), and denoted by $f \in C^\alpha(a, b)$.

In the following, we will prove an important property, that is the Lipschitz regularity closely correlates with the decay property of wavelet transform in scale.

Theorem 3.10 *Let $\alpha > 0$, and n be the largest integer which is less than α, $\psi \in L^2(\mathbb{R})$ satisfy $(1 + |x|)^\alpha \psi(x) \in L^1(\mathbb{R})$ and have vanishing moments of order n, i.e.,*

$$\int_{\mathbb{R}} t^k \psi(t)dt = 0, \quad (k = 0, 1, \cdots, n).$$

Then, there exists a constant $C > 0$, such that $\forall f \in L^2(\mathbb{R}) \cap C^\alpha(\mathbb{R})$, the following holds

$$|(W_\psi f(a,b)| \leq C|a|^{\alpha+\frac{1}{2}}, \quad (\forall a, b \in \mathbb{R}).$$

Proof $\forall a, b \in \mathbb{R}$, we have

$$\begin{aligned}
|(W_\psi f(a,b)| &= \left|\int_{\mathbb{R}} f(t)\frac{1}{|a|^{1/2}}\psi(\frac{t-x}{a})dt\right| \\
&= \left|\int_{\mathbb{R}} \left[f(t) - f(x) - f'(x)(t-x) - \cdots - \frac{f^{(n)}(x)}{n!}(t-x)^n\right]\right. \\
&\quad \left.\cdot \frac{1}{|a|^{1/2}}\psi(\frac{t-x}{a})dt\right| \\
&= \left|\int_{\mathbb{R}} \left[\frac{f^{(n)}(\xi)}{n!}(t-x)^n - \frac{f^{(n)}(x)}{n!}(t-x)^n\right]\right. \\
&\quad \left.\cdot \frac{1}{|a|^{1/2}}\psi(\frac{t-x}{a})dt\right| \quad (\xi \text{ is between } t \text{ and } x) \\
&\leq C\int_{\mathbb{R}} \frac{|t-x|^{\alpha-n}}{n!}|t-x|^n \frac{1}{|a|^{1/2}}\left|\psi(\frac{t-x}{a})\right| dt \\
&= \frac{C}{n!}|a|^{\alpha+\frac{1}{2}}\int_{\mathbb{R}} |t|^\alpha|\psi(t)|dt \\
&\leq C|a|^{\alpha+\frac{1}{2}}.
\end{aligned}$$

where C denotes a positive constant. The proof is complete. ∎

Theorem 3.11 *Let $0 < \alpha < 1$, and $\psi \in L^2(\mathbb{R})$ have vanishing moment of order 0 and satisfy:*

$$(1+|x|)^\alpha\psi(x) \in L^1(\mathbb{R}), \quad |\psi'(x)| = O(\frac{1}{1+|x|^\sigma}), \ (\exists\sigma > 1).$$

Then,

$$f \in C^\alpha(\mathbb{R}) \Longleftrightarrow \exists C > 0 : \ |W_\psi f(a,b)| \leq C|a|^{\alpha+\frac{1}{2}}, \quad (\forall a, b \in \mathbb{R}).$$

Proof Here we give only the proof of the sufficiency part. According to the note in Theorem 3.5, we have $f(x) \in C(\mathbb{R})$, and the following inverse wavelet transform holds:

$$f(x) = C_\psi^{-1}\int_{\mathbb{R}}\int_{\mathbb{R}}(W_\psi f)(a,b)\psi_{a,b}(x)\frac{dadb}{a^2}, \quad (\forall x \in \mathbb{R}).$$

Therefore

$$
\begin{aligned}
&C_\psi|f(x)-f(y)| \\
&\le \int_{\mathbb{R}}\int_{\mathbb{R}} |(W_\psi f)(a,b)| \frac{1}{|a|^{1/2}} \left|\psi(\frac{x-b}{a}) - \psi(\frac{y-b}{a})\right| \frac{dadb}{a^2} \\
&\le C \int_{|a|\le\delta}\int_{\mathbb{R}} |a|^{\alpha+\frac{1}{2}} \frac{1}{|a|^{1/2}} \left(\left|\psi(\frac{x-b}{a}\right| + \left|\psi(\frac{y-b}{a})\right| \right) \frac{dadb}{a^2} \\
&+C \int_{|a|>\delta}\int_{\mathbb{R}} |a|^{\alpha+\frac{1}{2}} \frac{1}{|a|^{1/2}} |\psi'(\xi)| \frac{|x-y|}{|a|} \frac{dadb}{a^2} \\
&\qquad\qquad\qquad (\text{where } \xi \text{ is between } \frac{x-b}{a} \text{ and } \frac{y-b}{a}) \\
&\le C \int_{|a|\le\delta} |a|^{\alpha-1} da \|\psi\|_1 + C \left(\int_{|a|>\delta}\int_{\mathbb{R}} |a|^{\alpha 3} \frac{1}{1+|\xi|^\sigma} dadb \right) |x-y|
\end{aligned}
$$

Since ξ is between $\frac{x-b}{a}$ and $\frac{y-b}{a}$, we have that (please refer to the note followed by this proof)

$$
\left|\frac{x-b}{a}\right| \le |\xi| + \left|\frac{x-b}{a} - \frac{y-b}{a}\right| = |\xi| + \left|\frac{x-y}{a}\right|.
$$

In particular, setting $\delta = |x-y|$, for $|a| > \delta$, we can deduce that

$$
\left|\frac{x-b}{a}\right| \le |\xi| + \frac{\delta}{|a|} \le |\xi| + 1.
$$

Therefore, we have

$$
1 + \left|\frac{x-b}{a}\right|^\sigma \le 1 + (|\xi|+1)^\sigma \le (1+2^\sigma)(|\xi|^\sigma + 1).
$$

Subsequently, we arrive at

$$
\frac{1}{1+|\xi|^\sigma} \le (1+2^\sigma) \frac{1}{1+\left|\frac{x-b}{a}\right|^\sigma},
$$

and hence, for $\delta = |x-y|$,

$$
\begin{aligned}
&C_\psi|f(x)-f(y)| \\
\le\ & C \int_{|a|\le\delta} |a|^{\alpha-1} da \|\psi\|_1 + C|x-y| \\
&\left(\int_{|a|>\delta}\int_{\mathbb{R}} |a|^{\alpha-3} (1+2^\sigma) \frac{1}{1+\left|\frac{x-b}{a}\right|^\sigma} dadb \right)
\end{aligned}
$$

$$
\begin{aligned}
&= C\frac{1}{\alpha}\delta^\alpha + C|x-y|\int_{|a|>\delta}|a|^{\alpha-2}da\int_{\mathbb{R}}\frac{db}{1+|b|^\sigma} \\
&\le C\frac{1}{\alpha}\delta^\alpha + C|x-y|\frac{1}{1-\alpha}\delta^{\alpha-1} \\
&\le C\left(\frac{1}{\alpha}+\frac{1}{1-\alpha}\right)|x-y|^\alpha.
\end{aligned}
$$

That means $f \in C^\alpha(\mathbb{R})$. The proof is complete. ■

Note: In the proof, we have used the following obvious results, namely, if ξ, a and b are three real numbers, ξ is between a and b, then $|a| \le |\xi|+|a-b|$. In fact, (1) If a, b have the different sign, then $|a| \le |a-b| \le |\xi| + |a-b|$; (2) If a, b have same sign, then $|a| \le |\xi| \le |b|$ or $|b| \le |\xi| \le |a|$. In the first case, $|a| \le |\xi| \le |\xi| + |a-b|$ already holds; In the second case, we have $|a| \le |b| + |a-b| \le |\xi| + |a-b|$. These are the results we wanted.

The above theorem can be extended to the case of $\alpha > 1$ as follows. We omit the proof here.

Theorem 3.12 *Let n be a no-negative integer and $\alpha \in \mathbb{R}$ satisfy $n < \alpha < n+1$. Suppose $\psi \in L^2(\mathbb{R})$ has vanishing moments of order n and satisfies:*

$$(1+|x|)^\alpha\psi(x),\ \ \psi, \psi', \cdots, \psi^{(n)} \in L^1(\mathbb{R}),$$

$$|\psi^{(n+1)}(x)| = O(\frac{1}{1+|x|^\sigma}),\quad (\exists\sigma > 1).$$

Then, $\forall f \in L^1(\mathbb{R})$, the following holds

$$f \in C^\alpha(\mathbb{R}) \Longleftrightarrow \exists C > 0 :\ |W_\psi f(a,b)| \le C|a|^{\alpha+\frac{1}{2}},\quad (\forall a, b \in \mathbb{R}).$$

3.1.4 *Some Examples of Basic Wavelets*

There are a great number of basic wavelets. Generally speaking, the derivative of a compact support function, which is continuous differentiable, is a basic wavelet. Several examples of basic wavelets and their basic properties are given in this section.

- **Gaussian Wavelets:**

$$\psi_\alpha(x) = -\frac{x}{4\alpha\sqrt{\pi\alpha}}e^{-\frac{x^2}{4\alpha}}.$$

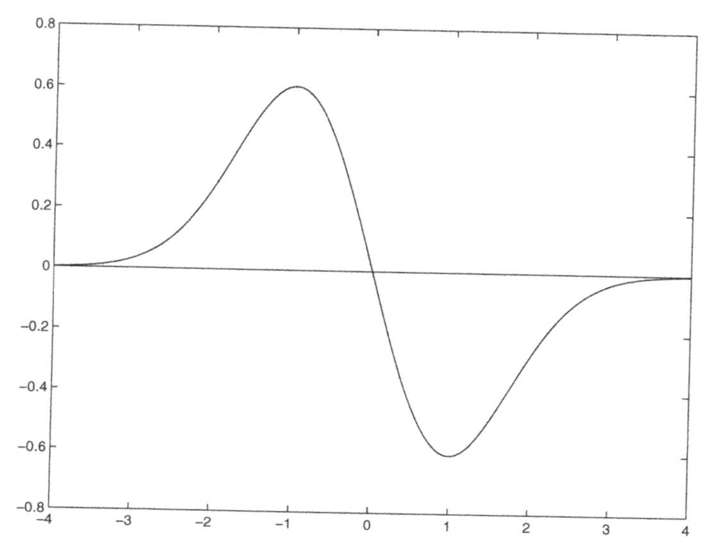

Fig. 3.3 Gaussian wavelet $f(x) = xe^{-\frac{x^2}{2}}$ and its Fourier transform.

It is the derivative of Gaussian function

$$g_\alpha(x) = \frac{1}{2\sqrt{\pi\alpha}} e^{-\frac{x^2}{4\alpha}},$$

i.e., $\psi_\alpha(x) = g'_\alpha(x)$. Its Fourier transform is

$$\hat{\psi}_\alpha(\xi) = i\xi e^{-\alpha\xi^2}.$$

The time-frequency window of ψ_α is

$$[-\sqrt{3\alpha},\ \sqrt{3\alpha}] \times \left[-\sqrt{\frac{3}{4\alpha}},\ \sqrt{\frac{3}{4\alpha}}\right].$$

Figure 3.3 graphically shows Gaussian wavelet with $\alpha = \frac{1}{2}$, and its Fourier transform. Both of them have the same shape. The difference between $\psi_\alpha(x)$ and its Fourier transform is only a constant.

- **Mexico hat-like wavelet**: When $\alpha = \frac{1}{2}$, the second derivative of Gaussian function $g_{\frac{1}{2}}$ is a wavelet, which is referred to as Mexico hat-like wavelet:

$$\psi_M(x) := -g''_{\frac{1}{2}}(x) = \frac{1}{\sqrt{2\pi}}(1-x^2)e^{-\frac{1}{2}x^2}.$$

Its Fourier transform is

$$\hat{\tilde{\psi}}_M(\xi) = \xi^2 e^{-\frac{1}{2}\xi^2}.$$

The time-frequency window is

$$\left[-\frac{\sqrt{42}}{6}, \frac{\sqrt{42}}{6}\right] \times \left[-\sqrt{\frac{8}{3\sqrt{\pi}}}, \sqrt{\frac{8}{3\sqrt{\pi}}}\right].$$

Mexico hat-like wavelet and its Fourier transform are shown in Figure 3.4.

- **Spline wavelet**: For $m \geq 1$, the B-spline of order m is defined by

$$N_m(x) = (\underbrace{N_1 * \cdots N_1}_{m})(x) = (N_{m-1} * N_1)(x) = \int_0^1 N_{m-1}(x-t)dt, \tag{3.15}$$

where $N_1(x)$ is a characteristic function of $[0,1)$, and can be written as

$$N_1(x) := \begin{cases} 1 & x \in [0,1); \\ 0 & \text{otherwise.} \end{cases}$$

Its Fourier transform is

$$\hat{N}_1(\xi) = \frac{2}{\xi} e^{-i\xi/2} \sin\frac{\xi}{2}.$$

We can easily prove that

$$N_m(x) = \frac{1}{(m-1)!} \sum_{k=0}^{m} (-1)^k \begin{pmatrix} m \\ k \end{pmatrix} (x-k)_+^{m-1} .$$

$N_m(x)$ has compact support $[0, m]$. For $m \geq 2$, the derivative of $N_m(x)$ is a basic wavelet, which is called the spline wavelet of order $m-1$ and can be represented as

$$\psi_{m-1}(x) := N'_m(x) = \frac{1}{(m-1)!} \sum_{k=0}^{m} (-1)^k \begin{pmatrix} m \\ k \end{pmatrix} (x-k)_+^{m-2} .$$

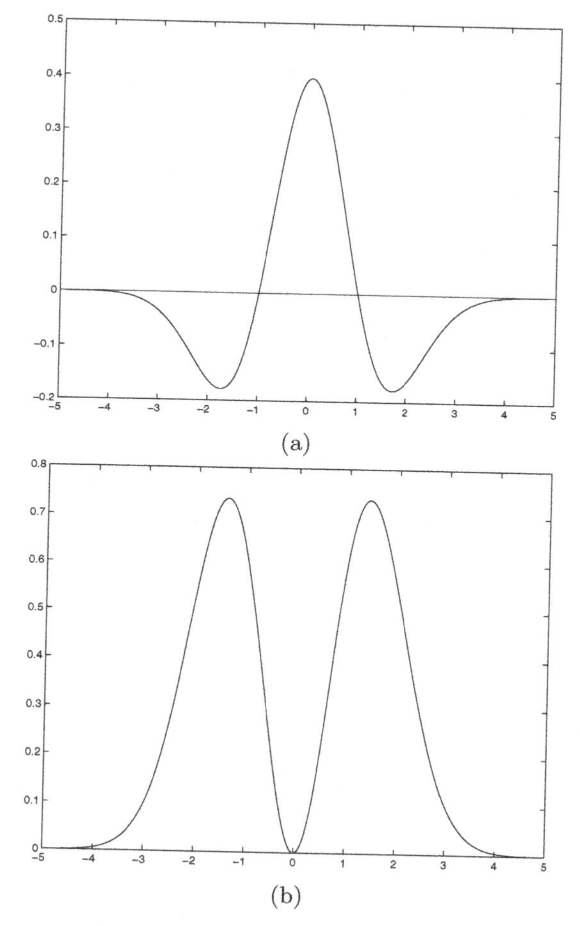

Fig. 3.4 (a): Mexico-hat wavelet ψ_M, and (b): its Fourier transform $\hat{\psi}_M$.

Its Fourier transform is

$$\begin{aligned}\hat{\psi}_{m-1}(\xi) &= (N_m')\hat{}(\xi) = i\xi\hat{N}_m(\xi) = i\xi(\hat{N}_1(\xi))^m \\ &= i\xi\left(\frac{2}{\xi}e^{-i\xi/2}\sin\frac{\xi}{2}\right)^m.\end{aligned}$$

In particular, we consider the following two cases, namely, $m = 2$ and $m = 4$:

(1). When $m = 2$, the first spline wavelet is

$$\psi_1(x) := \begin{cases} 1 & x \in (0,1) \\ -1 & x \in (1,2) \\ 0 & x \in \mathbb{R}\backslash[0,1] \end{cases}.$$

By taking dilation to the 1-D spline wavelet, Haar wavelet $h(x) := \psi_1(2x)$ can be achieved:

$$h(x) := \begin{cases} 1 & x \in (0,\frac{1}{2}) \\ -1 & x \in (\frac{1}{2},1) \\ 0 & x \in \mathbb{R}\backslash[0,1] \end{cases}.$$

Its Fourier transform is

$$\hat{h}(\xi) = \frac{1}{2}\hat{\psi}_1(\frac{\xi}{2}) = \frac{1}{4}i\xi e^{-i\xi/2}(\frac{4}{\xi}\sin\frac{\xi}{4})^2.$$

The time-frequency window is

$$[\frac{1}{2} - \frac{1}{2\sqrt{3}},\ \frac{1}{2} + \frac{1}{2\sqrt{3}}] \times (-\infty,\ \infty).$$

It is clearly that the width of its frequency window is infinite, Haar wavelet is good for the localized analysis in the time domain but in the frequency domain. Haar wavelet and its Fourier transform are graphically displayed in Figure 3.5.

(2). When $m = 4$, the quadratic spline wavelet is

$$\psi_3(x) := \frac{1}{6}\sum_{k=0}^{4}(-1)^k \left(\begin{array}{c} 4 \\ k \end{array}\right)(x-k)_+^2.$$

$\tilde{\psi}_3(x) := 4\psi_3(2x+2)$, the quadratic spline wavelet, is applied widely in signal processing. It is an odd function and has compact support $[-1,1]$. In $[0,\infty)$, it can be represented as

$$\tilde{\psi}_3(x) := \begin{cases} 8(3x^2-2x) & x \in [0,\frac{1}{2}) \\ -8(x-1)^2 & x \in [\frac{1}{2},1) \\ 0 & x \geq 1 \end{cases}.$$

Its Fourier transform can be written as

$$\hat{\tilde{\psi}}_3(\xi) = i\xi\left(\frac{4}{\xi}\sin\frac{\xi}{4}\right)^4.$$

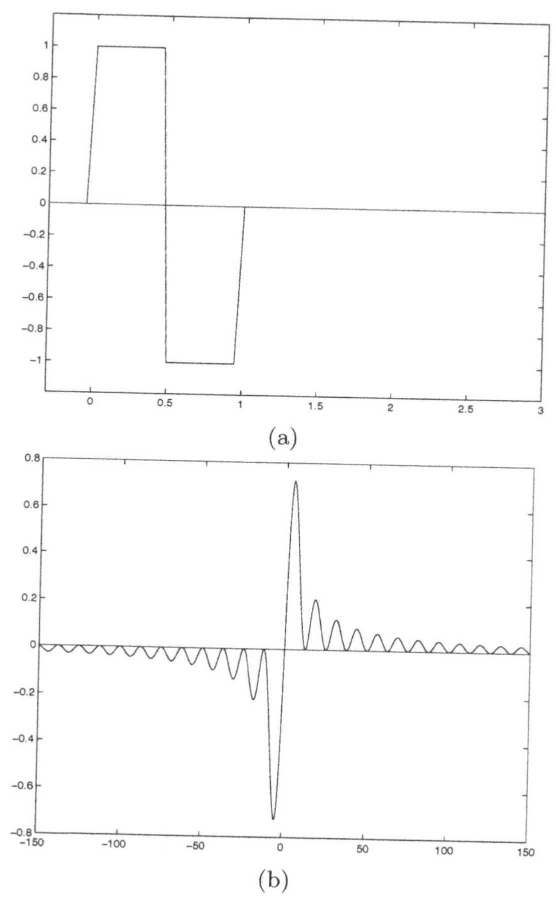

Fig. 3.5 (a): Haar wavelet $h(x)$ and (b): its Fourier transform $-ie^{i\xi/2}\hat{h}(\xi) = \frac{4}{\xi}(\sin\frac{\xi}{4})^2$.

The time-frequency window is

$$[-\frac{1}{\sqrt{7}}, \frac{1}{\sqrt{7}}] \times [-4, 4].$$

Quadratic spline wavelet wavelet and its Fourier transform are graphically displayed in Figure 3.6. Note: The width 4 of the frequency window was approximately computed with Matlab system, however, a little error maybe exists.

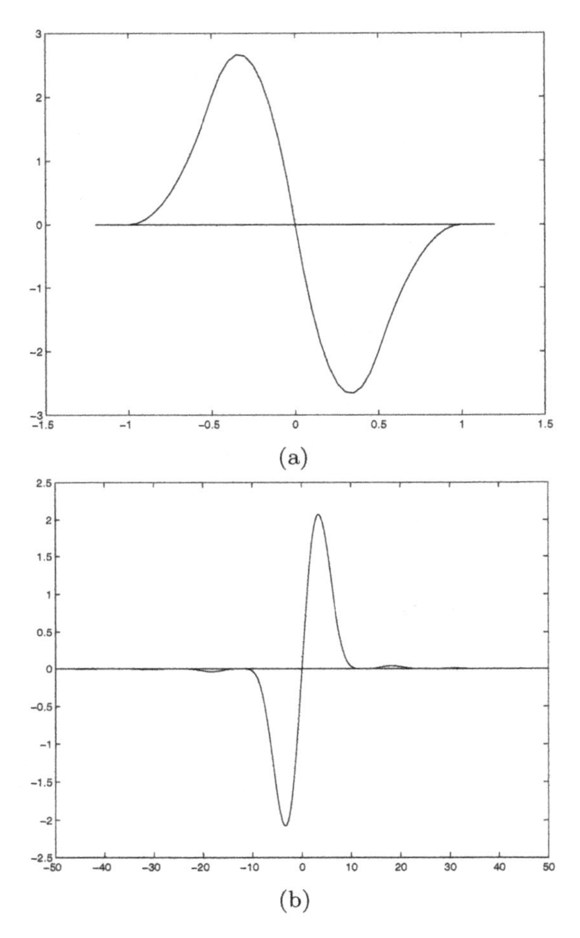

Fig. 3.6 (a): Quadratic Spline wavelet $\tilde{\psi}_3$ and (b): its Fourier transform $-i\hat{\tilde{\psi}}_3$.

3.2 Multiresolution Analysis (MRA) and Wavelet Bases

3.2.1 *Multiresolution Analysis*

3.2.1.1 *Basic Concept of MRA*

It was very difficult to construct a wavelet basis of $L^2(\mathbb{R})$ early on in the history of the wavelet development. From the point of view of function analysis, it is also a non-trival task to find such a function ψ with good regularity and localization to ensure that $\{\psi_{j,k}(x) := 2^{j/2}\psi(2^jx - k)\}_{j\in\mathbb{Z},\, k\in\mathbb{Z}}$ to be an orthonormal basis of $L^2(\mathbb{R})$. For a long time, people doubted the existence of this function. Fortunately, in 1980s, such functions were found, and a standard scheme to construct wavelet bases was set up as well. It has been proved theoretically that almost all useful wavelet bases can be constructed using the standard scheme, namely, Multiresolution Analysis (MRA). MRA was first published in 1989 by Mallat and Meyer [Mallat, 1989b; Meyer, 1990]. Since that time, the multiresolution analysis has become a very important tool in signal processing, image processing, pattern recognition, and other related fields. In this section, we will introduce the intuitive meaning of wavelet and multiresolution analysis. Our purpose is to find a function ψ to ensure that $\{\psi_{j,k}(x) := 2^{j/2}\psi(2^jx - k)\}_{j\in\mathbb{Z},\, k\in\mathbb{Z}}$ is an orthonormal bases of $L^2(\mathbb{R})$. Two variables are embedded in them, namely the translation factor $k \in \mathbb{Z}$, and the dilation factor $j \in \mathbb{Z}$. For $\{\psi_{j,k} | k \in \mathbb{Z}\}$, if j is fixed, there will be a fixed bandwidth:

$$|(\psi_{j,k})\hat{}(\xi)| = 2^{-j/2}|\hat{\psi}(\frac{\xi}{2^j})|.$$

Thus, the decomposition of $L^2(\mathbb{R})$ in frequency is as follows

$$W_j := \overline{span\{\psi_{j,k} |\ k \in \mathbb{Z}\}},$$

where the $\{\psi_{j,k}\}_{k\in\mathbb{Z}}$ denotes the orthomormal bases of W_j. We can write

$$L^2(\mathbb{R}) = \cdots \oplus W_{j-1} \oplus W_j \oplus W_{j+1} \oplus \cdots$$

in which, from left to right, the frequency of W_j changes from low to high. We denote that

$$V_j = \cdots \oplus W_{j-1} \oplus W_j, \qquad (j \in \mathbb{Z}),$$

where V_j refers to the function set with lower frequency, and satisfies

- $\cdots \subset V_{-1} \subset V_0 \subset V_1 \subset \cdots$
- $\bigcap_{j\in\mathbb{Z}} V_j = \{0\}, \quad \overline{\bigcup_{j\in\mathbb{Z}} V_j} = L^2(\mathbb{R})$
- $f(x) \in V_j \iff f(2x) \in V_{j+1}$

We can also write $W_j = V_{j+1} \ominus V_j$. Therefore $\{W_j\}_{j\in\mathbb{Z}}$ can also be represented by $\{V_j\}_{j\in\mathbb{Z}}$. According to the above definitions, $L^2(\mathbb{R})$ can be divided into the conjoint "concentric annulus" or the expanding "concentric balls" $\{W_j\}_{j\in\mathbb{Z}}$ by $\{V_j\}_{j\in\mathbb{Z}}$. Now, we consider such a question, namely, is there a function ϕ with low frequency such that $\{\phi(\cdot - k)\}_{k\in\mathbb{Z}}$ is an othonormal or a Riesz basis of V_0? The answer is affirmative. In this way, we can find ψ in from

$$\psi \in V_1 \ominus V_0.$$

By $\psi \in V_0 \subset V_1$, we obtain the two-scale equation as follows:

$$\phi(x) = 2\sum_{k\in\mathbb{Z}} h_k \phi(2x - k), \qquad \exists\{h_k\} \in l^2(\mathbb{Z}).$$

This is a famous framework referred to as the multiresolution analysis(MRA), by which wavelet can be constructed. In the following, we will show how to construct wavelet bases and the biorthonormal bases by using MRA. We will give the formal mathematical definitions of MRA.

Definition 3.7 Let H be a Hilbert space. Then a sequence in H, $\{e_j\}_{j=1}^{\infty}$ is said to be a Riesz basis of H, if the following conditions are satisfied:

(1) $\overline{span}\{e_j\}_{j=1}^{\infty} = H$, i. e., $\forall x \in H$ and $\forall \varepsilon > 0$, there exists $\sum_{j=1}^{n} c_j e_j$, such that $\|x - \sum_{j=1}^{n} c_j e_j\| < \varepsilon$;

(2) Constants A and B exist, such that

$$A\sum_{j=1}^{\infty} |c_j|^2 \le \|\sum_{j=1}^{\infty} c_j e_j\|^2 \le B\sum_{j=1}^{\infty} |c_j|^2, \qquad \forall\{c_j\}_{j=1}^{\infty} \in l^2.$$

A and B are called the lower and upper bounds of the Riesz basis, respectively.

Particularly, if $A = B = 1$, $\{e_j\}_{j=1}^{\infty}$ is said to be an orthonormal basis.

Definition 3.8 An closed subspace $\{V_j\}_{-\infty}^{\infty}$ of $L^2(\mathbb{R})$ is sail to be a multiresolution analysis (MRA), if

(1). $\cdots \subset V_{-1} \subset V_0 \subset V_1 \subset \cdots$

(2). $\bigcap_{j\in\mathbb{Z}} V_j = \{0\}$, $\quad \overline{\bigcup_{j\in\mathbb{Z}} V_j} = L^2(\mathbb{R})$

(3). $f(x) \in V_j \iff f(2x) \in V_{j+1}$

(4). $\exists\phi \in V_0$, such that $\{\phi(\cdot - k)\}_{k\in\mathbb{Z}}$ is a Riesz basis of space V_0. ϕ is called the scaling function of the MRA.

Especially, if the basis of (4) is orthonormal, the MRA is then called orthonormal multiresolution analysis. Furthermore, if $\{\phi(\cdot - k)\}_{k\in\mathbb{Z}}$ is the Riesz basis of V_0 with both upper bound B and low bound A, we call that ϕ generates a MRA with upper bound B and low bound A.

Note that, certainly, V_j is translation-invariant with $2^{-j}\mathbb{Z}$. Meanwhile, $\{\phi_{j,k}\}_{k\in\mathbb{Z}}$ is a Riesz basis of V_j (the upper and lower bounds are constants for any $j \in \mathbb{Z}$), here $\phi_{j,k}(x) := 2^{j/2}\phi(2^j x - k)$.

In the multiresolution analysis, the key point is how the scale function ϕ can be constructed. Because $\phi \in V_0 \subset V_1$, and $\{\phi(2\cdot -k)|k \in \mathbb{Z}\}$ is the Riesz basis of V_1, hence, $\{h_k\} \in l^2(\mathbb{Z})$ exists, such that

$$\phi(x) = 2\sum_{k\in\mathbb{Z}} h_k\phi(2x - k). \tag{3.16}$$

Applying Fourier transform to both sides of the above equation, we can find that it is equivalent to the following fact: There exists $m_0(\xi) = \sum_{k\in\mathbb{Z}} h_k e^{-ik\xi} \in L^2(\mathbb{T})$, such that

$$\hat{\phi}(\xi) = m_0(\frac{\xi}{2})\hat{\phi}(\frac{\xi}{2}). \tag{3.17}$$

Therefore, we can obtain

$$\hat{\phi}(\xi) = \left(\prod_{j=1}^{n} m_0(\frac{\xi}{2^j})\right)\hat{\phi}(\frac{\xi}{2^n}), \qquad (\forall n \in \mathbb{N}).$$

It means that if $\hat{\phi}(\xi)$ is continuous at $\xi = 0$ and $\phi(0) \neq 0$, we can infer $\prod_{j=1}^{\infty} m_0(2^{-j}\xi)$ is convergent, and

$$\hat{\phi}(\xi) = \prod_{j=1}^{\infty} m_0(\frac{\xi}{2^j}). \tag{3.18}$$

The construction of ϕ is then equivalent to looking for the function m_0 with $2\pi\mathbb{Z}$-period. The $\{h_k\} \in l^2(\mathbb{Z})$ can be regarded as the mask of the two-scale equation, and the $m_0(\xi) = \sum_{k\in\mathbb{Z}} h_k e^{-ik\xi}$ can be considered to be the filter function.

As in the above discussion, most of wavelet bases can be constructed by multiresolution analysis, and the key point of the MRA is how to construct ϕ. Moreover, ϕ can be constructed by solving the two-scale equation Eq. (3.16) or Eq. (3.17). Therefore, one of the most important problems in MRA is:

- The existence of the solution for the two-scale equation Eq. (3.16) or Eq. (3.17).

From Eq. (3.17), we can obtain $\hat{\phi}$ which is represented by Eq. (3.18). However, it is only the formal solution, in fact, it is very difficult to find the analytical expression of ϕ for the following reasons: First, we do not know whether Eq. (3.18) is convergent. Second, we do not know whether $\hat{\phi}$ belongs to $L^2(\mathbb{R})$. Actually, the analytical expressions do not exist for most of the scaling functions, which are solutions of Eq. (3.16). A treatment to handle these problems is to study the convergence of the partial product $\prod_{j=1}^{n} m_0(2^{-j}\xi)$, and thereafter, to check if it converges in $L^2(\mathbb{R})$.

Although the two-scale equation has a solution ϕ in $L^2(\mathbb{R})$, it cannot guarantee the generation of MRA from ϕ. The main reason is that $\{\phi(\cdot - k)|k \in \mathbb{Z}\}$ must be a Riesz basis of V_0. Consequently, the second important problem is:

- Whether the solutions of two-scale equation Eq. (3.16) can generate a MRA.

Proving that $\{\phi(\cdot - k)|k \in \mathbb{Z}\}$ is a Riesz basis of V_0 is equivalent to the proof of the $[\hat{\phi}, \hat{\phi}]$ has positive upper and lower bounds.

The orthonormality and biorthonormality are very important in multiresolution analysis and wavelet theory. Thus, the third important problem is:

- The orthonormality and biorthonormality of the solutions of two-scale equation.

The orthonormality and biorthonormality can strictly be defined below:

Definition 3.9 Suppose that $\tilde{\phi},\ \phi \in L^2(\mathbb{R})$.

(1). If

$$\langle \tilde{\phi}(\cdot - k), \phi(\cdot)\rangle = \delta_{0,k} \quad (\forall k \in \mathbb{Z}),$$

then $\{\tilde{\phi}, \phi\}$ is said to be biorthonormal; furthermore, if $\{\phi, \phi\}$ is biorthonormal, we can say ϕ is orthonormal.

(2). If $\{\tilde{\phi}, \phi\}$ is biorthonormal, and

$$\tilde{V}_j := \{\tilde{\phi}_{j,k}|k \in \mathbb{Z}\}, \quad V_j := \{\phi_{j,k}|k \in \mathbb{Z}\}, \qquad (j \in \mathbb{Z}),$$

form MRAs in $L^2(\mathbb{R})$ with scale functions $\tilde{\phi}$ and ϕ respectively, we say that $\{\tilde{\phi}, \phi\}$ generate a pair of biorthonormal MRAs; Particularly, if $\{\phi, \phi\}$ forms a pair of orthonormal MRAs, ϕ is said to generate an orthonormal MRA.

3.2.1.2 *The Solution of Two-Scale Equation*

• The General Solution of Two-Scale Equation

As mentioned above, the formal solution of the two-scale equation is

$$\prod_{j=1}^{\infty} m_0(\xi/2^j).$$

The following is a sufficient condition for its convergence:

Theorem 3.13 *Let $m_0(\xi) \in C(\mathbb{T})$, and satisfy*

$$\exists \varepsilon > 0, \ \textit{when } |\xi| \textit{ is small enough}: \ m_0(\xi) = 1 + O(|\xi|^{\varepsilon}).$$

then, $\prod_{j=1}^{\infty} m_0(\xi/2^j)$ is uniformly convergent on any compact subset of $\mathbb{R}$, consequently, a continuous function can be defined in $C(\mathbb{R})$. Furthermore, there are positive constants C, τ and δ such that the following holds:

$$\left|\prod_{j=1}^{n} m_0(2^{-j}\xi)\right| \leq \begin{cases} \exp(\tau|\xi|^{\varepsilon}), & \textit{when } |\xi| \leq \delta; \\ C|\xi|^{\sigma}, & \textit{when } |\xi| > \delta, \end{cases} \qquad (\forall n \in \mathbb{N} \cup \{+\infty\}),$$

where $\sigma := \log_2 \|m_0\|_{C(\mathbb{T})}$.

Proof According to the condition, there must exist positive constants M and δ, such that

$$|m_0(\xi) - 1| \leq M|\xi|^{\varepsilon} < 1, \qquad \forall |\xi| \leq \delta.$$

For each compact subset K of $\mathbb{R}$, we choose $J \in \mathbb{N}$ satisfying

$$|2^{-J}\xi| < \delta, \qquad \forall \xi \in K.$$

Then, $\forall J_2 \geq J$, we have

$$\begin{aligned} \ln\left(\prod_{j=J}^{J_2} m_0(2^{-j}\xi)\right) &= \sum_{j=J}^{J_2} \ln m_0(2^{-j}\xi) \\ &= \sum_{j=J}^{J_2} \ln\left[1 + (m_0(2^{-j}\xi) - 1)\right] \\ &\leq \sum_{j=J}^{J_2} [m_0(2^{-j}\xi) - 1] \end{aligned}$$

$$\leq M\sum_{j=J}^{J_2}|2^{-j}\xi|^{\varepsilon}$$

$$\leq M\frac{2^{-\varepsilon(J-1)}}{2^{\varepsilon}-1}|\xi|^{\varepsilon}. \qquad (*)$$

Therefore, when J is large enough, we can write

$$\ln\left(\prod_{j=J}^{J_2} m_0(2^{-j}\xi)\right) = O(2^{-J\varepsilon}) \to 0, \qquad (\forall \xi \in K,\ J \to \infty).$$

Consequently, $\prod_{j=1}^{\infty} m_0(\xi/2^j)$ is uniformly convergent on any compact subset of $\mathbb{R}$. Since $m_0 \in C(\mathbb{T})$, it defines a continuous function.

For the inequality, we only focus on $n \in \mathbb{N}$. It is obvious, for $n = \infty$.

$\forall \xi \in \mathbb{R}$ with $\xi \neq 0$, let $J_1 \in \mathbb{Z}$ satisfy

$$\frac{1}{2}\delta < |2^{-J_1}\xi| \leq \delta.$$

Now, we denote $J := \max(1, J_1)$, then $\forall J_2 \geq J$, according to the formula (*), we have

$$\prod_{j=J}^{J_2} m_0(2^{-j}\xi) \leq \exp\left(M\frac{2^{-\varepsilon(J-1)}}{2^{\varepsilon}-1}|\xi|^{\varepsilon}\right).$$

If $0 < |\xi| \leq \delta$, then $J_1 \leq 0,\ J = 1$, and $\forall n \in \mathbb{N}$, we denote $J_2 = n$, and obtain

$$\left|\prod_{j=1}^{m} m_0(2^{-j}\xi)\right| = \prod_{j=J}^{m} m_0(2^{-j}\xi) \leq \exp\left(\frac{M}{2^{\varepsilon}-1}|\xi|^{\varepsilon}\right) = \exp\left(\tau|\xi|^{\varepsilon}\right),$$

where $\tau := \frac{M}{2^{\varepsilon}-1} > 0$.

When $\xi = 0$, because $m_0(0) = 1$, it satisfies the inequality.

If $|\xi| > \delta$, then $J_1 \geq 1,\ J = J_1$, therefore, when $n \geq J_1$, we can deduce that

$$\begin{aligned}\left|\prod_{j=1}^{m} m_0(2^{-j}\xi)\right| &= \left|\prod_{j=1}^{J_1-1} m_0(2^{-j}\xi)\right| \prod_{j=J}^{m} m_0(2^{-j}\xi) \\ &\leq \left(\|m_0\|_{C(\mathbb{T})}\right)^{J_1-1} \exp\left(M\frac{2^{-\varepsilon(J-1)}}{2^{\varepsilon}-1}|\xi|^{\varepsilon}\right)\end{aligned}$$

$$\begin{aligned}
&= 2^{(J_1-1)\log_2 \|m_0\|_{C(\mathbb{T})}} \exp\left(M\frac{2^\varepsilon}{2^\varepsilon-1}|2^{-J_1}\xi|^\varepsilon\right)\\
&\le \left(\frac{|\xi|}{\delta}\right)^\sigma \exp\left(M\frac{2^\varepsilon}{2^\varepsilon-1}\delta^\varepsilon\right)\\
&= C|\xi|^\sigma,
\end{aligned}$$

where,

$$\sigma := \log_2 \|m_0\|_{C(\mathbb{T})}; \qquad C := \delta^{-\sigma}\exp\left(M\frac{2^\varepsilon}{2^\varepsilon-1}\delta^\varepsilon\right).$$

For $n < J_1$, by $\|m_0\|_{C(\mathbb{T})} \ge |m_0(0)| = 1$, we have

$$\left|\prod_{j=1}^m m_0(2^{-j}\xi)\right| \le \left(\|m_0\|_{C(\mathbb{T})}\right)^m \le \left(\|m_0\|_{C(\mathbb{T})}\right)^{J_1-1} \le \left(\frac{|\xi|}{\delta}\right)^\sigma \le C|\xi|^\sigma.$$

This concludes our proof. ■

Note: Especially, if m_0 is a trigonometric polynomial and $m_0(0) = 1$, then, the conditions of this theorem are satisfied.

We hope that $\prod_{j=1}^\infty m_0(\xi/2^j) \in L^2(\mathbb{R})$, so that the inverse Fourier transform of ϕ can be also in $L^2(\mathbb{R})$. This means that ϕ is a solution of equation Eq. (3.16). Unfortunately, under the conditions of (3.13), $\prod_{j=1}^\infty m_0(\xi/2^j)$ may not be in $L^2(\mathbb{R})$, which can be shown by a simple example, namely, let $m_0(\xi) \equiv 1$, then $\prod_{j=1}^\infty m_0(\xi/2^j) \equiv 1 \notin L^2(\mathbb{R})$. We can handle this problem in two ways: First, we can enhance the conditions for $m_0(\xi)$, so as to ensure $\prod_{j=1}^\infty m_0(\xi/2^j) \in L^2(\mathbb{R})$. Second, we can discuss the solution of the equation Eq. (3.16) in a larger space over $L^2(\mathbb{R})$. We will now focus on the second way. In the conditions of (3.13), $\prod_{j=1}^\infty m_0(\xi/2^j)$ is continuous and increase at most polynomially fast at infinite. It belongs to φ', the space of all the slowly increasing generalized functions, where the operation of Fourier transform is very convenient. We will discuss the solution of two-scale equation in φ'.

As we know, the space φ' consistes of all the continuous linear functionals on φ. φ is the space of all the fast decreasing C^∞ functions, whose definition is as below:

$$\varphi := \{\varphi \in C^\infty(\mathbb{R}) \mid \sup_{x\in\mathbb{R}}(1+|x|^2)^{\frac{k}{2}}|\partial^\alpha\varphi(x)| \le M_{k,\alpha} < \infty\,(k,\ |\alpha| = 0, 1, \cdots)\}.$$

By equiping countable semi-norms as follows:

$$\|\varphi\|_m := \sup_{\substack{|\alpha| \leq m \\ x \in \mathbb{R}}} (1+|x|^2)^{\frac{m}{2}} |\partial^\alpha \varphi(x)|, \quad (m = 0, 1, \cdots),$$

φ becomes a countable-norm space, i.e., a B_0^* space (see [Zhang, 1986]).

For $1 \leq p \leq \infty$, we define

$$PL^p := PL^p(\mathbb{R}) := \{f \mid \exists m \in \mathbb{N} : (1+|x|)^{-m} f(x) \in L^p(\mathbb{R})\}. \tag{3.19}$$

Let $f \in PL^1$, and we define

$$\langle f, g \rangle := \int_{\mathbb{R}} f(x) g(x) dx \qquad (\forall g \in \varphi).$$

It is obvious that

$$PL^\infty \subset PL^p \subset PL^1 \subset \varphi' \quad (1 \leq p \leq \infty).$$

Now, we will prove that the following two equations are equivalent to each other:

$$\phi = 2 \sum_{k \in \mathbb{Z}} h_k \phi(2 \cdot -k) \quad (\text{In } \varphi', \text{ the right term is convergent}) \tag{3.20}$$

$$\hat{\phi} = m_0(\frac{\cdot}{2}) \hat{\phi}(\frac{\cdot}{2}) \qquad (\text{In } \varphi') \tag{3.21}$$

Theorem 3.14 *Let $m_0(\xi) = \sum_{k \in \mathbb{Z}} h_k e^{-ik\xi} \in C(\mathbb{T})$, and $\phi \in \varphi'$. If one of the following two conditions is satisfied:*

(1). $\hat{\phi} \in PL^1$ and $\sum_{k \in \mathbb{Z}} |h_k| < \infty$;

(2). $\sum_{k \in \mathbb{Z}} |h_k| |k|^n < \infty$ $(\forall n \in \mathbb{N})$,

then, ϕ satisfies Eq. (3.20) if and only if ϕ satisfies Eq. (3.21).

Proof $\forall \phi \in \varphi'$, $g \in \varphi$, we denote

$$I_n := \lim_{n \to \infty} \langle (\sum_{|k| \leq n} h_k e^{-ik \frac{\cdot}{2}}) \hat{\phi}(\frac{\cdot}{2}) - m_0(\frac{\cdot}{2}) \hat{\phi}(\frac{\cdot}{2}), \ g \rangle.$$

The satisfaction of the condition (1) produces

$$|I_n| = \left| \int_{\mathbb{R}} \hat{\phi}(\frac{\xi}{2}) (\sum_{|k| > n} h_k e^{-ik\frac{\xi}{2}}) g(\xi) d\xi \right|$$

$$\begin{aligned} &\leq \Big(\sum_{|k|>n} |h_k|\Big) \int_{\mathbb{R}} \left|\hat{\phi}(\frac{\xi}{2}) g(\xi)\right| d\xi \\ &\to 0 \quad (n \to \infty). \end{aligned}$$

The satisfaction of the condition (2) yields

$$g_n(\xi) := \Big(\sum_{|k|>n} h_k e^{-ik\frac{\xi}{2}}\Big) g(\xi) \in \varphi.$$

Consequently, we have, $\forall m \in \mathbb{Z}^+$,

$$\begin{aligned} & \|g_n\|_m \\ = & \sup_{\xi\in\mathbb{R},|\alpha|\leq m} |(1+|\xi|^2)^{\frac{m}{2}} D^\alpha g_n(\xi)| \\ = & \sup_{\xi\in\mathbb{R},|\alpha|\leq m} \left|(1+|\xi|^2)^{\frac{m}{2}} \sum_{\beta\leq\alpha} \binom{\alpha}{\beta} \Big(\sum_{|k|>n} h_k(-i\frac{k}{2})^\beta e^{-ik\frac{\xi}{2}}\Big) D^{\alpha-\beta} g(\xi)\right| \\ \leq & \sup_{\xi\in\mathbb{R},|\alpha|\leq m} |(1+|\xi|^2)^{\frac{m}{2}} \sum_{\beta\leq\alpha} \binom{\alpha}{\beta} \Big(\sum_{|k|>n} |h_k||k|^{|\beta|}\Big) |D^{\alpha-\beta} g(\xi)| \\ \leq & C\Big(\sum_{|k|>n} |h_k||k|^m\Big) \sup_{\xi\in\mathbb{R},|\alpha|\leq m} |(1+|\xi|^2)^{\frac{m}{2}} |D^\alpha g(\xi)| \\ = & C\Big(\sum_{|k|>n} |h_k||k|^m\Big) \|g\|_m \\ \to & 0 \quad (n\to\infty), \end{aligned}$$

where, C stands for an constant which depends only on m. Thus, we can obtain $g_n \to 0$ (in φ as $n \to 0$). Therefore, we have

$$|I_n| = \left|\langle \hat{\phi}(\frac{\cdot}{2}),\ \Big(\sum_{|k|>n} h_k e^{-ik\frac{\cdot}{2}}\Big) g\rangle\right| = \left|\langle \hat{\phi}(\frac{\cdot}{2}),\ g_n\rangle\right| \to 0 \quad (n \to 0).$$

It implies that

$$\begin{aligned} \phi \text{ satisfies Eq. (3.20)} &\iff \hat{\phi} = 2\sum_{k\in\mathbb{Z}} h_k (\phi(2\cdot -k))\hat{} \\ &\iff \hat{\phi} = \lim_{n\to\infty} \sum_{|k|\leq n} h_k e^{-ik\frac{\cdot}{2}} \hat{\phi}(\frac{\cdot}{2}) \end{aligned}$$

$$\iff \langle \hat{\phi},\ g\rangle = \lim_{n\to\infty}\langle (\sum_{|k|\leq n} h_k e^{-ik\frac{\cdot}{2}})\hat{\phi}(\frac{\cdot}{2}),\ g\rangle \quad (\forall g\in\varphi)$$

$$\iff \langle \hat{\phi} - m_0(\frac{\cdot}{2})\hat{\phi}(\frac{\cdot}{2}),\ g\rangle$$

$$= \lim_{n\to\infty}\langle [\sum_{|k|\leq n} h_k e^{-ik\frac{\cdot}{2}} - m_0(\frac{\cdot}{2})]\hat{\phi}(\frac{\cdot}{2}),\ g\rangle \quad (\forall g\in\varphi)$$

$$\iff \langle \hat{\phi} - m_0(\frac{\cdot}{2})\hat{\phi}(\frac{\cdot}{2}),\ g\rangle = \lim_{n\to\infty} I_n = 0 \quad (\forall g\in\varphi)$$

$$\iff \phi \text{ satisfy Eq. (3.21).}$$

This establishes our proof. ■

As for the solution of two-scale equation in φ', we have the following results.

Theorem 3.15 *Let $m_0(\xi) = \sum_{k\in\mathbb{Z}} h_k e^{-ik\xi} \in C(\mathbb{T})$, such that $\prod_{j=1}^{\infty} m_0(\frac{\xi}{2^j})$ converges pointwise to $M(\xi)\in PL^1$. Then, $\phi := \check{M}\in\varphi'$ is a solution of Eq. (3.21). Futhermore, if $M(\xi)$ is continuous at $\xi = 0$ and $M(0) = 1$, then ϕ is the unique solution of Eq. (3.21) in φ_0' ($\varphi_0'\subset\varphi$) which is defined by:*

$$\varphi_0' := \{\phi\in\varphi' \mid \hat{\phi}\in PL^1,\ \hat{\phi}(\xi) \text{ is continuous at } \xi = 0\ ,\ \hat{\phi}(0) = 1\} \quad (3.22)$$

Proof It is clear that

$$M(\xi) = m_0(\frac{\xi}{2})M(\frac{\xi}{2}), \qquad a.e.\ \xi\in\mathbb{R}.$$

Therefore, as a generalized function in φ', we have $M(\cdot) = m_0(\frac{\cdot}{2})M(\frac{\cdot}{2})$, that is, $\phi := \check{M}\in\varphi'$ satisfies

$$\hat{\phi}(\cdot) = m_0(\frac{\cdot}{2})\hat{\phi}(\frac{\cdot}{2}) \quad \text{in } \varphi'.$$

Consequetly, ϕ is the solution of Eq. (3.21).

If $M(\xi)$ is continuous at $\xi = 0$, and $M(0) = 1$, it is obvious that, $\phi\in\varphi_0'$, i.e. $\phi := \check{M}$ is the solution of Eq. (3.21) in φ_0'.

Now, we will prove the unicity of the solution. Suppose $\phi\in\varphi_0'$ is a solution of Eq. (3.21), hence, as a generalized function of φ', it satisfies

$$\hat{\phi} = m_0(\frac{\cdot}{2})\hat{\phi}(\frac{\cdot}{2}) \quad (\text{in } \varphi').$$

As $\hat{\phi} \in PL^1$, the above result is equivalent to that of ordinary function, thus, we obtain

$$\hat{\phi}(\xi) = m_0(\frac{\xi}{2})\hat{\phi}(\frac{\xi}{2}) \quad a.\ e.\ \xi \in \mathbb{R}.$$

Therefore,

$$\hat{\phi}(\xi) = m_0(\frac{\xi}{2})\hat{\phi}(\frac{\xi}{2}) = \cdots = (\prod_{j=1}^{n} m_0(\frac{\xi}{2^j}))\hat{\phi}(\frac{\xi}{2^n}) \quad a.\ e.\ \xi \in \mathbb{R}.$$

Let $n \to \infty$, we have

$$\hat{\phi}(\xi) = \prod_{j=1}^{\infty} m_0(\frac{\xi}{2^j}) = M(\xi) \quad a.\ e.\ \xi \in \mathbb{R}.$$

As a conclusion, $\phi = \check{M}$ holds. The proof is complete. ■

Corollary 3.1 *Let $m_0(\xi) = \sum_{k\in\mathbb{Z}} h_k e^{-ik\xi} \in C(\mathbb{T})$, such that*

$$\sum_{k\in\mathbb{Z}} |h_k| < \infty,$$

and

$$\exists \varepsilon > 0, \text{ when } |\xi| \text{ is small enough, } m_0(\xi) = 1 + O(|\xi|^{\varepsilon}).$$

Then, Eqs. (3.20) and (3.21) have the same solution in φ_0', which is just the inverse Fourier transform of $\prod_{j=1}^{\infty} m_0(\frac{\cdot}{2^j})$.

Proof According to threorem 3.14, it is known that the two-scale equations Eqs. (3.20) and (3.21) have the same solution in φ'. By Theorems 3.13 and 3.15, a unique solution exists in φ_0', which is just the inverse Fourier transform of $\prod_{j=1}^{\infty} m_0(\frac{\cdot}{2^j})$. ■

This result is enough for wavelet analysis in $L^2(\mathbb{R})$, although the solution of two-scale equation is in φ_0'. To solve the unique solution of two-scale equation in $L^2(\mathbb{R})$ is much more difficult than that in φ'. Thus, we discussed the case of φ' first. Meanwhile, from the above analysis, we found that the unicity of the solution in φ_0' implicates the unicity in $L^2(\mathbb{R})$. Consequently, our task will be changed to investigate the conditions which can guarantee that the solution in φ_0' belongs to $L^2(\mathbb{R})$.

The Cascade Algorithm to Solve Two-Scale Equation

It is assumed that $m_0(\xi) = \sum_{k\in\mathbb{Z}} h_k e^{-ik\xi} \in C(\mathbb{R})$, we take $\eta_0 \in L^\infty(\mathbb{R}) \cap L^1(\mathbb{R})$, such that $\hat{\eta}_0(0) = 1$. Hence, the general scheme of the Cascade algorithm is

$$\eta_n(x) := 2\sum_{k\in\mathbb{Z}} h_k \eta_{n-1}(2x-k), \qquad (n = 1, 2, \cdots). \tag{3.23}$$

In order to guarantee the convergence of the series in the Cascade algorithm, we suppose that $\sum_{j=1}^{\infty} |h_k| < \infty$. It is easy to see that the right hand side of the Cascade algorithm is absolutely convergent, $\eta_n \in L^\infty(\mathbb{R}) \cap L^1(\mathbb{R})$ $(n = 1, 2, \cdots)$, and

$$\|\eta_n\|_\infty \leq (2\sum_{k\in\mathbb{Z}} |h_k|)\|\eta_{n-1}\|_\infty, \qquad \|\eta_n\|_1 \leq (\sum_{k\in\mathbb{Z}} |h_k|)\|\eta_{n-1}\|_1,$$

for $n = 1, 2, \cdots$. Since

$$\hat{\eta}_n(\xi) = (\prod_{j=1}^{n} m_0(\frac{\xi}{2^j}))\hat{\eta}_0(\frac{\xi}{2^n}),$$

we have, under the condition of Theorem 3.13, that

$$\hat{\eta}_n(\xi) \to \prod_{j=1}^{\infty} m_0(\frac{\xi}{2^j}) \quad a.e.\ \xi \in \mathbb{R},$$

thus, positive constants C and σ exist, such that

$$|\hat{\eta}_n(\xi)| \leq C(1 + |\xi|)^\sigma, \qquad (\forall \xi \in \mathbb{R}).$$

Therefore, $\forall g \in \varphi$, using Lebesgue's dominated convergence theorem, we conclude that

$$\langle \hat{\eta}_n,\ g\rangle = \int_{\mathbb{R}} \hat{\eta}_n(\xi) g(\xi) d\xi \to 0, \quad (n \to \infty),$$

that is

$$\hat{\eta}_n \to (\prod_{j=1}^{\infty} m_0(\frac{\xi}{2^j})) \quad \text{in } \varphi'.$$

Therefore,

$$\eta_n \to (\prod_{j=1}^{\infty} m_0(\frac{\xi}{2^j}))^\vee \quad \text{in } \varphi'.$$

Consequently, the limit of $\{\eta_n\}$ in φ_0' is just the unique solution of Eq. (3.20) in φ'. In summary, we have the following theorem.

Theorem 3.16 *Let $m_0(\xi) = \sum_{k\in\mathbb{Z}} h_k e^{-ik\xi} \in C(\mathbb{T})$ satisfy $\sum_{k\in\mathbb{Z}} |h_k| < \infty$, and*

$$\exists \varepsilon > 0, \text{ when } |\xi| \text{is small enough}, \ m_0(\xi) = 1 + O(|\xi|^{\varepsilon}).$$

Let $\eta_0 \in L^\infty(\mathbb{R})$ be a function with compact support $[A, B]$, and satisfy $\hat{\eta}_0(0) = 1$. Then the function sequence $\{\eta_n\}$ defined by Cascade algorithm Eq. (3.23) is convergent in φ'. The limit is the unique solution of Eq. (3.20) in φ_0'.

If $\{h_k\}_{k\in\mathbb{Z}}$ has finite length, that is, the nonzero items in $\{h_k\}_{k\in\mathbb{Z}}$ is finit, it can be proved that the solution of Eq. (3.20) has a compact support in φ_0'. The support of $\phi \in \varphi'$ is defined as

$$\text{supp}\phi := \{x \in \mathbb{R} | \ \forall \text{ open set } O_x \ni x, \ \exists g \in C_c^\infty(O_x) : \ \langle \phi, \ g \rangle \neq 0\}, \quad (3.24)$$

where $C_c^\infty(O_x)$ denotes the space of all the C^∞ functions which is compactly supported in O_x.

Obviously, the support defined in the above is a closed set in $\mathbb{R}$. When ϕ degenerates to an ordinary locally integrable function, the definition of its support is the same as that of the ordinary meaning. The definition of the support of an ordinary function is defined by

$$\text{supp}\phi := \{x \in \mathbb{R}^d | \ \forall O_x \ni x, \text{ always has} : \ |O_x \cap \{\phi(x) \neq 0\}| > 0\}. \quad (3.25)$$

where, O_x denotes an open set containing x, and $|O_x \cap \{\phi(x) \neq 0\}|$ denotes the Lebesgue measure of the set $O_x \cap \{\phi(x) \neq 0\}$. In order to prove that the unique solution of the equation Eq. (3.20) in φ_0' is compactly supported when $\{h_k\}_{k\in\mathbb{Z}}$ has finite length, we first consider the following lemma:

Lemma 3.1 *Let $\{h_k\}_{k\in\mathbb{Z}}$ be a complex sequence supported in $[N, M]$ in the meaning that $h_k = 0$ if $k \notin [N, M]$. Let $\eta_0 \in L^\infty(\mathbb{R})$ is compactly supported on $[A, B]$. Then, for the function sequence $\{\eta_n\}$ which is given by Cascade algorithm Eq. (3.23), we have*

$$\text{supp}\eta_n \subset [A^{(n)}, B^{(n)}], \qquad (n = 0, 1, \cdots),$$

where $A^{(n)}$, $B^{(n)}$ *are defined by*

$$\begin{cases} A^{(0)} = A \\ B^{(0)} = B \end{cases} \qquad \begin{cases} A^{(n)} = \frac{1}{2}(N + A^{(n-1)}) \\ B_i^{(n)} = \frac{1}{2}(M + B^{(n-1)}) \end{cases}, \qquad (n = 1, 2, \cdots),$$

and satisfy

$\{A^{(n)}\}_{n=1}^{\infty}$ *is a monotonous sequence between* N *and* A, *and* $A^{(n)} \to N$ $(n \to \infty)$;

$\{B^{(n)}\}_{n=1}^{\infty}$ *is a monotonous sequence between* M *and* B, *and* $B^{(n)} \to M$ $(n \to \infty)$.

Proof We use induction for n to prove $\text{supp}\eta_n \subset [A^{(n)}, B^{(n)}]$. Obviously, for $n = 0$, our conclusion is established. Suppose it is right for $n-1$. As for $\text{supp}\eta_n$, let $\eta_n(x) \neq 0$, then, $\exists k \in [N, M]$, such that $2x - k \in \text{supp}\eta_{n-1} \subset [A^{(n-1)}, B^{(n-1)}]$. This means $A^{(n-1)} \le 2x - k \le B^{(n-1)}$ or $\frac{1}{2}(A^{(n-1)} + k) \le x \le \frac{1}{2}(B^{(n-1)} + k)$. Therefore, $\frac{1}{2}(A^{(n-1)} + N) \le x \le \frac{1}{2}(B^{(n-1)} + M)$. So far, we proved that

$$\text{supp}\eta_n \subset \left[\frac{(A^{n-1)} + N}{2}, \frac{B^{(n-1)} + M}{2}\right] = [A^{(n)}, B^{(n)}].$$

By induction, the result holds for $n \in \mathbb{Z}_+$.

We now discuss the monotonicity of $\{A^{(n)}\}_{n=0}^{\infty}$, in two cases:

If $A^{(0)} = A \ge N$, then, by $A^{(n)} = \frac{1}{2}(N + A^{(n-1)})$, it is easy to prove that $A^{(n)} \ge N$, so $A^{(n)} \le A^{(n-1)} \le \cdots \le A^{(0)}$. i. e. $\{A^{(n)}\}_{n=1}^{\infty}$ is a monotonous sequence valued between N and A.

If $A^{(0)} = A \le N$, the same conclusion can be proved in the same way.

Leting $n \to \infty$ on the both sides of $A^{(n)} = \frac{1}{2}(N + A^{(n-1)})$, we obtain that $A^{(n)} \to N$.

We can give the proof in the same way for $B^{(n)}$.

The proof of the Lemma completes. ■

Theorem 3.17 *Let* $\{h_k\}_{k \in \mathbb{Z}}$ *be a complex sequence which is supported in* $[N, M]$, $\sum_{k \in \mathbb{Z}} h_k = 1$. *Then, the unique solution* ϕ *of equation Eq. (3.20) in* φ_0' *is compactly supported on* $[N, M]$.

Proof Let $\eta_0 \in L^\infty(\mathbb{R})$ satisfy $\text{supp}\eta_0 \subset [A, B]$, and $\eta_0(0) = 1$. Then, for the function sequence $\{\eta_n\}$ derivated by the cascade agorithm, we have $\eta_n \to \phi$ (in φ'), where ϕ is the unique solution of Eq. (3.20) in φ_0', and

$$\text{supp}\eta_n \subset [A^{(n)}, B^{(n)}] \qquad (n = 0, 1, \cdots),$$

where $A^{(n)}$, $B^{(n)}$ satisfy

$$A^{(n)} \to N\ , \quad B^{(n)} \to M, \qquad (n \to \infty).$$

$\forall x \notin [N, M]$, there exists a neighborhood O_x of x, such that the distance between $[N, M]$ and O_x is positive. Then, when n is large enough, we have

$$([A^{(n)}, B^{(n)}]) \bigcap O_x = \text{empty set}.$$

For any $g \in C_c^\infty(O_x)$, we have

$$\langle \phi,\ g\rangle = \lim_{n\to\infty} \langle \eta_n,\ g\rangle = 0.$$

Therefore, $x \notin \text{supp}\phi$. Now that x is an arbitrary point of $\mathbb{R}\backslash[N, M]$, we have

$$\mathbb{R}\backslash[N, M] \subset \mathbb{R}\backslash\text{supp}\phi,$$

i.e., $\text{supp}\phi \subset [N, M]$.

This ends the proof. ∎

3.2.2 *The Construction of MRAs*

To facilitate the discussion, we assume the $2\pi\mathbb{Z}$-periodic function, $m_0(\xi)$, is continuous on $\mathbb{R}$. We will study how MRAs can be constructed from such $m_0(\xi)$ that satisfies some basic conditions as follows.

Definition 3.10 Suppose $m_0(\xi)$ is $2\pi\mathbb{Z}$-periodic measurable function on $\mathbb{R}$.

- m_0 is said to satisfy the basic condition I , if $m_0(\xi) \in C(\mathbb{T})$, $m_0(0) = 1$ and $m_0(\pi) = 0$;
- m_0 is said to satisfy the basic condition II , if $\prod_{j=1}^{\infty} m_0(\xi/2^j)$ *a.e.* converges to a non-zero function, which is denoted by $\hat{\phi}$ and is called the corresponding limit function, (where $\hat{\phi}$ denotes a function instead of the Fourier transform of a function temporarily.)

To study the sufficient conditions such that the solution of the two-scale equation belongs to $L^2(\mathbb{R})$, the transition operators will be introduced first.

Let $\tilde{m}_0$, m_0 be $2\pi\mathbb{Z}-$periodic functions on $\mathbb{R}$, we define:

$$Tf(\xi) := \left|m_0(\frac{\xi}{2})\right|^2 f(\frac{\xi}{2}) + \left|m_0(\frac{\xi}{2}+\pi)\right|^2 f(\frac{\xi}{2}+\pi) \tag{3.26}$$

$$\tilde{T}f(\xi) := \left|\tilde{m}_0(\frac{\xi}{2})\right|^2 f(\frac{\xi}{2}) + \left|\tilde{m}_0(\frac{\xi}{2}+\pi)\right|^2 f(\frac{\xi}{2}+\pi) \tag{3.27}$$

$$Sf(\xi) := \tilde{m}_0(\frac{\xi}{2})\bar{m}_0(\frac{\xi}{2})f(\frac{\xi}{2}) + \tilde{m}_0(\frac{\xi}{2}+\pi)\bar{m}_0(\frac{\xi}{2}+\pi)f(\frac{\xi}{2}+\pi) \tag{3.28}$$

$$|S|f(\xi) := \left|\tilde{m}_0(\frac{\xi}{2})\bar{m}_0(\frac{\xi}{2})\right| f(\frac{\xi}{2}) + \left|\tilde{m}_0(\frac{\xi}{2}+\pi)\bar{m}_0(\frac{\xi}{2}+\pi)\right| f(\frac{\xi}{2}+\pi) \tag{3.29}$$

The following lemma is important.

Lemma 3.2 *Let $m_0 \in C(\mathbb{T})$ satisfy the basic condition II and the corresponding limit function $\hat{\phi} \in C(\mathbb{R})$. Let X be a closed subspace of $C(\mathbb{T})$ and satisfy both of the following conditions:*

(1). $T(X) \subset X$ and the spectral radius $r_\sigma(T|_X)$ of $T|_X$, which is the restriction of the transition operator T on X, satisfies $r_\sigma(T|_X) < 1$;

(2). There exists a non-negative function $f \in X$ which has at most one zero point $\xi = 0$ in $[-\pi, \pi]$.

Then $\hat{\phi} \in L^2(\mathbb{R})$ and $\|\hat{\phi}_n - \hat{\phi}\|_2 \to 0$ $(n \to \infty)$, where $\hat{\phi}_n$ is defined by

$$\hat{\phi}_n(\xi) := \prod_{j=1}^{n} m_0(2^{-j}\xi)\chi_{2^n[-\pi,\pi]}(\xi). \tag{3.30}$$

Proof It can easily be deduced that $\hat{\phi}(0) = 1$. Let $\delta:\ 0 < \delta < \pi$ satisfy

$$\frac{1}{2} \le |\hat{\phi}(\xi)| \le 2, \qquad \forall \xi \in [-\delta, \delta].$$

Since f is continuous and positive on $[-\pi,\pi]\backslash(-\frac{1}{2}\delta, \frac{1}{2}\delta)$, a constant $C > 0$ must exist, such that

$$C^{-1} \le f(\xi) \le C, \qquad \forall \xi \in [-\pi,\pi]\backslash(-\frac{1}{2}\delta, \frac{1}{2}\delta).$$

We denote

$$E_n := [-2^n\delta, 2^n\delta], \qquad (n = 0, 1, \cdots).$$

Then $2^{-n}\xi \in E_0 = [-\delta, \delta]$ for any $\xi \in E_n$. Therefore,

$$|\hat{\phi}(\xi)| = |\hat{\phi}_n(\xi)\hat{\phi}(2^n\xi)| \le |\hat{\phi}_n(\xi)| \sup_{\xi \in E_0} |\hat{\phi}(\xi)| \le 2|\hat{\phi}_n(\xi)|.$$

Hence we have

$$\begin{aligned}
\int_{E_n\backslash E_{n-1}} |\hat{\phi}(\xi)|^2 d\xi &\leq 4\int_{E_n\backslash E_{n-1}} |\hat{\phi}_n(\xi)|^2 d\xi \\
&\leq 4\int_{[-2^n\pi,2^n\pi]\backslash E_{n-1}} |\hat{\phi}_n(\xi)|^2 d\xi \\
&\leq 4C\int_{[-2^n\pi,2^n\pi]\backslash E_{n-1}} |\hat{\phi}_n(\xi)|^2 f(2^{-n}\xi) d\xi \\
&\leq 4C\int_{\mathbb{R}} |\hat{\phi}_n(\xi)|^2 f(2^{-n}\xi) d\xi \\
&= 4C\int_{\mathbb{T}} (T^n f)(\xi) d\xi \\
&= 4C\int_{\mathbb{T}} ((T|_X)^n f)(\xi) d\xi \\
&\leq 4C(2\pi)\|f\|_{C(\mathbb{T})}\|(T|_X)^n\|.
\end{aligned}$$

Let ρ satisfy $r_\sigma(T|_X) < \rho < 1$. According to the spectral radius formula, we deduce that $\lim_{n\to\infty}\|(T|_X)^n\|^{1/n} = r_\sigma(T|_X)$. Thus there exists $N > 0$ such that $\|(T|_X)^n\| \leq \rho^n$ for $n > N$, which concludes that

$$\int_{E_n\backslash E_{n-1}} |\hat{\phi}(\xi)|^2 d\xi \leq 4\int_{[-2^n\pi,2^n\pi]\backslash E_{n-1}} |\hat{\phi}_n(\xi)|^2 d\xi \leq 4C(2\pi)\|f\|_{C(\mathbb{T})}\rho^n.$$

Consequently,

$$\begin{aligned}
\int_{\mathbb{R}} |\hat{\phi}(\xi)|^2 d\xi &= \int_{E_N} |\hat{\phi}(\xi)|^2 d\xi + \sum_{n=N+1}^{\infty}\int_{E_n\backslash E_{n-1}} |\hat{\phi}(\xi)|^2 d\xi \\
&\leq (2^{N+1}\delta)\|\hat{\phi}\|_{C(E_N)} + 4C(2\pi)\|f\|_{C(\mathbb{T})}\sum_{n=N+1}^{\infty}\rho^n \\
&< \infty.
\end{aligned}$$

$\hat{\phi} \in L^2(\mathbb{R})$ is proved.

To prove $\|\hat{\phi}_n - \hat{\phi}\|_2 \to 0$ $(n \to \infty)$. We deduce that

$$\begin{aligned}
\|\hat{\phi}_n\|_2^2 &= \int_{\mathbb{R}} |\hat{\phi}_n(\xi)|^2 d\xi = \int_{[-2^n\pi,2^n\pi]} |\hat{\phi}_n(\xi)|^2 d\xi \\
&= \int_{[-2^n\pi,2^n\pi]\backslash E_{n-1}} |\hat{\phi}_n(\xi)|^2 d\xi + \int_{E_{n-1}} |\hat{\phi}_n(\xi)|^2 d\xi.
\end{aligned}$$

Since $2^{-n}\xi \in [-\frac{1}{2}\delta, \frac{1}{2}\delta] \subset [-\delta, \delta]$ for any $\xi \in E_{n-1}$, we have

$$|\hat{\phi}(\xi)| = |\hat{\phi}_n(\xi)\hat{\phi}(2^{-n}\xi)| \geq \frac{1}{2}|\hat{\phi}_n(\xi)|.$$

Hence

$$|\hat{\phi}_n(\xi)|^2 \chi_{E_{n-1}}(\xi) \leq 4|\hat{\phi}(\xi)|^2, \qquad \forall \xi \in \mathbb{R}.$$

Now that $\hat{\phi}_n(\xi)\chi_{E_{n-1}}(\xi) \to \hat{\phi}(\xi)$ $a.e.$ $\xi \in \mathbb{R}$, the Lebesgue dominated convergence theorem deduces that

$$\int_{E_{n-1}} |\hat{\phi}_n(\xi)|^2 d\xi = \int_{\mathbb{R}} |\hat{\phi}_n(\xi)|^2 \chi_{E_{n-1}}(\xi) d\xi \to \int_{\mathbb{R}} |\hat{\phi}(\xi)|^2 d\xi = \|\hat{\phi}\|_2^2,$$

which together with the following result

$$\int_{[-2^n\pi, 2^n\pi]\setminus E_{n-1}} |\hat{\phi}_n(\xi)|^2 d\xi \leq C(2\pi)\|f\|_{C(\mathbb{T})}\rho^n \to 0 \quad (n \to 0)$$

concludes that $\|\hat{\phi}_n\|_2 \to \|\hat{\phi}\|_2$ $(n \to \infty)$. Note that $\hat{\phi}_n(\xi) \to \hat{\phi}(\xi)$ $a.e.$ $\xi \in \mathbb{R}$ $(n \to \infty)$, we obtain $\|\hat{\phi}_n - \hat{\phi}\|_2 \to 0$ $(n \to \infty)$. Thus, the proof of the lemma is complete. ■

Particularly, if m_0 satisfies the basic conditions I and II, we set

$$X = \dot{C}(\mathbb{T}) := \{f \in C(\mathbb{T}) | f(0) = 0\} \quad \text{or} \quad X = \dot{\mathcal{P}}_N := \{f \in \mathcal{P}_N | f(0) = 0\}$$

corresponding to $m_0 \in C(\mathbb{T})$ or $m_0 \in \mathcal{P}_N^+$ respectively, where

$$\mathcal{P}_N := \left\{ \sum_{n=-N}^{N} c_n e^{-in\xi} \mid \forall \text{ sequence } \{c_n\}_{n\in\mathbb{Z}} \right\},$$

$$\mathcal{P}_N^+ := \left\{ \sum_{n=0}^{N} c_n e^{-in\xi} \mid \forall \text{ sequence } \{c_n\}_{n\in\mathbb{Z}} \right\}.$$

Then we have $T(X) \subset X$. According to the above lemma, we obtain the following theorem.

Theorem 3.18 *Suppose m_0 meets the basic conditions I and II and its limit function satisfies $\hat{\phi} \in C(\mathbb{R})$. If the spectral radius $r_\sigma(T|_{\dot{C}(\mathbb{T})})$ of $T|_X$, the restriction of T defined by Eq. (3.26), satisfies $r_\sigma(T|_{\dot{C}(\mathbb{T})}) < 1$, then $\hat{\phi} \in L^2(\mathbb{R})$ and $\|\hat{\phi}_n - \hat{\phi}\|_2 \to 0$ $(n \to \infty)$, where $\hat{\phi}_n$ is defined by Eq. (3.30).*

Proof According to the above lemma, the theorem holds obviously if we set $X = \dot{C}(\mathbb{T})$ and $f(\xi) := (1 - \cos\xi)^2$. ■

Now, we give a theorem for the case that m_0 is a trigonometric polynomial. The readers can refer to [Long, 1995] for the proof.

Theorem 3.19 *Let $m_0 \in \mathcal{P}_N^+$ ($N \neq 0$), $m_0(0) = 1, m_0(\pi) = 0$ and $\hat{\phi}$ be its limit function. We denote*

$$\dot{\mathcal{P}}_N := \{f \in \mathcal{P}_N | f(0) = 0\}.$$

Then the following three conditions are equivalent to each other:

(1). $\lambda = 1$ is the simple eigenvalue of the restriction $T|_{\mathcal{P}_N}$ of T on $\mathcal{P}_N$, and each of its other eigenvalues λ satisfies $|\lambda| < 1$, where T is defined by Eq. (3.26).

(2). each eigenvalue λ of the restriction $T|_{\dot{\mathcal{P}}_N}$ of T on $\dot{\mathcal{P}}_N$ satisfies $|\lambda| < 1$.

(3). $\hat{\phi} \in L^2(\mathbb{R})$ and $\|\hat{\phi}_n - \hat{\phi}\|_2 \to 0$ $(n \to \infty)$, where $\hat{\phi}_n$ is defined by Eq. (3.30)

Furthermore, each of them implies the following results:

(i). $\Phi := [\hat{\phi}, \hat{\phi}] \in \mathcal{P}_N$ and Φ is the eigenvetor of $T|_{\mathcal{P}_N}$ corresponding to the simple eigenvalue $\lambda = 1$

(ii). There exist constants $0 < \varepsilon < 2$ and $C > 0$ such that

$$\sum_{k\in\mathbb{Z}} |\hat{\phi}(\xi + 2k\pi)|^{2-\varepsilon} \leq C, \quad |\hat{\phi}(\xi)| \leq C(1 + |\xi|)^{-\varepsilon}, \qquad (\forall \xi \in \mathbb{R}),$$

where

$$\Phi(\xi) := [\hat{\phi}, \hat{\phi}] := \sum_{k\in\mathbb{Z}} |\hat{\phi}(\xi + 2\pi k)|^2. \tag{3.31}$$

We consider whether ϕ generates an orthonormal MRA or a biorthogonal MRA. The estimation of the lower and upper bounds of $\Phi = [\hat{\phi}, \hat{\phi}]$ is key to this question. In mathematics, it is difficult since Φ is defined by an infinite sum and not easily to be expressed analytically in general. The above theorem, however, tells us that under certain conditions, Φ is just the eigenvector of $T|_{\mathcal{P}_N}$ corresponding to the simple eigenvalue $\lambda = 1$. The latter can be solved with a computer.

A sufficient condition such that $\hat{\phi} \in L^2(\mathbb{R})$ and $\|\hat{\phi}_n - \hat{\phi}\|_2 \to 0$ $(n \to \infty)$ is given as follows. Its proof is omitted here and can be found in some references such as [Daubechies, 1992; Long, 1995].

Theorem 3.20 *Suppose* m_0 *can be written as*

$$m_0(\xi) = (\frac{1+e^{-i\xi}}{2})^L M_0(\xi)$$

where $L \in \mathbb{Z}_+$ *and* $M_0 \in C(\mathbb{T})$ *satisfies:*

(1). $M_0(\pi) \neq 0$;

(2). There exists a constant $0 < \delta < 1$ *satisfying* $M_0(\xi) = 1 + O(|\xi|^\delta)$ *in some neighborhood of* $\xi = 0$;

(3). There exists $k \in \mathbb{N}$ *such that*

$$B_k := \max_{\xi \in \mathbb{R}} |M_0(\xi) \cdots M_0(2^{k-1}\xi)|^{1/k} < 2^{L-\frac{1}{2}}.$$

Then, $\hat{\phi} \in L^2(\mathbb{R})$, $\|\hat{\phi}_n - \hat{\phi}\|_2 \to 0 \quad (n \to \infty)$ *and a constant* $C > 0$ *exists, such that*

$$|\hat{\phi}(\xi)| \leq \frac{C}{(1+|\xi|)^{\frac{1}{2}+\varepsilon}}, \qquad (\forall \xi \in \mathbb{R}),$$

where $\varepsilon := L - \frac{1}{2} - \log B_k > 0$ *and* $\hat{\phi}_n$ *is defined by Eq. (3.30).*

The following is a sufficient and necessary condition, which can guarantee that ϕ generates a MRA.

Theorem 3.21 *Let* m_0 *satisfy the basic conditions I and II and* $\hat{\phi} \in C(\mathbb{R})$ *be its limit function. Then* ϕ *generates MRA if and only if both of the following two conditions hold:*

(1). $\hat{\phi} \in L^2(\mathbb{R})$ *and* $\|\hat{\phi}_n - \hat{\phi}\|_2 \to 0$ $(n \to \infty)$, *where* $\hat{\phi}_n$ *is defined by Eq. (3.30).*

(2). Two positive constants A and B exist, such that $A \leq (T^n 1)(\xi) \leq B$ *a.e.* $\xi \in \mathbb{T}$ *for any* $n \in \mathbb{N}$, *where T is the transition operator defined by Eq. (3.26).*

Proof To prove the necessity we assume that ϕ generates a MRA of $L^2(\mathbb{R})$. Two constants A and B exist, such that Φ defined by Eq. (3.31) satisfies

$$A \leq \Phi(\xi) \leq B \qquad a.e.\ \xi \in \mathbb{T},$$

which concludes (1).

Now, we prove (2). Since T^n is a positive linear operator for any $n \in \mathbb{N}$, we deduce that $T^n f(\xi) \leq T^n g(\xi)$ *a.e.* ξ for any $f(\xi) \leq g(\xi)$ *a.e.* ξ. Using

$T^n\Phi = \Phi$, we have:

$$\begin{aligned} A \leq \Phi(\xi) &= T^n\Phi(\xi) \leq (T^nB)(\xi) = B(T^n1)(\xi), \quad a.e.\ \xi \in \mathbb{T}; \\ B \geq \Phi(\xi) &= T^n\Phi(\xi) \geq (T^nA)(\xi) = A(T^n1)(\xi), \quad a.e.\ \xi \in \mathbb{T}. \end{aligned}$$

Therefore, $\frac{A}{B} \leq (T^n1)(\xi) \leq \frac{B}{A}$ *a.e.* $\xi \in \mathbb{T}$. The proof of (2) is complete.

To prove the sufficiency, we suppose (1) and (2) hold. For any function g which is $2\pi\mathbb{Z}$-periodic, bounded, non-negative and measurable, the following holds:

$$\int_{\mathbb{R}} |\hat{\phi}_n(\xi)|^2 g(\xi)d\xi = \int_{\mathbb{T}} (T^n1)(\xi)g(\xi)d\xi.$$

Hence

$$A\int_T g(\xi)d\xi \leq \int_{\mathbb{R}} |\hat{\phi}_n(\xi)|^2 g(\xi)d\xi \leq B\int_T g(\xi)d\xi.$$

Let $n \to \infty$, we have

$$A\int_T g(\xi)d\xi \leq \int_{\mathbb{R}} |\hat{\phi}(\xi)|^2 g(\xi)d\xi \leq B\int_T g(\xi)d\xi,$$

i.e.

$$A\int_T g(\xi)d\xi \leq \int_{\mathbb{T}} \Phi(\xi)g(\xi)d\xi \leq B\int_{\mathbb{T}} g(\xi)d\xi.$$

Thus, $A \leq \Phi(\xi) \leq B$ *a.e.*$\xi \in \mathbb{T}$ which implies consequently that $\{\phi(\cdot-k)|k \in \mathbb{Z}\}$ constitute a Riesz basis of $V_0 := \overline{span}\{\phi(\cdot - k)|k \in \mathbb{Z}\}$. By the fact of $\hat{\phi}(0) = m_0(0) = 1$ we deduce that $\overline{\cup_{j\in\mathbb{Z}}V_j} = L^2(\mathbb{R})$ for $V_j := \{f(2^j\cdot)|f \in V_0\}$ $(j \in \mathbb{Z})$. Therefore, ϕ generates a MRA of $L^2(\mathbb{R})$. This ends the proof of the theorem. ■

For the case that m_0 is trigonometric polynomial, we further have following result.

Corollary 3.2 *Let $m_0 \in \mathcal{P}_N^+$ (where $N \in \mathbb{Z}_+$, $N \neq 0$). Then $\hat{\phi}(\xi) := \prod_{j=1}^{\infty} m_0(\xi/2^j)$ converges a.e. $\xi \in \mathbb{R}$, belongs to $L^2(\mathbb{R})$ and ϕ generates a MRA, if and only if the following three conditions hold:*

(1). $m_0(0) = 1$ and $m_0(\pi) = 0$,

(2). $\lambda = 1$ is the simple eigenvalue of $T|_{\mathcal{P}_N}$ which is the restriction of T defined by Eq. (3.26) on $\mathcal{P}_N$, and each of its other eigenvalues λ satisfies $|\lambda| < 1$.

(3). The eigenvector $g \in \mathcal{P}_N$ of $T|_{\mathcal{P}_N}$ corresponding to the eigenvalue 1 *has positive lower bound if* $g(0) = 1$, *that is, there is a constant* $C > 0$ *such that* $g(\xi) \geq C$ $(\forall \xi \in \mathbb{T})$.

Proof To prove the necessity, we choose $\xi_0 \in \mathbb{R}$ such that $\prod_{j=1}^{\infty} m_0(\xi_0/2^j)$ converges. Then $\lim_{j\to\infty} m_0(\xi_0/2^j) = 1$, i.e. $m_0(0) = 1$.

Since $\hat{\phi} \in L^2(\mathbb{R})$, and $m_0 \in \mathcal{P}_N^+$ implies that ϕ is compactly supported, we conclude that $\Phi := [\hat{\phi}, \hat{\phi}]$ is a trigonometric polynomial. According to the fact that ϕ generates a MRA, it can be shown easily that Φ has positive upper and lower bounds on $\mathbb{T}$. Therefore, (1) holds.

By Theorems 3.19 and 3.21, it can be concluded that (2) holds and $\Phi = [\hat{\phi}, \hat{\phi}]$ is the eigenvector of $T|_{\mathcal{P}_N}$ corresponding to the eigenvalue 1. Since 1 is the simple eigenvalue of $T|_{\mathcal{P}_N}$, we deduce that $g \in \mathcal{P}_N$ is just some constant times of Φ. Now, $g(0) = \Phi(0) = 1$, we further have $g \equiv \Phi$. Hence g has positive lower bound and (3) is proved.

To prove the sufficiency, we deduce that $\hat{\phi} \in L^2(\mathbb{R})$, $\Phi \in \mathcal{P}_N$ and Φ has positive lower bound according to Theorem 3.19 and condition (3). On the other hand, $\Phi \in \mathcal{P}_N$ implies that Φ has positive upper bound. Hence, Φ has positive upper and lower bounds, which together with the equality $\hat{\phi}(0) = 1$ concludes that ϕ generates a MRA. The proof is complete. ■

This corollary is more convenient to be applied in practice. It can be easily implemented with a computer.

3.2.2.1 *The Biorthonormal MRA*

The purpose of this section is to construct such $\tilde{\phi}$, $\phi \in L^2(\mathbb{R})$ that generate a pair of biorthonormal MRA. The first task is to construct the biorthonormal functions $\{\tilde{\phi}, \phi\}$, or equivalently, to find out the conditions for masks $\tilde{m}_0$ and m_0, such that following two-scale relation:

$$\hat{\phi}(\xi) = m_0(\frac{\xi}{2})\hat{\phi}(\frac{\xi}{2}) \tag{3.32}$$

$$\hat{\tilde{\phi}}(\xi) = \tilde{m}_0(\frac{\xi}{2})\hat{\tilde{\phi}}(\frac{\xi}{2}) \tag{3.33}$$

Generally, it is very difficult to find out the sufficient and necessary conditions. To avoid such difficulty, we will first study some basic necessary conditions for $\{\tilde{m}_0, m_0\}$, Thereafter, by enhancing the conditions until we meet the sufficient ones.

Let $\{\tilde{\phi}, \phi\}$ generate a pair of biorthonormal MRA. There exists a constant $C > 0$ such that

$$C^{-1} \leq \tilde{\Phi}(\xi),\ \Phi(\xi) \leq C \qquad a.e.\ \xi \in \mathbb{R},$$

where $\Phi(\xi)$ is defined by Eq. (3.31), and

$$\tilde{\Phi}(\xi) := [\hat{\tilde{\phi}}, \hat{\tilde{\phi}}] := \sum_{k\in\mathbb{Z}} |\hat{\tilde{\phi}}(\xi + 2k\pi)|^2. \tag{3.34}$$

Enhance the above inequality by letting

$$C^{-1} \leq \tilde{\Phi}(\xi),\ \Phi(\xi) \leq C \quad (\forall \xi \in \mathbb{R}).$$

Then it can be deduced that

$$m_0(0) = 1,\ m_0(\pi) = 0.$$

In order to define $\tilde{\phi}$, ϕ from $\tilde{m}_0$, m_0 based on the two-scale relation, we assume that $\prod_{j=1}^{\infty} \tilde{m}_0(\xi/2^j)$ and $\prod_{j=1}^{\infty} m_0(\xi/2^j)$ are pointwise convergent. This assumption is necessary if $\hat{\tilde{\phi}}(\xi)$ and $\hat{\phi}(\xi)$ are expected to be continuous at $\xi = 0$ and are not zero functions.

According to the biorthonormal property, we have (see Theorem 3.23):

$$\tilde{m}_0(\xi)\bar{m}_0(\xi) + \tilde{m}_0(\xi + \pi)\bar{m}_0(\xi + \pi) = 1 \qquad a.e.\ \xi \in \mathbb{R}.$$

To facilitate the discussion, we assume that $2\pi\mathbb{Z}$-period functions $\tilde{m}_0(\xi)$ and $m_0(\xi)$ are continuous and satisfy the following basic conditions. We will study such sufficient conditions $\{\tilde{m}_0(\xi),\ m_0(\xi)\}$ satisfy, that $\{\tilde{\phi}(\xi),\ \phi(\xi)\}$ generate a pair of biorthonomal.

Definition 3.11 Let $\{\tilde{m}_0(\xi),\ m_0(\xi)\}$ be a pair of $2\pi\mathbb{Z}$-period measurable functions on $\mathbb{R}$. The following three conditions are called the basic conditions:

- Basic condition I: $\tilde{m}_0(\xi),\ m_0(\xi) \in C(\mathbb{T})$, $\tilde{m}_0(0) = m_0(0) = 1$, $\tilde{m}_0(\pi) = m_0(\pi) = 0$.
- Basic condition II: $\prod_{j=1}^{\infty} \tilde{m}_0(\xi/2^j)$ and $\prod_{j=1}^{\infty} m_0(\xi/2^j)$ are pointwise convergent to non-zero functions $\hat{\tilde{\phi}}$ and $\hat{\phi}$ respectively, which are called the corresponding limit functions. (Here $\hat{\tilde{\phi}}$, $\hat{\phi}$ denote two functions instead of the Fourier transforms of functions temporarily).

- Basic condition III: $\tilde{m}_0(\xi)\bar{m}_0(\xi)+\tilde{m}_0(\xi+\pi)\bar{m}_0(\xi+\pi) = 1,\ a.e.\ \xi \in \mathbb{R}$.

$\{\tilde{m}_0(\xi),\ m_0(\xi)\}$ is called to be basic if it satisfies the above three conditions.

Particularly, if $\tilde{m}_0 = m_0$, the corresponding basic conditions are defined below:

Definition 3.12 Let $m_0(\xi)$ be a $2\pi\mathbb{Z}$-period measurable functions on $\mathbb{R}$. The following three conditions are called the basic conditions:

- Basic condition I: $m_0(\xi) \in C(\mathbb{T})$ and $m_0(0) = 1, m_0(\pi) = 0$.
- Basic condition II: $\prod_{j=1}^{\infty} m_0(\xi/2^j)$ is pointwise convergent to a non-zero function $\hat{\phi}$, which is called the corresponding limit functions. (Here $\hat{\phi}$ denotes a function instead of the Fourier transform of a function temporarily).
- Basic condition III: $|m_0(\xi)|^2 + |m_0(\xi+\pi)|^2 = 1,\ a.e.\ \xi \in \mathbb{R}$.

$m_0(\xi)$ is called to be basic if it satisfies the above three conditions.

Note 1: if $m_0(\xi)$ is basic and its limit function $\hat{\phi} \in L^2(\mathbb{R}) \cap C(\mathbb{R})$, then $\hat{\phi}(2\pi\alpha) = \delta_{0,\alpha}$ $(\forall \alpha \in \mathbb{Z})$.

Note 2: Differing from the case of function pair in Definition 3.11, the conditions $m_0(0) = 1$ and $m_0(\pi) = 0$ in the basic condition I of Definition 3.12 is implied in the basic conditons II and III of Definition 3.12. Furthermore, $\hat{\phi} \in L^2(\mathbb{R})$ can also be concluded by them. The details are included in the following theorem.

Theorem 3.22 *Assume that $\{\tilde{m}_0, m_0\} \subset C(\mathbb{R})$ satisfies the basic condition II, $\hat{\tilde{\phi}}$, $\hat{\phi}$ are the corresponding limit functions, then*

(1). if there is a constant $B > 0$, such that operator $|S|$, which is defined by Eq. (3.29), satisfies $|S|^n 1(\xi) \le B$ (a.e. $\xi \in \mathbb{T}$, for all $n \in N$), then, $\hat{\tilde{\phi}}\bar{\hat{\phi}} \in L^1(\mathbb{R})$.

(2). if m_0 satisfies the basic condition III, then $\hat{\phi} \in L^2(\mathbb{R})$, $m_0(0) = 1, and\ m_0(\pi) = 0$.

Proof We will prove (1) first. Similar to Eq. (3.30), we denote

$$\hat{\tilde{\phi}}_n(\xi) := \prod_{j=1}^{n} \tilde{m}_0(2^{-j}\xi)\chi_{2^n[-\pi,\pi]}(\xi),) \tag{3.35}$$

then it is easy to see that

$$\int_{\mathbb{R}} |\hat{\tilde{\phi}}_n(x)\bar{\hat{\phi}}_n(x)|dx = \int_{\mathbb{R}} (|S|^n 1)(\xi)d\xi \le B(2\pi)\ (\forall n \in \mathbb{N}).$$

Since

$$\hat{\tilde{\phi}}_n(\xi) \to \hat{\tilde{\phi}}(\xi) \; \hat{\phi}_n(\xi) \to \hat{\phi}(\xi) \qquad a.e.\ \xi \in \mathbb{R}.$$

By Fatou lemma, letting $n \to \infty$, we have

$$\int_{\mathbb{R}} |\hat{\tilde{\phi}}(\xi)\bar{\hat{\phi}}(\xi)| d\xi \leq B2\pi.$$

This establishes $\hat{\tilde{\phi}}\bar{\hat{\phi}} \in L^1(\mathbb{R})$.

To prove (2), let $\hat{\tilde{\phi}} = \hat{\phi}$. It is clear that the corresponding operator $|S|$ satisfies $|S|^n 1(\xi) \equiv 1 \quad (a.e.\ \xi \in \mathbb{T},\ \forall n \in \mathbb{N})$, therefore, $\hat{\phi}\bar{\hat{\phi}} \in L^1(\mathbb{R})$, or equivalently, $\hat{\phi} \in L^2(\mathbb{R})$. Since

$$\hat{\phi}_{n+1}(\xi) = m_0(\xi/2^{n+1})\hat{\phi}_n(\xi/2), \quad (\forall \xi \in 2^n[-\pi, \pi)),$$

by letting $n \to \infty$, we have

$$\hat{\phi}(\xi) = m_0(0)\hat{\phi}(\xi) \qquad a.e.\ \xi \in \mathbb{R}.$$

It can be observed easily that $\hat{\phi}(0) = 1$, therefore $m_0(0) = 1$. Furthermore, since m_0 satisfies the basic condition Ⅲ, (2) is proven. ■

Theorem 3.23 *Let $\tilde{\phi},\ \phi \in L^2(\mathbb{R})$. Then,*

$$\begin{aligned}\{\tilde{\phi},\ \phi\} \text{ is biorthonormal} \quad &\Longleftrightarrow \quad F(\xi) = 1 \qquad a.e.\ \xi \in \mathbb{R} \\ &\Longrightarrow \quad \exists C > 0: \ |F|(\xi) \geq C \qquad a.e.\ \xi \in \mathbb{R} \\ &\Longrightarrow \quad \exists C > 0: \ \Phi(\xi)\tilde{\Phi}(\xi) \geq C \quad a.e.\ \xi \in \mathbb{T},\end{aligned}$$

where

$$F(\xi) := \sum_{k \in \mathbb{Z}} \hat{\tilde{\phi}}(\xi + 2k\pi)\bar{\hat{\phi}}(\xi + 2k\pi). \tag{3.36}$$

Furthermore,

(1). If there are $2\pi\mathbb{Z}$-periodic measurable functions $\tilde{m}_0(\xi)$ and $m_0(\xi)$ such that two-scale equations Eq. (3.32) and Eq. (3.33) hold, then

$$\begin{aligned}&\{\tilde{\phi},\ \phi\} \text{ is biorthonormal} \\ \Longrightarrow\ &\tilde{m}_0(\xi)\bar{m}_0(\xi) + \tilde{m}_0(\xi + \pi)\bar{m}_0(\xi + \pi) = 1 \quad a.e.\ \xi \in \mathbb{R}.\end{aligned}$$

(2). If $\tilde{\Phi},\ \Phi \in L^\infty(\mathbb{T})$, then

$\{\tilde{\phi},\ \phi\}$ *is biorthonormal* $\Longrightarrow$ *There is a constant* $C > 0$, *such that:*

$$C^{-1} \le \tilde{\Phi}(\xi),\ \Phi(\xi) \le C \qquad a.e.\ \xi \in \mathbb{T}.$$

Proof By

$$\int_{\mathbb{R}} \tilde{\phi}(x-k)\bar{\phi}(x)dx = \left(\frac{1}{2\pi}\right)\int_{\mathbb{R}} \hat{\tilde{\phi}}(\xi)\bar{\hat{\phi}}(\xi)e^{-ik\xi}d\xi$$

$$= \left(\frac{1}{2\pi}\right)\int_{\mathbb{T}} F(\xi)e^{-ik\xi}d\xi\ (\forall k \in \mathbb{Z}),$$

we have

$$\int_{\mathbb{R}} \tilde{\phi}(x-k)\bar{\phi}(x)dx = \delta_{0,k}\ (\forall k \in \mathbb{Z}) \iff F(\xi) = 1 \qquad a.e.\ \xi \in \mathbb{T}.$$

(1). If there are $2\pi\mathbb{Z}$-periodic measurable functions $\tilde{m}_0(\xi)$ and $m_0(\xi)$ such that two-scale equations Eq. (3.32) and Eq. (3.33) hold, by

$$F(2\xi) = \tilde{m}_0(\xi)\bar{m}_0(\xi)F(\xi) + \tilde{m}_0(\xi+\pi)\bar{m}_0(\xi+\pi)F(\xi+\pi),$$

we conclude that

$$F(\xi) = 1 \qquad a.e.\ \xi \in \mathbb{R}$$

implies

$$\tilde{m}_0(\xi)\bar{m}_0(\xi) + \tilde{m}_0(\xi+\pi)\bar{m}_0(\xi+\pi) = 1, \qquad a.e.\ \xi \in \mathbb{R}.$$

(2). If $\tilde{\Phi},\ \Phi \in L^\infty(\mathbb{T})$, then

$$\begin{aligned}
&\{\tilde{\phi},\ \phi\} && \text{is biorthonormal} \\
&\Longrightarrow && \exists C > 0:\ \Phi(\xi)\tilde{\Phi}(\xi) \ge C \qquad a.e.\ \xi \in \mathbb{T} \\
&\Longrightarrow && \exists C > 0:\ \tilde{\Phi}(\xi),\ \Phi(\xi) \ge C \qquad a.e.\ \xi \in \mathbb{T} \\
&\Longrightarrow && \exists C > 0:\ C^{-1} \le \tilde{\Phi}(\xi),\ \Phi(\xi) \le C\ \ a.e.\ \xi \in \mathbb{T}.
\end{aligned}$$

This finishes our proof. ∎

Note: According to the above discussion, $\forall \tilde{\phi},\ \phi \in L^2(\mathbb{R})$, if $\{\tilde{\phi},\ \phi\}$ is biorthonormal and $\tilde{\Phi},\ \Phi \in L^\infty(\mathbb{T})$, then $\{\tilde{\phi}(\cdot - k)|k \in \mathbb{Z}\}$ is a Riesz basis of $\tilde{V}_0 := \overline{span}\{\tilde{\phi}(\cdot - k)|k \in \mathbb{Z}\}$ and $\{\phi(\cdot - k)|k \in \mathbb{Z}\}$ is a Riesz basis of $V_0 := \overline{span}\{\phi(\cdot - k)|k \in \mathbb{Z}\}$. Furthermore, if $\tilde{\phi},\ \phi \in L(\mathbb{R})$, then

$\hat{\tilde{\phi}}(0) = \hat{\phi}(0) = 1$. Consequently, biorthonormal MRA $\{\tilde{V}_j\}$, $\{V_j\}$ can be generated from $\{\tilde{\phi},\ \phi\}$, here,

$$\tilde{V}_j := \{f(2^j\cdot)|f \in \tilde{V}_0\}, \qquad V_j := \{f(2^j\cdot)|f \in \tilde{V}_0\}\ (\forall j \in \mathbb{Z}).$$

Theorem 3.24 *Let* $\{\tilde{m}_0(\xi),\ m_0(\xi)\}$ *be basic,* $\{\hat{\tilde{\phi}},\ \hat{\phi}\} \subset L^2(\mathbb{R}) \cap C(\mathbb{R})$ *be their limit functions. Suppose* $\{\hat{\phi}_n\}$ *is defined by Eq. (3.30) and* $\{\hat{\tilde{\phi}}_n\}$ *is defined by Eq. (3.35). Then,*

$$\begin{aligned}
\{\tilde{\phi}, \phi\} \textit{ is biorthonormal} \quad &\Longleftrightarrow \quad F(\xi) = 1 \qquad a.e.\ \xi \in \mathbb{R} \\
&\Longleftrightarrow \quad \exists C > 0:\ |F|(\xi) \geq C \qquad a.e.\ \xi \in \mathbb{R} \\
&\Longleftrightarrow \quad \exists C > 0:\ \Phi(\xi)\tilde{\Phi}(\xi) \geq C \quad a.e.\ \xi \in \mathbb{T} \\
&\Longleftrightarrow \quad \|\hat{\tilde{\phi}}_n\overline{\hat{\phi}}_n - \hat{\tilde{\phi}}\overline{\hat{\phi}}\|_1 \to 0\ (n \to \infty).
\end{aligned}$$

Furthermore, if $\tilde{\Phi},\ \Phi \in L^\infty(\mathbb{T})$*, then*

$$\begin{aligned}
\{\tilde{\phi}, \phi\} \quad \textit{can} \quad &\textit{generate a pair of biorthonormal MRAs} \\
\Longleftrightarrow \quad &F(\xi) = 1 \qquad a.e.\ \xi \in \mathbb{R} \\
\Longleftrightarrow \quad &\exists C > 0:\ |F|(\xi) \geq C \quad a.e.\ \xi \in \mathbb{R} \\
\Longleftrightarrow \quad &\exists C > 0:\ \Phi(\xi)\tilde{\Phi}(\xi) \geq C \quad a.e.\ \xi \in \mathbb{T} \\
\Longleftrightarrow \quad &\|\hat{\tilde{\phi}}_n\overline{\hat{\phi}}_n - \hat{\tilde{\phi}}\overline{\hat{\phi}}\|_1 \to 0\ (n \to \infty).
\end{aligned}$$

Proof According to Theorem 3.23, we only need to prove that

$$\exists C > 0:\ \Phi(\xi)\tilde{\Phi}(\xi) \geq C\ a.e.\xi \in \mathbb{T} \quad \Longrightarrow \quad \|\hat{\tilde{\phi}}_n\overline{\hat{\phi}}_n - \hat{\tilde{\phi}}\overline{\hat{\phi}}\|_1 \to 0\ (n \to \infty); \tag{3.37}$$

and

$$\|\hat{\tilde{\phi}}_n\overline{\hat{\phi}}_n - \hat{\tilde{\phi}}\overline{\hat{\phi}}\|_1 \to 0\ (n \to \infty) \quad \Longrightarrow \quad \{\tilde{\phi}, \phi\} \text{ is biorthonormal.} \tag{3.38}$$

It is known that

$$\begin{aligned}
\int_{\mathbb{R}} \tilde{\phi}(x)\overline{\phi}(x-k)dx &= (\frac{1}{2\pi})\int_{\mathbb{R}} \hat{\tilde{\phi}}(\xi)\overline{\hat{\phi}}(\xi)e^{ik\xi}d\xi \\
&= \lim_{n\to\infty}(\frac{1}{2\pi})\int_{\mathbb{R}} \hat{\tilde{\phi}}_n(\xi)\overline{\hat{\phi}}_n(\xi)e^{ik\xi}d\xi \\
&= \delta_{0,k}.
\end{aligned}$$

Therefore, the second implication is proven. We will ignore the proof of the first implication for short (see [Long, 1995]).

When $\tilde{\Phi}$, $\Phi \in L^\infty(\mathbb{T})$, the equivalent equations in the theorem can easily be proved based on the note of Theorem 3.23. ■

A key problem of the theory of the wavelet construction is to find out the sufficient conditions for $\{\tilde{m}_0, m_0\}$ to ensure that $\{\tilde{\phi}, \phi\}$ is biorthonormal. Two important results are given below.

Theorem 3.25 *Let m_0 satisfy the conditions in Theorem 3.20, and*

$$|m_0(\xi)|^2 + |m_0(\xi + \pi)|^2 = 1, \ \forall \xi \in \mathbb{R}.$$

Then, $\prod_{j=1}^{\infty} m_0(\xi/2^j)$ is pointwise convergent to $\hat{\phi} \in L^2(\mathbb{R}) \cap C(\mathbb{R})$, and ϕ can generate an orthonormal MRA.

Proof By Theorems 3.20 and 3.24, it is clear that the theorem holds. ■

The famous Daubechies wavelet is based on this construction method. The key problem is to ensure that m_0 satisfies both Theorem 3.20 and the basic condition Ⅲ. This means to construct the M_0 in Theorem 3.20, and ensure that m_0 satisfies the basic condition Ⅲ (see [Daubechies, 1992]).

Theorem 3.26 *Let $\{\tilde{m}_0, m_0\} \subset \mathcal{P}_N^+$ $(N \neq 0)$ be basic; $\{\hat{\tilde{\phi}}, \hat{\phi}\}$ be their limit functions. Then, the following two statements are equivalent:*

(1). $\{\tilde{\phi}, \phi\}$ can generate a pair of biorthonormal MRAs.

(2). Each eigenvalue λ of $\tilde{T}|_{\dot{\mathcal{P}}_N}$ and $T|_{\dot{\mathcal{P}}_N}$, which are the restrictions of the translation operators $\tilde{T}$ and T in $\dot{\mathcal{P}}_N$, satisfies $|\lambda| < 1$.

Corollary 3.3 *Let $\{\tilde{m}_0, m_0\} \subset \mathcal{P}_N$ $(N \neq 0)$ satisfy the basic condition I; $\{\hat{\tilde{\phi}}, \hat{\phi}\}$ be their limit functions. Then, the following two statements are equivalent:*

(1). $\hat{\tilde{\phi}}$, $\hat{\phi} \in L^2(\mathbb{R})$, and $\{\tilde{\phi}, \phi\}$ can generate a pair of biorthonormal MRAs.

(2). $\tilde{m}_0(\xi)\bar{m}_0(\xi) + \tilde{m}_0(\xi + \pi)\bar{m}_0(\xi + \pi) = 1$ a.e. $\xi \in \mathbb{R}$, and each eigenvalue λ of $\tilde{T}|_{\dot{\mathcal{P}}_N}$ and $T|_{\dot{\mathcal{P}}_N}$, which are the restrictions of the translation operators $\tilde{T}$ and T in $\dot{\mathcal{P}}_N$, satisfies $|\lambda| < 1$.

Proof Obviously, $\{\tilde{m}_0, m_0\}$ satisfies the basic condition Ⅱ.

"(1)⇒(2)": By (1) in Theorem 3.23, we obtain

$$\tilde{m}_0(\xi)\bar{m}_0(\xi) + \tilde{m}_0(\xi+\pi)\bar{m}_0(\xi+\pi) = 1 \qquad a.e.\ \xi \in \mathbb{R},$$

i.e., $\{\tilde{m}_0, m_0\}$ satisfies the basic condition III. Thus, $\{\tilde{m}_0,\ m_0\}$ is basic. By Theorem 3.26, (2) is proven.

"(2)⇒(1)": Since $\{\tilde{m}_0,\ m_0\}$ is basic, according to Theorem 3.26, (1) is true. ■

3.2.2.2 *Examples of Constructing MRA*

Examples of constructing MRAs based on Theorem 3.3 will be discussed in this subsection.

First of all, the basic condition I can be equivalently illustrated by masks. That is:

Theorem 3.27 *Let $m_0(\xi) := \sum_{k\in\mathbb{Z}} h_k e^{-ik\xi}$, where $\{h_k\} \in l^1(\mathbb{Z})$. Then*

$$m_0(0) = 1 \text{ and } m_0(\pi) = 0 \iff \sum_{k\in\mathbb{Z}} h_{2k} = \sum_{k\in\mathbb{Z}} h_{2k+1} = \frac{1}{2}.$$

Proof For $\mu = 0$ or 1, we have

$$\begin{aligned} m_0(\pi\mu) &= \sum_{k\in\mathbb{Z}} h_k e^{-i\pi k\mu} \\ &= (\sum_{k\in\mathbb{Z}} h_{2k}) + (\sum_{k\in\mathbb{Z}} h_{2k+1})e^{-i\pi\mu}, \end{aligned}$$

or,

$$\left\{ \begin{array}{l} m_0(0) = \sum_{k\in\mathbb{Z}} h_{2k} + \sum_{k\in\mathbb{Z}} h_{2k+1} \\ m_0(\pi) = \sum_{k\in\mathbb{Z}} h_{2k} - \sum_{k\in\mathbb{Z}} h_{2k+1} \end{array} \right.$$

Therefore,

$$\left\{ \begin{array}{l} m_0(0) = 1 \\ m_0(\pi) = 0 \end{array} \right. \iff \left\{ \begin{array}{l} \sum_{k\in\mathbb{Z}} h_{2k} + \sum_{k\in\mathbb{Z}} h_{2k+1} = 1 \\ \sum_{k\in\mathbb{Z}} h_{2k} - \sum_{k\in\mathbb{Z}} h_{2k+1} = 0 \end{array} \right.$$

$$\iff \sum_{k\in\mathbb{Z}} h_{2k} = \sum_{k\in\mathbb{Z}} h_{2k+1} = \frac{1}{2}.$$

This finishes our proof. ■

For $m_0 \in \mathcal{P}_N^+$ $(N \neq 0)$, by corollary 3.2, we see that whether ϕ, which is defined by m_0, can generates a MRA, depends on the properties of the eigenvalues and eigenvectors of $T|_{\mathcal{P}_N}$. It is easy to see that $T|_{\mathcal{P}_N}$ can be represented by a $2N+1$-order matrix. Since the eigenvalues and eigenvectors of a matrix can be calculated easily with a computer, the construction of MRAs in this case is always feasible.

Let $m_0(\xi) = \sum_{k\in\mathbb{Z}} h_k e^{-ik\xi} \in \mathcal{P}_N^+$ $(N \neq 0)$. $\{e^{-ik\xi} | k \in \mathbb{Z} \cap [-N, N]\}$ be a set of basis of $\mathcal{P}_N$. Our first step is to get the matrix of $T|_{\mathcal{P}_N}$ based on this basis. We have that

$$\begin{aligned}
(Te^{-ilx})(\xi) &= |m_0(\frac{\xi}{2})|^2 e^{-il\frac{\xi}{2}} + |m_0(\frac{\xi}{2}\pi)|^2 e^{-il\frac{\xi}{2}} e^{-il\pi} \\
&= \left(|m_0(\frac{\xi}{2})|^2 + |m_0(\frac{\xi}{2}+\pi)|^2 e^{-il\pi}\right) e^{-il\frac{\xi}{2}} \\
&= \sum_{n\in\mathbb{Z}}\sum_{k\in\mathbb{Z}} h_n \bar{h}_k e^{-i(n-k)\frac{\xi}{2}} [1 + e^{-i(n-k+l)\pi}] e^{-il\frac{\xi}{2}} \\
&= \sum_{n\in\mathbb{Z}}\sum_{k\in\mathbb{Z}} h_n \bar{h}_k e^{-i(n-k+l)\frac{\xi}{2}} [1 + e^{-i(n-k+l)\pi}] \\
&= \sum_{n\in\mathbb{Z}}\sum_{k\in\mathbb{Z}} h_n \bar{h}_{n+l-k} e^{-ik\frac{\xi}{2}} [1 +^{-ik\pi}] \\
&= \sum_{k\in\mathbb{Z}} (2 \sum_{n\in\mathbb{Z}} h_n \bar{h}_{n+l-2k}) e^{-ik\xi} ,
\end{aligned}$$

Let $\{r_1, \cdots, r_s\}$ $(s := 2N+1)$ be a permutation of the set $\mathbb{Z} \cap [-N, N]$, and

$$(Te^{-ir_1\xi}, \cdots, Te^{-ir_s\xi}) = (e^{-ir_1\xi}, \cdots, e^{-ir_s\xi})A,$$

where A, the corresponding matrix of $T|_{\mathcal{P}_N}$ based on the basis $\{e^{-ir_1\xi}, \cdots, e^{-ir_s\xi}\}$, is a $2N+1$-order matrix:

$$A = \begin{pmatrix} a_{1,1} & a_{1,2} & \cdots & a_{1,s} \\ a_{2,1} & a_{2,2} & \cdots & a_{2,s} \\ \vdots & \vdots & \vdots & \vdots \\ a_{s,1} & a_{s,2} & \cdots & a_{s,s} \end{pmatrix}.$$

The (k, l)-th element of A is:

$$a_{k,l} = 2\sum_{n\in\mathbb{Z}} h_n \bar{h}_{n+r_l-2r_k} \ (k, l = 1, \cdots, s).$$

By corollary 3.2, we have

Theorem 3.28 *Let* $m_0(\xi) = \sum_{k\in\mathbb{Z}} h_k e^{-ik\xi} \in \mathcal{P}_N^+$, $(N \neq 0)$, $\{r_1, \cdots, r_s\}$ *be a permutation of* $\{-N, -N+1, \cdots, N\}(s := 2N+1)$, *and the* (k, l)*-th* *element of A is:*

$$a_{k,l} = 2\sum_{n=0}^{N} h_n \bar{h}_{n+r_l-2r_k} \ (k,l = 1,\cdots,s).$$

Then, $\hat{\phi}(\xi) := \prod_{j=1}^{\infty} m_0(\xi/2^j)$ is pointwise convergent on $\mathbb{R}$ and ϕ generates a MRA of $L^2(\mathbb{R})$, if and only if the following three conditions are satisfied:

(1). $\sum_{k\in\mathbb{Z}} h_{2k} = \sum_{k\in\mathbb{Z}} h_{2k} = \frac{1}{2}$;

(2). 1 is the simple eigenvalue of matrix A, and each of the other eigenvalues λ satisfies $|\lambda| < 1$;

(3). The eigenpolynomial , $g(\xi) := \sum_{k=1}^{s} b_k e^{-ir_k\xi}$, has a positive lower-bound, where $(b_1,\cdots,b_s)^T$ is the eigenvector of matrix A, which corresponds to the eigenvalue 1 and satisfies $\sum_{k=1}^{s} b_k = 1$.

Proof Due to that A is the corresponding matrix of $T|_{\mathcal{P}_N}$ based on the basis $\{e^{-ir_1\xi},\cdots,e^{-ir_s\xi}\}$ of $\mathcal{P}_N$, it is clear that (1) and (2) in the theorem are equivalent to (1) and (2) in corollary 3.2 respecteively. We conclude that $g(\xi) := \sum_{k=1}^{s} b_k e^{-ir_k\xi}$ is the eigenvector of $T|_{\mathcal{P}_N}$ corresponding to the eigenvalue 1, and satisfies $g(0) = 1$, if and only if, $(b_1,\cdots,b_s)^T$ is the eigenvector of A corresponding to the eigenvalue 1 and satisfies $\sum_{k=1}^{s} b_k = 1$. In fact, it is easy to see that

$$\begin{aligned}
Tg(\xi) &= g(\xi) \\
\iff & (Te^{-ir_1\xi},\cdots,Te^{-ir_s\xi})\begin{pmatrix} b_1 \\ \vdots \\ b_s \end{pmatrix} = (e^{-ir_1\xi},\cdots,e^{-ir_s\xi})\begin{pmatrix} b_1 \\ \vdots \\ b_s \end{pmatrix} \\
\iff & (e^{-ir_1\xi},\cdots,e^{-ir_s\xi})A\begin{pmatrix} b_1 \\ \vdots \\ b_s \end{pmatrix} = (e^{-ir_1\xi},\cdots,e^{-ir_s\xi})\begin{pmatrix} b_1 \\ \vdots \\ b_s \end{pmatrix} \\
\iff & A\begin{pmatrix} b_1 \\ \vdots \\ b_s \end{pmatrix} = \begin{pmatrix} b_1 \\ \vdots \\ b_s \end{pmatrix}.
\end{aligned}$$

Obviously, $g(0) = 1 \iff \sum_{k=1}^{s} b_k = 1$. By Corollary 3.2, our proof is complete. ■

For a natural permutation of $\mathbb{Z}\cap[-N,N] = \{-N,-N+1,\cdots,N-1,N\}$:

$$r_1 = -N,\ \cdots,\ r_k = -N-1,\ \cdots,\ r_{2N+1} = N.$$

The corresponding matrix A is a $2N+1$-order matrix whose (k,l)-th elements is:

$$a_{k,l} = 2\sum_{n=0}^{N} h_n \bar{h}_{n+N+1+l-2k}, \qquad (k,l = 1,\cdots,2N+1),$$

where $h_k = 0$ for $k < 0$ and $k > N$.

To construct MRA, a necessary condition that $\{h_0,\cdots,h_N\}$ should satisfy is

$$\sum_{k\in\mathbb{Z},0\leq 2k\leq N} h_{2k} = \sum_{k\in\mathbb{Z},0\leq 2k+1\leq N} h_{2k+1} = \frac{1}{2}.$$

Now, for $N = 1, 2$, we discuss the conditions that $\{h_0,\cdots,h_N\}$ should satisfy so that it can generate a MRA. We will focus on the three conditions in Theorem 3.28.

1). For $N = 1$, we have $h_0 = h_1 = \frac{1}{2}$. A is a 3-order matrix whose elements are:

$$a_{k,l} = 2(h_0\bar{h}_{2+l-2k} + h_1\bar{h}_{3+l-2k}) = h_{2+l-2k} + h_{3+l-2k}, \quad (k,l = 1,2,3).$$

Thus, A can be represented as:

$$A = \begin{pmatrix} \frac{1}{2} & 0 & 0 \\ \frac{1}{2} & 1 & \frac{1}{2} \\ 0 & 0 & \frac{1}{2} \end{pmatrix}.$$

The eigenvalues of A are 1, $\frac{1}{2}$. The eigenvector corresponding to 1 is $(b_1,b_2,b_3)^t = (0,1,0)^t$, and the sum of its components equals to 1. Hence, the eigenpolynomial

$$g(\xi) = e^{-ir_2\xi} = e^{-i0\xi} \equiv 1$$

has a positive lower-bound. This tells us that there is only one MRA when $N = 1$. Now, we intend to find out its corresponding scale function. By

$$m_0(\xi) = e^{-i\frac{1}{2}\xi} cos\frac{\xi}{2},$$

we have

$$\hat{\phi}(\xi) \;=\; \lim_{n\to\infty}\prod_{j=1}^{n} m_0(\frac{\xi}{2^j})$$

$$
\begin{aligned}
&= \lim_{n\to\infty} \prod_{j=1}^{n} (e^{-i\frac{\xi}{2^{j+1}}} \cos \frac{\xi}{2^{j+1}}) \\
&= e^{-i\frac{\xi}{2}} \lim_{n\to\infty} \prod_{j=1}^{n} \cos \frac{\xi}{2^{j+1}} \\
&= e^{-i\frac{\xi}{2}} \lim_{n\to\infty} \frac{\sin \frac{\xi}{2}}{2^n \sin \frac{\xi}{2^{n+1}}} \\
&= e^{-i\frac{\xi}{2}} \frac{2}{\xi} \sin \frac{\xi}{2}
\end{aligned}
$$

This is just the Fourier transform of the characteristic function $\chi_{[0,1]}(x)$ of interval $[0, 1]$. Therefore,

$$
\phi(x) = \chi_{[0,1]}(x) := \begin{cases} 1 & \text{when } x \in [0,1] \\ 0 & \text{otherwise} \end{cases}
$$

As a conclusion, for $N = 1$, there is only one MRA, which is generated by $\chi_{[0,1]}(x)$.

2). For $N = 2$, we have

$$
h_0 + h_2 = \frac{1}{2},\ h_1 = \frac{1}{2}.
$$

Therefore, A is a 5-order matrix whose (k, l)-th elements is:

$$
a_{k,l} = 2(h_0\bar{h}_{3+l-2k} + \frac{1}{2}\bar{h}_{4+l-2k} + h_2\bar{h}_{5+l-2k}), \quad (k, l = 1, \cdots, 5).
$$

Thus,

$$
\begin{cases}
a_{1,l} = 2h_0\bar{h}_{1+l} \\
a_{2,l} = 2(h_0\bar{h}_{l-1} + \frac{1}{2}\bar{h}_l + h_2\bar{h}_{l+1}) \\
a_{3,l} = 2(h_0\bar{h}_{l-3} + \frac{1}{2}\bar{h}_{l-2} + h_2\bar{h}_{l-1}) \\
a_{4,l} = 2(h_0\bar{h}_{l-5} + \frac{1}{2}\bar{h}_{l-4} + h_2\bar{h}_{l-3}) \\
a_{5,l} = 2h_2\bar{h}_{l-5}
\end{cases}
$$

Hence, matrix A can be written as:

$$
\begin{pmatrix}
h_0 - 2|h_0|^2 & 0 & 0 & 0 & 0 \\
1 - h_0 - \bar{h}_0 + 4|h_0|^2 & \frac{1}{2} + h_0 - \bar{h}_0 & h_0 - 2|h_0|^2 & 0 & 0 \\
\bar{h}_0 - 2|h_0|^2 & \frac{1}{2} + \bar{h}_0 - h_0 & 1 - h_0 - \bar{h}_0 + 4|h_0|^2 & \frac{1}{2} + h_0 - \bar{h}_0 & h_0 - 2|h_0|^2 \\
0 & 0 & \bar{h}_0 - 2|h_0|^2 & \frac{1}{2} + \bar{h}_0 - h_0 & 1 - h_0 - \bar{h}_0 + 4|h_0|^2 \\
0 & 0 & 0 & 0 & \bar{h}_0 - 2|h_0|^2
\end{pmatrix}.
$$

The eigenpolynomial of A is

$$|\lambda I - A| = (\lambda - h_0 + 2|h_0|^2)(\lambda - \bar{h}_0 + 2|h_0|^2)$$

$$\cdot \begin{vmatrix} \lambda - \frac{1}{2} - h_0 + \bar{h}_0 & -h_0 + 2|h_0|^2 & 0 \\ -\frac{1}{2} - \bar{h}_0 + h_0 & \lambda - 1 + h_0 + \bar{h}_0 - 4|h_0|^2 & -\frac{1}{2} - h_0 + \bar{h}_0 \\ 0 & -\bar{h}_0 + 2|h_0|^2 & \lambda - \frac{1}{2} - \bar{h}_0 + h_0 \end{vmatrix}$$

$$= (\lambda - h_0 + 2|h_0|^2)(\lambda - \bar{h}_0 + 2|h_0|^2)$$

$$\cdot \begin{vmatrix} \lambda - \frac{1}{2} - h_0 + \bar{h}_0 & -h_0 + 2|h_0|^2 & 0 \\ \lambda - 1 & \lambda - 1 & \lambda - 1 \\ 0 & -\bar{h}_0 + 2|h_0|^2 & \lambda - \frac{1}{2} - \bar{h}_0 + h_0 \end{vmatrix}$$

$$= (\lambda - h_0 + 2|h_0|^2)(\lambda - \bar{h}_0 + 2|h_0|^2)(\lambda - 1)$$

$$\cdot \begin{vmatrix} \lambda - \frac{1}{2} - h_0 + \bar{h}_0 & -h_0 + 2|h_0|^2 & 0 \\ 1 & 1 & 1 \\ 0 & -\bar{h}_0 + 2|h_0|^2 & \lambda - \frac{1}{2} - \bar{h}_0 + h_0 \end{vmatrix}$$

$$= (\lambda - h_0 + 2|h_0|^2)(\lambda - \bar{h}_0 + 2|h_0|^2)(\lambda - 1)$$

$$\cdot \begin{vmatrix} \lambda - \frac{1}{2} - h_0 + \bar{h}_0 & 1 & 0 \\ -h_0 + 2|h_0|^2 & 1 & -\bar{h}_0 + 2|h_0|^2 \\ 0 & 1 & \lambda - \frac{1}{2} - \bar{h}_0 + h_0 \end{vmatrix}$$

$$= -(\lambda - h_0 + 2|h_0|^2)(\lambda - \bar{h}_0 + 2|h_0|^2)(\lambda - 1)$$

$$\cdot \begin{vmatrix} 1 & \lambda - \frac{1}{2} - h_0 + \bar{h}_0 & 0 \\ 1 & -h_0 + 2|h_0|^2 & -\bar{h}_0 + 2|h_0|^2 \\ 1 & 0 & \lambda - \frac{1}{2} - \bar{h}_0 + h_0 \end{vmatrix}$$

$$= -(\lambda - h_0 + 2|h_0|^2)(\lambda - \bar{h}_0 + 2|h_0|^2)(\lambda - 1)$$

$$\cdot \begin{vmatrix} 1 & \lambda - \frac{1}{2} - h_0 + \bar{h}_0 & 0 \\ 0 & -\lambda + \frac{1}{2} + 2|h_0|^2 - \bar{h}_0 & 2|h_0|^2 - \bar{h}_0 \\ 0 & h_0 - 2|h_0|^2 & \lambda - \frac{1}{2} + h_0 - 2|h_0|^2 \end{vmatrix}$$

$$= -(\lambda - h_0 + 2|h_0|^2)(\lambda - \bar{h}_0 + 2|h_0|^2)(\lambda - 1)$$

$$\cdot \begin{vmatrix} -\lambda + \frac{1}{2} + 2|h_0|^2 - \bar{h}_0 & 2|h_0|^2 - \bar{h}_0 \\ h_0 - 2|h_0|^2 & \lambda - \frac{1}{2} + h_0 - 2|h_0|^2 \end{vmatrix}$$

$$= -(\lambda - h_0 + 2|h_0|^2)(\lambda - \bar{h}_0 + 2|h_0|^2)(\lambda - 1)$$

$$\cdot \begin{vmatrix} -\lambda + \frac{1}{2} & 2|h_0|^2 - \bar{h}_0 \\ -\lambda + \frac{1}{2} & \lambda - \frac{1}{2} + h_0 - 2|h_0|^2 \end{vmatrix}$$

$$= (\lambda - h_0 + 2|h_0|^2)(\lambda - \bar{h}_0 + 2|h_0|^2)(\lambda - 1)(\lambda - \frac{1}{2})$$

$$\cdot \begin{vmatrix} 1 & 2|h_0|^2 - \bar{h}_0 \\ 1 & \lambda - \frac{1}{2} + h_0 - 2|h_0|^2 \end{vmatrix}$$

$$= (\lambda - h_0 + 2|h_0|^2)(\lambda - \bar{h}_0 + 2|h_0|^2)(\lambda - 1)(\lambda - \frac{1}{2})$$

$$\cdot \begin{vmatrix} 1 & 2|h_0|^2 - \bar{h}_0 \\ 0 & \lambda - \frac{1}{2} + h_0 + \bar{h}_0 - 4|h_0|^2 \end{vmatrix}$$

$$= (\lambda - h_0 + 2|h_0|^2)(\lambda - \bar{h}_0 + 2|h_0|^2)(\lambda - 1)(\lambda - \frac{1}{2})$$

$$\cdot(\lambda - \frac{1}{2} + h_0 + \bar{h}_0 - 4|h_0|^2).$$

Therefore, the eigenvalues of A are:

$$1,\ \frac{1}{2},\ h_0 - 2|h_0|^2,\ \bar{h}_0 - 2|h_0|^2,\ \frac{1}{2} + 4|h_0|^2 - h_0 - \bar{h}_0.$$

In order to generate MRA, the moduli of the above eigenvalues, except 1, must be smaller than 1, that is,

$$\begin{cases} |h_0 - 2|h_0|^2| < 1 \\ |\frac{1}{2} + 4|h_0|^2 - h_0 - \bar{h}_0| < 1 \end{cases}.$$

We now simplify these conditions as follows. By

$$\frac{1}{2}|h_0|^2 - h_0 - \bar{h}_0 = \frac{1}{2}{}_0\bar{h}_0 - h_0 - \bar{h}_0 = \frac{1}{4}|h_0 - \frac{1}{4}|^2 > 0.$$

we get

$$|\frac{1}{2}+4|h_0|^2-h_0-\bar{h}_0|<1 \iff |h_0-\frac{1}{4}|<\frac{\sqrt{3}}{4}.$$

For $|h_0-\frac{1}{4}|<\frac{\sqrt{3}}{4}$, we have:

$$\begin{aligned}
\left|h_0-2|h_0|^2\right| &= \sqrt{|2|h_0|^2-Re(h_0)|^2+|Im(h_0)|^2} \\
&= \sqrt{|2|h_0|^2-\frac{1}{2}(h_0+\bar{h}_0)|^2+|Im(h_0)|^2} \\
&= \sqrt{\frac{1}{4}\left|4|h_0|^2-h_0-\bar{h}_0\right|^2+|Im(h_0)|^2} \\
&= \sqrt{\frac{1}{4}\left|4|h_0-\frac{1}{4}|^2-\frac{1}{4}\right|^2+|Im(h_0)|^2} \\
&< \sqrt{\frac{1}{4}\left(\frac{1}{2}\right)^2+|Im(h_0)|^2} \\
&= \frac{1}{2}\sqrt{\frac{1}{4}+4|Im(h_0)|^2}.
\end{aligned}$$

And by

$$|\frac{1}{2}-h_0+\bar{h}_0|=|\frac{1}{2}-2iIm(h_0)|=\sqrt{\frac{1}{4}+4|Im(h_0)|^2},$$

we have

$$\begin{aligned}
\left|h_0-2|h_0|^2\right| &< \frac{1}{2}\left|\frac{1}{2}-h_0+\bar{h}_0\right| \\
&= \frac{1}{2}\left|\frac{1}{2}-(h_0-\frac{1}{4})+(\bar{h}_0-\frac{1}{2})\right| \\
&< \frac{1}{2}\left(\frac{1}{2}+\frac{\sqrt{3}}{4}+\frac{\sqrt{3}}{4}\right) \\
&= \frac{1+\sqrt{3}}{4}<1,
\end{aligned}$$

Hence

$$\begin{cases} |h_0-2|h_0|^2|<1 \\ |\frac{1}{2}+4|h_0|^2-h_0-\bar{h}_0|<1 \end{cases} \iff |h_0-\frac{1}{4}|<\frac{\sqrt{3}}{4}.$$

Now we suppose $|h_0 - \frac{1}{4}| < \frac{\sqrt{3}}{4}$. To solve the eigenvector corresponding to 1, such that the sum of whose components equals to 1, we apply the elementary row transformation to matrix $I - A$ as follows:

$$
I - A \Longrightarrow
\begin{pmatrix}
1-h_0+2|h_0|^2 & 0 & 0 & 0 & 0 \\
h_0+\bar{h}_0-1-4|h_0|^2 & \frac{1}{2}-h_0+\bar{h}_0 & 2|h_0|^2-h_0 & 0 & 0 \\
2|h_0|^2-\bar{h}_0 & h_0-\bar{h}_0-\frac{1}{2} & -4|h_0|^2+h_0+\bar{h}_0 & \bar{h}_0-h_0-\frac{1}{2} & 2|h_0|^2-h_0 \\
0 & 0 & 2|h_0|^2-\bar{h}_0 & \frac{1}{2}+h_0-\bar{h}_0 & h_0+\bar{h}_0-1-4|h_0|^2 \\
0 & 0 & 0 & 0 & 1+2|h_0|^2-\bar{h}_0
\end{pmatrix}
$$

$$
\Rightarrow
\begin{pmatrix}
1-h_0+2|h_0|^2 & 0 & 0 & 0 & 0 \\
h_0+\bar{h}_0-1-4|h_0|^2 & \frac{1}{2}-h_0+\bar{h}_0 & 2|h_0|^2-h_0 & 0 & 0 \\
0 & 0 & 0 & 0 & 0 \\
0 & 0 & 2|h_0|^2-\bar{h}_0 & \frac{1}{2}{}_0+h_0-\bar{h}_0 & h_0+\bar{h}_0-1-4|h_0|^2 \\
0 & 0 & 0 & 0 & 1+2|h_0|^2-\bar{h}_0
\end{pmatrix}
$$

$$
\Rightarrow
\begin{pmatrix}
1 & 0 & 0 & 0 & 0 \\
0 & \frac{1}{2}-h_0+\bar{h}_0 & 2|h_0|^2-h_0 & 0 & 0 \\
0 & 0 & 2|h_0|^2-\bar{h}_0 & \frac{1}{2}+h_0-\bar{h}_0 & 0 \\
0 & 0 & 0 & 0 & 1 \\
0 & 0 & 0 & 0 & 0
\end{pmatrix}.
$$

By solving the following linear system:

$$
\begin{cases}
\begin{pmatrix}
1 & 0 & 0 & 0 & 0 \\
0 & \frac{1}{2}-h_0+\bar{h}_0 & 2|h_0|^2-h_0 & 0 & 0 \\
0 & 0 & 2|h_0|^2-\bar{h}_0 & \frac{1}{2}+h_0-\bar{h}_0 & 0 \\
0 & 0 & 0 & 0 & 1 \\
0 & 0 & 0 & 0 & 0
\end{pmatrix}
\begin{pmatrix} b_1 \\ b_2 \\ b_3 \\ b_4 \\ b_5 \end{pmatrix} = 0 \\
b_1+b_2+b_3+b_4+b_5 = 1
\end{cases},
$$

we obtain the eigenvector corresponding to the eigenvalue 1, so that the sum of its components equals to 1. It is

$$
\begin{pmatrix} b_1 \\ b_2 \\ b_3 \\ b_4 \\ b_5 \end{pmatrix}
= \frac{8|\frac{1}{2}-h_0+\bar{h}_0|^2}{3-16|h_0-\frac{1}{4}|^2}
\begin{pmatrix} 0 \\ \frac{h_0-2|h_0|^2}{\frac{1}{2}-h_0+\bar{h}_0} \\ 1 \\ \frac{\bar{h}_0-2|h_0|^2}{\frac{1}{2}-\bar{h}_0+h_0} \\ 0 \end{pmatrix}
$$

Thus, the eigenpolynomial is:

$$
g(\xi) \;=\; b_1e^{i2\xi} + b_2e^{i\xi} + b_3 + b_4e^{-i\xi} + b_5e^{-i2\xi}
$$

$$
\begin{aligned}
&= \frac{8|\frac{1}{2}-h_0+\bar{h}_0|^2}{3-16|h_0-\frac{1}{4}|^2}\left(\frac{h_0-2|h_0|^2}{\frac{1}{2}-h_0\bar{h}_0}e^{i\xi}+1+\frac{\bar{h}_0-2|h_0|^2}{\frac{1}{2}-\bar{h}_0+h_0}e^{-i\xi}\right)\\
&= \frac{8|\frac{1}{2}-h_0+\bar{h}_0|^2}{3-16|h_0-\frac{1}{4}|^2}\left(1+2Re(\frac{h_0-2|h_0|^2}{\frac{1}{2}-h_0+\bar{h}_0}e^{i\xi})\right)\\
&\geq \frac{8|\frac{1}{2}-h_0+\bar{h}_0|^2}{3-16|h_0-\frac{1}{4}|^2}\left(1-2\left|\frac{h_0-2|h_0|^2}{\frac{1}{2}-h_0+\bar{h}_0}\right|\right)\\
&> 0,
\end{aligned}
$$

which shows that $g(\xi)$ has a positive lower-bound.

All the discussion above shows that, for $N=2$, $m_0(\xi) := h_0 + h_1 e^{-i\xi} + h_2 e^{-i2\xi}$ can generate a MRA based on Theorem 3.28 if and only if:

$$
\begin{cases} h_0 \neq 0,\ h_0 \neq \frac{1}{2} \text{ and } |h_0-\frac{1}{4}| < \frac{\sqrt{3}}{4} \\ h_1 = \frac{1}{2} \\ h_2 = \frac{1}{2} - h_0 \end{cases} \tag{3.39}
$$

There are infinite bank of $\{h_0, h_1, h_2\}$ which satisfy the above conditions. Generally speaking, it is very difficult to find the corresponding scale function ϕ. Now, we consider a particular case.

For $h_0 = \frac{1}{4}$, we have

$$
\begin{aligned}
m_0(\xi) &= \frac{1}{4}+\frac{1}{2}e^{-i\xi}+\frac{1}{4}e^{-2i\xi}\\
&= \frac{1}{4}(1+e^{-i\xi})^2\\
&= \left(e^{-i\xi/2}\cos(\frac{1}{2}\xi)\right)^2.
\end{aligned}
$$

Obviously, it is just the square of the filter function m_0 for $N=1$. Therefore,

$$
\hat{\phi}(\xi) = \left(\hat{\chi}_{[0,1)}(\xi)\right)^2 = (\chi_{[0,1)} * \chi_{[0,1)})\hat{}(\xi),
$$

i.e.,

$$
\phi(x) = \chi_{[0,1)} * \chi_{[0,1)}(x) = \begin{cases} x & x\in[0,1) \\ 2-x & x\in[1,2) \\ 0 & \text{otherwise} \end{cases}.
$$

This is the well-known first order B-spline function.

3.2.3 The Construction of Biorthonormal Wavelet Bases

Multiresolution Analysis (MRA), since it was proposed by S. Mallat and Y. Meyer in late 1980's, has become the standard scheme for construction of wavelet bases. It has shown that almost all the wavelet bases with usual properties can be constructed from MRAs. In this section, we will study how to construct a general (orthonormal or not) wavelet basis $\{\psi_{j,k}\}$ from an MRA.

Definition 3.13 $\phi \in L^2(\mathbb{R})$ is said to satisfy the basic smooth condition if there exist positive numbers $\delta < 2$ and C such that

$$\begin{cases} \sum_{k\in\mathbb{Z}} |\hat{\phi}(\xi + 2\pi k)|^{2-\delta} \leq C & \text{a. e. } \xi \in \mathbb{T} \\ \sum_{j=0}^{\infty} |\hat{\phi}(2^j\xi)|^{\delta} \leq C & \text{a. e. } 1 \leq |\xi| \leq 2. \end{cases}$$

Note 1: The condition in the definition is called the basic smooth condition since the smoothness of ϕ corresponds to the decreasing of its Fourier transform.

Note 2: It can be proved that ϕ defined by $\hat{\phi}(\xi) := \prod_{j=1}^{\infty} m_0(\xi/2^j)$ satisfies the basic smooth condition if m_0 satisfies either the following conditions:

(1) the conditions of Theorem 3.20;
(2) $m_0 \in \mathcal{P}_N^+ \ (N \neq 0), m_0(0) = 1, m_0(\pi) = 0$ and any eigenvalue λ of $T_{\dot{\mathcal{P}}_N}$, the restriction of transition operator T on $\dot{\mathcal{P}}_N$, satisfies $|\lambda| < 1$.

Theorem 3.29 *Suppose $\tilde{\phi}, \phi \in L^2(\mathbb{R})$, generate biorthonormal MRA $\{\tilde{V}_j\}, \{V_j\}$, satisfy the basic smooth condition and the corresponding filter function $\tilde{m}_0, m_0 \in L^\infty(\mathbb{T})$ satisfy*

$$|m_0(\xi)|^2 + |m_0(\xi + \pi)|^2 \neq 0,$$

and

$$\tilde{m}_0(\xi)m_0(\xi) + \tilde{m}_0(\xi + \pi)m_0(\xi + \pi) = 1, \quad a.e. \xi \in \mathbb{R}.$$

Then, for

$$m_1(\xi) = a(2\xi)\bar{\tilde{m}}_0(\xi + \pi)e^{-i\xi}, \quad \tilde{m}_1(\xi) = \overline{a^{-1}(2\xi)}\bar{m}_0(\xi + \pi)e^{-i\xi},$$

where $a(\xi)$ denotes a 2π-periodic function bounded by positive numbers from both the above and below, $\{\tilde{\psi}_{j,k},\ \psi_{j,k}\}_{j,k\in\mathbb{Z}}$ which is defined by

$$\hat{\tilde{\psi}}(\xi) := \tilde{m}_1(\frac{\xi}{2})\hat{\tilde{\phi}}(\frac{\xi}{2}), \quad \hat{\psi}(\xi) := m_1(\frac{\xi}{2})\hat{\phi}(\frac{\xi}{2}) \tag{3.40}$$

constitutes a pair of biorthonaoral wavelet bases of $L^2(\mathbb{R})$, that is, both $\{\tilde{\psi}_{j,k}\}_{j,k\in\mathbb{Z}}$ and $\{\psi_{j,k}\}_{j,k\in\mathbb{Z}}$ constitute a Riesz basis of $L^2(\mathbb{R})$ and they are biorthonoral each other, i.e.

$$\langle\tilde{\psi}_{j,k}\ ,\psi_{j',k'}\rangle = \delta_{j,j'}\delta_{k,k'} \qquad (\forall j,\ j'\in\mathbb{Z},\ k,\ k'\in\mathbb{Z}).$$

Further, the following results hold:
(1). The operators defined by

$$\tilde{P}_j f := \sum_{k\in\mathbb{Z}}\langle f,\phi_{j,k}\rangle\tilde{\phi}_{j,k} \qquad P_j f := \sum_{k\in\mathbb{Z}}\langle f,\tilde{\phi}_{j,k}\rangle\phi_{j,k} \tag{3.41}$$

$$\tilde{Q}_j f := \sum_{k\in\mathbb{Z}}\langle f,\psi_{j,k}\rangle\tilde{\psi}_{j,k} \qquad Q_j f := \sum_{k\in\mathbb{Z}}\langle f,\tilde{\psi}_{j,k}\rangle\psi_{j,k} \tag{3.42}$$

satisfy that, $\forall j\in\mathbb{Z}$,

$$\begin{cases} \tilde{P}_{j+1} = \tilde{P}_j + \tilde{Q}_j, \qquad P_{j+1} = P_j + Q_j, \\ \lim_{j\to-\infty}\|\tilde{P}_j f\|_2 = \lim_{j\to-\infty}\|P_j f\|_2 = 0 \qquad \forall f\in L^2(\mathbb{R}), \\ \lim_{j\to+\infty}\|\tilde{P}_j f - f\|_2 = \lim_{j\to+\infty}\|P_j f - f\|_2 = 0 \qquad \forall f\in L^2(\mathbb{R}). \end{cases}$$

(2). If we denote

$$\tilde{W}_j := \overline{span}\{\tilde{\psi}_{j,k}|k\in\mathbb{Z}\}, \quad W_j := \overline{span}\{\psi_{j,k}|k\in\mathbb{Z}\}, \tag{3.43}$$

then, for any $j\in\mathbb{Z}$, operators $\tilde{P}_j$, P_j, $\tilde{Q}_j$, Q_j are the projectors from $L^2(\mathbb{R})$ to $\tilde{V}_j$, V_j, $\tilde{W}_j$ and W_j respectively satisfying:

$$\begin{cases} V_j\bigcap\tilde{V}_j^{\perp} = \tilde{V}_j\bigcap V_j^{\perp} = \{0\} \\ W_j\bigcap\tilde{W}_j^{\perp} = \tilde{W}_j\bigcap W_j^{\perp} = \{0\} \end{cases}$$

$$\begin{cases} \tilde{W}_j = \tilde{V}_{j+1}\bigcap V_j^{\perp} \\ W_j = V_{j+1}\bigcap\tilde{V}_j^{\perp} \end{cases} \qquad \begin{cases} \tilde{V}_{j+1} = \tilde{V}_j\dot{+}\tilde{W}_j \\ V_{j+1} = V_j\dot{+}W_j \end{cases},$$

where $\dot{+}$ denotes the direct sum.

We omit the proof of the theorem because it is complicated and refers to some mathematical analysis which is not included in this book. The reader can obtain the details from [Daubechies, 1992; Long, 1995] and other related references.

Sometimes, $V_j \perp W_j$ is useful in practice. We will discuss this question here. A lemma is given first.

Lemma 3.3 *Let m_0 be a 2π-periodic measurable function on $\mathbb{R}$ satisfying $|m_0(\xi)|^2 + |m_0(\xi+\pi)|^2 \neq 0$ a.e. $\xi \in \mathbb{R}$. Then, a 2π-periodic measurable function m_1 satisfies*

$$m_0(\xi)m_1(\xi) + m_0(\xi+\pi)m_1(\xi+\pi) = 0, \qquad a.e.\ \xi \in \mathbb{R},$$

if and only if a 2π-periodic measurable function ν exists such that

$$m_1(\xi) = e^{-i\xi}\nu(2\xi)m_0(\xi+\pi), \qquad a.e.\ \xi \in \mathbb{R}.$$

Proof The part of the sufficiency holds obviously. We need to prove only the necessity, and we assume without losing generality that, $\forall \xi \in \mathbb{R}$,

$$\begin{cases} m_0(\xi) = m_0(\xi+2\pi), \quad m_1(\xi) = m_1(\xi+2\pi) \\ |m_0(\xi)|^2 + |m_0(\xi+\pi)|^2 \neq 0 \\ m_0(\xi)m_1(\xi) + m_0(\xi+\pi)m_1(\xi+\pi) = 0 \end{cases}.$$

It is easy to see that

$$\lambda(\xi) := \begin{cases} -\frac{m_1(\xi+\pi)}{m_0(\xi)} & \text{for } m_0(\xi) \neq 0 \\ \frac{m_1(\xi)}{m_0(\xi+\pi)} & \text{for } m_0(\xi) = 0. \end{cases}$$

is a 2π-periodic measurable function. We claim that

$$\lambda(\xi) + \lambda(\xi+\pi) = 0, \qquad \forall \xi \in \mathbb{T}.$$

In fact,

(1). If $m_0(\xi) = 0$, it is clear that $m_0(\xi+\pi) \neq 0$, therefore,

$$\lambda(\xi) + \lambda(\xi+\pi) = \frac{m_1(\xi)}{m_0(\xi+\pi)} - \frac{m_1(\xi+2\pi)}{m_0(\xi+\pi)} = 0;$$

(2). If $m_0(\xi+\pi) = 0$, it is still easy to deduce that $m_0(\xi) \neq 0$. By the result of (1), we have $\lambda(\xi+\pi) + \lambda(\xi+2\pi) = 0$, i.e.

$$\lambda(\xi) + \lambda(\xi+\pi) = 0;$$

(3). At last, if $m_0(\xi) \neq 0$ and $m_0(\xi+\pi) \neq 0$, we can obtain

$$\begin{aligned} \lambda(\xi)+\lambda(\xi+\pi) &= -\frac{m_1(\xi+\pi)}{m_0(\xi)} - \frac{m_1(\xi)}{m_0(\xi+\pi)} \\ &= -\frac{m_1(\xi)m_0(\xi)+m_1(\xi+\pi)m_0(\xi+\pi)}{m_0(\xi)m_0(\xi+\pi)} \\ &= 0. \end{aligned}$$

Our claim is proved.

If $m_0(\xi)=0$, we have

$$m_1(\xi)=\lambda(\xi)m_0(\xi+\pi).$$

If $m_0(\xi) \neq 0$, it is obvious that

$$m_1(\xi+\pi)=-\lambda(\xi)m_0(\xi),$$

which concludes that

$$\begin{aligned} m_1(\xi)m_0(\xi) &= -m_1(\xi+\pi)m_0(\xi+\pi) \\ &= \lambda(\xi)m_0(\xi)m_0(\xi+\pi). \end{aligned}$$

Hence, we also have

$$m_1(\xi)=\lambda(\xi)m_0(\xi+\pi).$$

Obviously, $\lambda(\xi)+\lambda(\xi+\pi)=0$ implies that $\lambda(\xi)e^{-i\xi}$ is π-periodic. We denote

$$\nu(\xi) := \lambda(\frac{\xi}{2})e^{-i\frac{1}{2}\xi},$$

then $\nu(\xi)$ is a 2π-periodic measurable function and $\lambda(\xi)=\nu(2\xi)e^{i\xi}$. Therefore,

$$m_1(\xi)=e^{i\xi}\nu(2\xi)m_0(\xi+\pi).$$

This ends the proof. ■

According to the above lemma, the conditions of $V_j \perp W_j$ can be characterized as follows:

Theorem 3.30 *Let m_0 and ϕ be the filter function and scale function of a MRA $\{V_j\}_{j\in\mathbb{Z}}$ respectively, $m_1 \in L^\infty(\mathbb{T})$. We denote*

$$\hat{\psi}(\xi) := m_1(\frac{\xi}{2})\hat{\phi}(\frac{\xi}{2}),$$

$$W_j := \overline{span}\{\psi(2^j \cdot -k) | k \in \mathbb{Z}\}, \qquad (\forall j \in \mathbb{Z}).$$

Then $V_j \perp W_j$ $(\forall j \in \mathbb{Z})$ *if and only if there is a* $2\pi-$*periodic measurable function* $\nu(\xi)$ *such that*

$$m_1(\xi) = e^{-i\xi}\nu(2\xi)\bar{m}_0(\xi+\pi)\Phi(\xi+\pi),$$

where $\Phi := [\hat{\phi}, \hat{\phi}]$.

Proof It is easy to see that

$$\begin{aligned} V_j \perp W_j \ (\forall j \in \mathbb{Z}) \quad &\Longleftrightarrow \quad V_0 \perp W_0 \\ &\Longleftrightarrow \quad \langle \psi(\cdot - k), \phi(\cdot) \rangle = 0 \quad (\forall k \in \mathbb{Z}) \\ &\Longleftrightarrow \quad [\hat{\psi}, \hat{\phi}](\xi) = 0 \quad a.e.\ \xi \in \mathbb{T}. \end{aligned}$$

Since

$$\begin{aligned} [\hat{\psi}, \hat{\phi}](\xi) &= \sum_{k \in \mathbb{Z}} \hat{\psi}(\xi + 2\pi k)\bar{\hat{\phi}}(\xi + 2\pi k) \\ &= \sum_{k \in \mathbb{Z}} m_1(\frac{\xi}{2} + \pi k)\hat{\phi}(\frac{\xi}{2} + \pi k)\bar{m}_0(\frac{\xi}{2} + \pi k)\bar{\hat{\phi}}(\frac{\xi}{2} + \pi k) \\ &= \sum_{k \in \mathbb{Z}} m_1(\frac{\xi}{2} + \pi k)\bar{m}_0(\frac{\xi}{2} + \pi k)|\hat{\phi}(\frac{\xi}{2} + \pi k)|^2 \\ &= \sum_{\alpha \in \mathbb{Z}} \Big[m_1(\frac{\xi}{2})\bar{m}_0(\frac{\xi}{2}))|\hat{\phi}(\frac{\xi}{2} + 2\pi\alpha)|^2 \\ &\quad + m_1(\frac{\xi}{2} + \pi)\bar{m}_0(\frac{\xi}{2} + \pi))|\hat{\phi}(\frac{\xi}{2} + 2\pi\alpha + \pi)|^2 \Big] \\ &= m_1(\frac{\xi}{2})\bar{m}_0(\frac{\xi}{2})\Phi(\frac{\xi}{2}) + m_1(\frac{\xi}{2} + \pi)\bar{m}_0(\frac{\xi}{2} + \pi)\Phi(\frac{\xi}{2} + \pi), \end{aligned}$$

we conclude that

$$\begin{aligned} V_j \perp W_j \ (\forall j \in \mathbb{Z}) \quad \Longleftrightarrow \quad & m_1(\xi)\bar{m}_0(\xi)\Phi(\xi) \\ & + m_1(\xi + \pi)\bar{m}_0(\xi + \pi)\Phi(\xi + \pi) \\ & = 0 \ a.e.\ \xi \in \mathbb{T}. \end{aligned}$$

Set m_0 in Lemma 3.3 to be $\bar{m}_0(\xi)\Phi(\xi)$ here and notice that ϕ generates an MRA, it can be concluded easily that there exist positive numbers A and B such that

$$A \le |\bar{m}_0(\xi)\Phi(\xi)|^2 + |\bar{m}_0(\xi + \pi)\Phi(\xi + \pi)|^2 \le B, \qquad a.e.\ \xi \in \mathbb{R}.$$

By Lemma 3.3, our theorem is proved. ∎

3.2.4 *S. Mallat Algorithms*

Under the conditions of Theorem 3.29, the well-known S. Mallat algorithm can be deduced easily. This algorithm provides a recursive scheme to calculate the coefficients of the biorthonormal wavelet expansion of a signal from one layer to the next layer. It can be applied to image processing, pattern recognition and other related subjects effectively.

Theorem 3.31 *Suppose the conditions of Theorem 3.29 are satisfied and denote $\tilde{m}_0$, m_0, $\tilde{m}_1$, m_1 of Theorem 3.29 as follows:*

$$\tilde{m}_0(\xi) = \sum_{k\in\mathbb{Z}} \tilde{h}_k e^{-ik\xi}, \qquad m_0(\xi) = \sum_{k\in\mathbb{Z}} h_k e^{-ik\xi},$$
$$\tilde{m}_1(\xi) = \sum_{k\in\mathbb{Z}} \tilde{g}_k e^{-ik\xi}, \qquad m_1(\xi) = \sum_{k\in\mathbb{Z}} g_k e^{-ik\xi}.$$

Then, $\forall f \in L^2(\mathbb{R})$, the following S.Mallat decomposition algorithm holds:

$$\begin{cases} \langle f, \phi_{j,k}\rangle = \sqrt{2}\sum_{l\in\mathbb{Z}} h_l \langle f, \phi_{j+1,2k+l}\rangle \\ \langle f, \psi_{\mu,j,k}\rangle = \sqrt{2}\sum_{l\in\mathbb{Z}} g_l \langle f, \phi_{j+1,2k+l}\rangle \end{cases} \quad (\forall j \in \mathbb{Z}, k \in \mathbb{Z});$$

$$\begin{cases} \langle f, \tilde{\phi}_{j,k}\rangle = \sqrt{2}\sum_{l\in\mathbb{Z}} \tilde{h}_l \langle f, \tilde{\phi}_{j+1,2k+l}\rangle \\ \langle f, \tilde{\psi}_{j,k}\rangle = \sqrt{2}\sum_{l\in} \tilde{g}_l \langle f, \tilde{\phi}_{j+1,2k+l}\rangle \end{cases} \quad (\forall j \in \mathbb{Z}, k \in \mathbb{Z}).$$

And the corresponding reconstruction algorithm holds as follows:

$$\begin{aligned} \langle f, \phi_{j+1,k}\rangle &= \sqrt{2}\sum_{l\in\mathbb{Z}} \bar{\tilde{h}}_{k-2l}\langle f, \phi_{j,l}\rangle + \sqrt{2}\sum_{l\in\mathbb{Z}} \bar{\tilde{g}}_{k-2l}\langle f, \psi_{j,l}\rangle, \\ &\quad (\forall j \in \mathbb{Z},\ k \in \mathbb{Z}); \end{aligned}$$

$$\begin{aligned} \langle f, \tilde{\phi}_{j+1,k}\rangle &= \sqrt{2}\sum_{l\in\mathbb{Z}} \bar{h}_{k-2l}\langle f, \tilde{\phi}_{j,l}\rangle + \sqrt{2}\sum_{l\in\mathbb{Z}} \bar{g}_{k-2l}\langle f, \tilde{\psi}_{j,l}\rangle, \\ &\quad (\forall j \in \mathbb{Z},\ k \in \mathbb{Z}). \end{aligned}$$

Proof Using the two-scale relation and the expression of m_0, we have

$$\phi(x) = 2\sum_{l\in\mathbb{Z}} h_l \phi(2x - l),$$

which is equivalent to

$$\phi_{j,k}(x) = \sum_{l\in\mathbb{Z}} h_l \phi_{j+1,2k+l}(x).$$

Hence

$$\langle f, \phi_{j,k}\rangle = \sum_{l\in\mathbb{Z}} h_l \langle f, \phi_{j+1,2k+l}\rangle.$$

The other formulae of the decomposition algorithm can be proved similarly. Now we turn to the proof of the reconstruction algorithm.

By Theorem 3.29, we deduce that, $\forall j \in \mathbb{Z}, l \in \mathbb{Z}$,

$$\tilde{P}_{j+1}\tilde{\phi}_{j+1,l} = \tilde{P}_j \tilde{\phi}_{j+1,l} + \tilde{Q}_j \tilde{\phi}_{j+1,l},$$

i.e.

$$\begin{aligned}
\tilde{\phi}_{j+1,l} &= \sum_{k\in\mathbb{Z}} \langle \tilde{\phi}_{j+1,l}, \phi_{j,k}\rangle \tilde{\phi}_{j,k} + \sum_{k\in\mathbb{Z}} \langle \tilde{\phi}_{j+1,l}, \psi_{j,k}\rangle \tilde{\psi}_{j,k} \\
&= \sum_{k\in\mathbb{Z}} \langle \tilde{\phi}_{j+1,l}, \sum_{m\in\mathbb{Z}} h_m \phi_{j+1,2k+m}\rangle \tilde{\phi}_{j,k} \\
&\quad + \sum_{k\in\mathbb{Z}} \langle \tilde{\phi}_{j+1,l}, \sum_{m\in\mathbb{Z}} g_{\mu,m} \phi_{j+1,2k+m}\rangle \tilde{\psi}_{j,k} \\
&= \sum_{k\in\mathbb{Z}} \bar{h}_{l-2k} \tilde{\phi}_{j,k} + \sum_{k\in\mathbb{Z}} \bar{g}_{l-2k} \tilde{\psi}_{j,k}.
\end{aligned}$$

Therefore, we have

$$\langle f, \tilde{\phi}_{j+1,l}\rangle = \sum_{k\in\mathbb{Z}} \bar{h}_{l-2k} \langle f, \tilde{\phi}_{j,k}\rangle + \sum_{k\in\mathbb{Z}} \bar{g}_{l-2k} \langle f, \tilde{\psi}_{j,k}\rangle.$$

The another formula of the reconstruction algorithm can be shown similarly. The proof of this theorem is complete. ■

Note: S.Mallat decomposition algorithm and reconstruction algorithm can be illustrated in Table 3.1.

S. Mallat algorithm can be applied effectively to image processing, pattern recognition, fast computation of singular integrals and some other areas. It decomposes a signal from the lower frequency to the higher frequency locally both in time and frequency domains. In practical applications, we

Table 3.1 S. Mallat algorithm.

Decompostion:								
$\cdots$	$\longrightarrow$	$\langle f, \phi_{j+1,k}\rangle$	$\longrightarrow$	$\langle f, \phi_{j,k}\rangle$	$\longrightarrow$	$\langle f, \phi_{j-1,k}\rangle$	$\longrightarrow$	$\cdots$
	$\searrow$		$\searrow$		$\searrow$		$\searrow$	
		$\langle f, \psi_{j+1,k}\rangle$		$\langle f, \psi_{j,k}\rangle$		$\langle f, \psi_{j-1,k}\rangle$		$\cdots$
Reconstruction:								
$\cdots$	$\longrightarrow$	$\langle f, \phi_{j-1,k}\rangle$	$\longrightarrow$	$\langle f, \phi_{j,k}\rangle$	$\longrightarrow$	$\langle f, \phi_{j+1,k}\rangle$	$\longrightarrow$	$\cdots$
	$\nearrow$		$\nearrow$		$\nearrow$		$\nearrow$	
$\cdots$		$\langle f, \psi_{j-1,k}\rangle$		$\langle f, \psi_{j,k}\rangle$		$\langle f, \psi_{j+1,k}\rangle$		

usually need the discrete form of S. Mallat algorithm. Let

$$\begin{cases} s_k^j := 2^{j/2}\langle f, \phi_{j,k}\rangle \\ \tilde{s}_k^j := 2^{j/2}\langle f, \tilde{\phi}_{j,k}\rangle \end{cases}, \qquad \begin{cases} t_k^j := 2^{j/2}\langle f, \psi_{j,k}\rangle \\ \tilde{t}_k^j := 2^{j/2}\langle f, \tilde{\psi}_{j,k}\rangle. \end{cases} \tag{3.44}$$

Then, its discrete form can be written as follows: $\forall j \in \mathbb{Z}, k \in \mathbb{Z}$,

Decomposition Algorithm:

$$\begin{cases} s_k^j = \sum_{l\in\mathbb{Z}} h_l s_{2k+l}^{j+1} \\ t_k^j = \sum_{l\in\mathbb{Z}} g_l s_{2k+l}^{j+1} \end{cases} \qquad \begin{cases} \tilde{s}_k^j = \sum_{l\in\mathbb{Z}} \tilde{h}_l \tilde{s}_{2k+l}^{j+1} \\ \tilde{t}_k^j = \sum_{l\in\mathbb{Z}} \tilde{g}_l \tilde{s}_{2k+l}^{j+1}; \end{cases} \tag{3.45}$$

Reconstruction Algorithm:

$$\begin{cases} s_k^{j+1} = 2^d \sum_{l\in\mathbb{Z}} \bar{\bar{h}}_{k-2l} s_l^j + 2^d \sum_{l\in\mathbb{Z}} \bar{\bar{g}}_{k-2l} t_l^j \\ \tilde{s}_k^{j+1} = 2^d \sum_{l\in\mathbb{Z}} \bar{h}_{k-2l} \tilde{s}_l^j + 2^d \sum_{l\in\mathbb{Z}} \bar{g}_{k-2l} \tilde{s}_l^j. \end{cases} \tag{3.46}$$

The discrete S.Mallat algorithm is depicted in Table 3.2.

Table 3.2 Discrete form of S. Mallat algorithm.

	$\cdots$	$\longrightarrow$	s_k^{j+1}	$\longrightarrow$	s_k^j	$\longrightarrow$	s_k^{j-1}	$\longrightarrow$	$\cdots$
Decomposition:		$\searrow$		$\searrow$		$\searrow$		$\searrow$	
			t_k^{j+1}		t_k^j		t_k^{j-1}		$\cdots$
	$\cdots$	$\longrightarrow$	s_k^{j-1}	$\longrightarrow$	s_k^j	$\longrightarrow$	s_k^{j+1}	$\longrightarrow$	$\cdots$
Reconstruction:		$\nearrow$		$\nearrow$		$\nearrow$		$\nearrow$	
	$\cdots$		t_k^{j-1}		t_k^j		t_k^{j+1}		

The first formula of the decomposition algorithm Eq. (3.45) can be explained as follows: An input signal $\{s_k^{j+1}\}_{k\in\mathbb{Z}}$ is first filtered by a filter

$\{h_{-k}\}_{k\in\mathbb{Z}}$, which corresponds to a convolution in mathematics, then the output signal $\{\sum_{l\in\mathbb{Z}} h_l s_{k+l}^{j+1}\}$ is sampled alternately (that is, only the points with even indices are kept), which is called "downsample" and denoted by "$2\downarrow$" . At last, we get $\{s_k^j\}_{k\in\mathbb{Z}}$. The procedure is represented below:

$$\{s_k^{j+1}\}_{k\in\mathbb{Z}} \longrightarrow \boxed{\{h_{-k}\}_{k\in\mathbb{Z}}} \longrightarrow \boxed{\quad 2\downarrow \quad} \longrightarrow \{s_k^j\}_{k\in\mathbb{Z}}.$$

The other formulae of Eq. (3.45) can be illustrated similarly.

The implementation of the first part of the first formula of the reconstruction algorithm Eq. (3.46), $x_k^{j+1} := \sum_{l\in\mathbb{Z}} \bar{\bar{h}}_{k-2l} s_l^j$, is just in inverse order. At first, the input signal is upsampled. More precisely, a zero is placed in between every two consecutive terms of the input sequence $\{s_k^j\}_{k\in\mathbb{Z}}$, which is denoted by "$2\uparrow$". Then the upsampled sequence is filtered by $\{\bar{\bar{h}}_k\}_{k\in\mathbb{Z}}$ and at last the output signal $\{x_k^{j+1}\}_{k\in\mathbb{Z}}$ is obtained. The following is an illustration of the procedure.

$$\{s_k^j\}_{k\in\mathbb{Z}} \longrightarrow \boxed{\quad 2\uparrow \quad} \longrightarrow \boxed{\{\bar{\bar{h}}_k\}_{k\in\mathbb{Z}}} \longrightarrow \{x_k^{j+1}\}_{k\in\mathbb{Z}}$$

The other parts of Eq. (3.46) can be explained similarly.

Chapter 4

Document Analysis by Fractal Dimension

Document processing requires the extraction of features from regions of the document image, and processing of these features with a pattern recognition or image processing algorithm. Many of the features used in our applications tend to be local in nature, which means their calculation requires a connected region of the document, over which some significant features such as the average or other statistic properties are extracted. This chapter aims at exploring an approach using the fractal dimensions to the document analysis for automatic knowledge acquisition, i.e. extraction of the geometric (layout) structure from a document.

4.1 Introduction

A document is considered to have two structures: geometric structure and logical structure. They play a key role in the process of the knowledge acquisition, which can be viewed as a process of acquiring the above structures. Generally, document processing is divided into two phases: *document analysis* and *document understanding*. Extraction of the geometric structure from a document is defined as document analysis; mapping the geometric structure into a logical structure is defined as document understanding. Once the logical structure has been captured, its meaning can be decoded by AI or other techniques. In Figure 1.1 of Chapter 1, the relationships among the geometric structure, logical structure, document analysis and document understanding are depicted.

Document analysis is defined as the extraction of the geometric structure of a document. In this way, a document image is broken down into several blocks, which represent coherent components of a document, such as text lines, headlines, graphics, etc. with or without the knowledge regarding the specific format [Tang et al., 1999]. So far, many techniques have been developed, and all of them can be classified into two categories: Hierarchical and No-hierarchical methods.

- Hierarchical Methods: When we divide a page of document into blocks, we consider the geometric relationship among the blocks. In this way, we have two approaches, i.e.
  - Top-down approach
  - Bottom-up approach
- No-hierarchical Methods: When we break up a page of document into some pieces, we do not consider the geometric relationship among the blocks. This chapter will discuss this new approach.

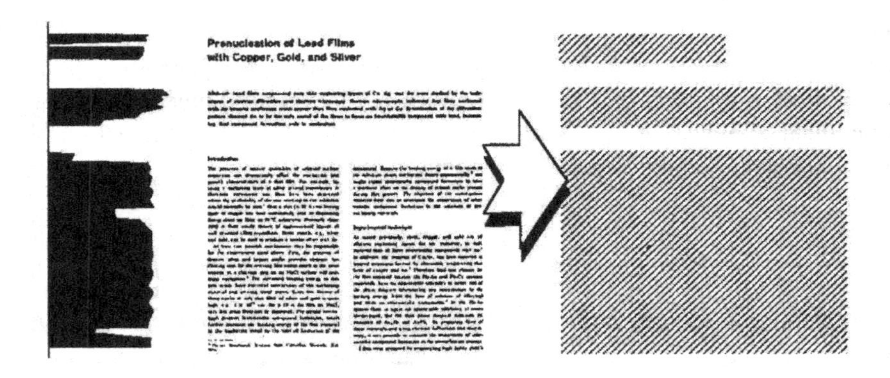
Prenucleation of Lead Films
with Copper, Gold, and Silver

Fig. 4.1 An example of the top-down approach of document analysis (1): A page of document is divided into 3 parts by the horizontal projection profiles.

Extracting the geometric structure from a document refers to document analysis. Traditionally, the top-down and bottom-up approaches have been used in document analysis [Parrish, 1989; Ramamoorthy and Wah, 1989; Tang et al., 1994].

Figures 4.1, 4.2 and 4.3 show an example of top-down approach using the horizontal and vertical projection profiles. In Figure 4.1, a page of document is divided into 3 parts by the horizontal projection profiles. In

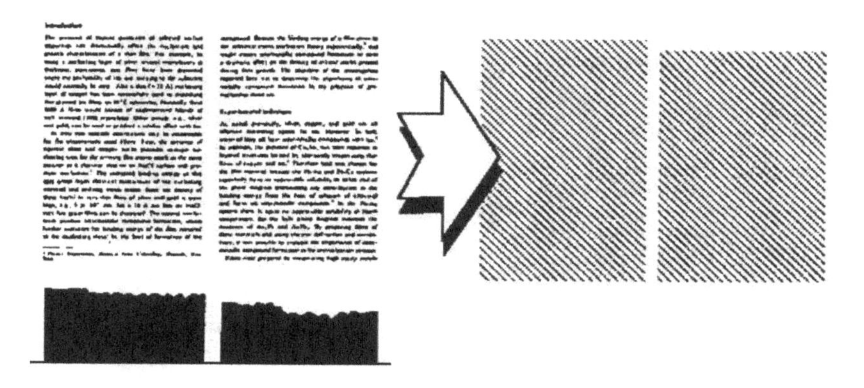

Fig. 4.2 An example of the top-down approach of document analysis (2): The third part from Figure 4.1 is broken up into two smaller parts (sub-parts) by the vertical projection profiles.

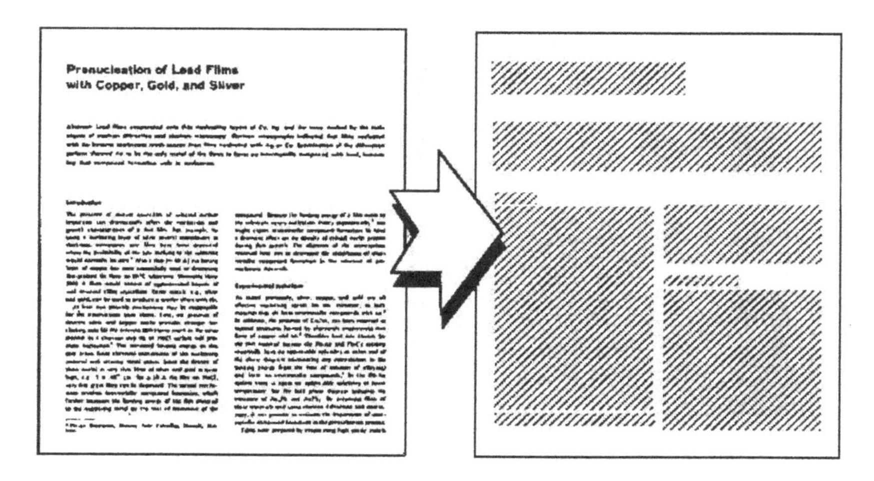

Fig. 4.3 An example of the top-down approach of document analysis (3): The above processes (Figure 4.1 and Figure 4.2) are continued in all sub-parts until the document is splited into the indivisible document blocks.

Figure 4.2, the third part in Figure 4.1 is broken up into two smaller parts which are called "sub-parts" by the vertical projection profiles. The above processes (Figure 4.1 and Figure 4.2) are continued in all sub-parts until the document is splited into the indivisible document blocks, which are illustrated in Figure 4.3.

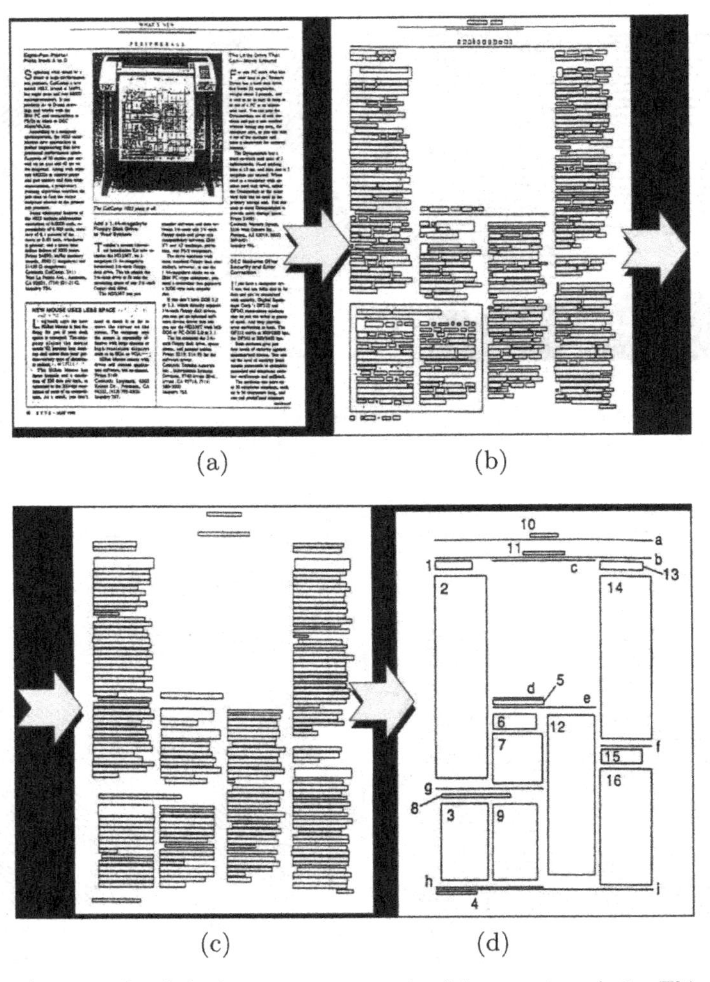

Fig. 4.4 An example of the bottom-up approach of document analysis: This method starts with the smallest components of document, i.e. English letters. Then, several letters go to a word. Thereafter, several words merge to a text lines. Next, the text lines get together and become a text block. Finally, a page of document is divided into text blocks.

Figure 4.4 shows an example of bottom-up approach. In this approach, firstly, the English letters are merged into English words, this process can be illustrated in Figure 4.4(a) to Figure 4.4(b). Secondly, several English words are combined into a text line, this process can be illustrated in Figure 4.4(b) to Figure 4.4(c). Finally, several text lines are united into a text block, which can be found in Figure 4.4(c) to Figure 4.4(d). Thus, this method starts with the smallest components of document, i.e. English letters. Then, several letters go to a word. Thereafter, several words merge to a text lines. Next, the text lines get together and become a text block.

However, both approaches have their weaknesses:

- They are not effective for processing documents with high geometrical complexity. Specifically, the top-down approach can process only simple documents, which have specific format or contain some a priori information. It fails to process the documents, which have complicated geometric structures as shown in Figure 4.5.
- To extract the geometric (layout) structure of a document, the top-down approach needs iterative operations to break the document into several blocks while the bottom-up approach needs to merge small components into large ones iteratively. Consequently, both approaches are time consuming.

Fractal behavior and structure can be intuitively appreciated in a variety of ways. Fractal is mathematical sets with a high degree of geometrical complexity, which can model many classes of time-series data as well as images. The fractal dimension is an important characteristic of fractals. It contains information about their geometrical structure. As the interest in fractal geometry rises, the applications are getting more and more numerous in many domains. Since Mandelbrot [Mandelbrot, 1983] proposed this technique, the subject of fractal dimension has drawn a great deal of attention from mathematicians, physicists, chemists, biologists, geologists, electrical and computer engineers in various disciplines. Specifically, in the area of image processing, the fractal dimension has been used for image compression, texture analysis, image encoding, etc. providing a novel technique for achieving compression ratios of 10,000 to 1 - or even higher [Barnsley and Sloan, 1988]. Earlier results on texture analysis using fractal techniques were reported by Nguyen and Quinqueton [Nguyen and Quinqueton, 1988] in which one-dimensional fractal analysis along a space filling curve was used. A full two-dimensional analysis was performed by Pentland [Pentland, 1984], and the statistics of differences of gray level between pairs of pixels at varying distances were used as indicators of the fractal

pinned by political and community consensus." At least one industry representative sees the WMI as a catalyst that will help launch a period of renewed commitment.

"We haven't made the progress we wanted to this year," says Bob Kelly of Newfoundland-based Iron Ore Company of Canada, and member of the International Organization for Standardization (ISO) technical committee on iron ores (ISO/TC 102). "But we slowed to wait for the mining industry."

Now, with the signing of the WMI accord and the impending environmental management standards coming from the Canadian Standards Association (CSA) and ISO, Mr. Kelly anticipates there will be a greater movement toward sustainable development in the mining sector.

Agriculture responds to public concerns

Canada's agricultural land base is far more limited than most Canadians appreciate. While occupying the second largest territorial area of any country in the world, Canada has little arable land. Only 11 per cent of the land is capable of any form of agriculture and less than five per cent can produce crops. There are no vast agricultural reserves remaining — virtually all of the climatically-favored land that hasn't been built upon or paved over, is being used.

The move toward achieving sustainability in agriculture has not been smooth. Some of the first steps were taken to conserve soil resources during the Great Depression of the 1930s. Despite the gains that have been made, the soils on which our agriculture depends are continuing to erode and lose their fertility. According to Statistics Canada's *Human Activity and the Environment 1994*, soil degradation is estimated to cost Canadians more than $1 billion annually.

Federal and provincial governments have developed programs to lessen the negative impact of agriculture practices on the environment. The federal *Green Plan* includes a commitment to encourage environmentally-sound on-farm practices. Agriculture and Agri-Food Canada has established a National Agriculture Environment Committee to advise it on a national strategy for sustainable agriculture. And the North American Waterfowl Management plan has been successfully restoring.

It is widely believed that fertilizer and pesticides have been responsible for allowing fewer farmers to manage larger farms. While it is true that the use of chemicals in farming grew steadily from the mid 1930s, the evidence suggests that their use peaked in the mid 1980s and has been declining ever since. Statistics Canada reports the percentage of Canadian farms applying commercial fertilizer has dropped from 66 per cent in 1985 to 59 per cent in 1990, while 40 per cent of farms reported using herbicides in 1990.

Despite these statistics and international recognition of Canada as having one of the safest and most efficient food systems in the world, a major concern of Canadian consumers continues to be the use of chemicals in food production. This has led to the creation of a market for organically-produced products. There have also been efforts to develop national standards for growing and certifying organic products, but Bruce Dodd, senior program manager for standards at the Canadian General Standards Board (CGSB) stakeholders have been unable.

There are currently over 20 different organizations certifying organic produce and one provincial standards body, the Certified Organic Association of British Columbia. The various standards currently in place may include restrictions on fertilizer and pesticide use, buffer zones (strips of plants such as trees or hedges) surrounding fields, record-keeping, farm waste disposal and other factors not easily tested for. The fact that Agriculture and Agri-Food Canada cannot detect what is an organically-grown product and what isn't, poses another roadblock.

On January 1, 1994, working with Agriculture and Agri-Food Canada, the Standards Council of Canada (SCC) launched its Laboratory Accreditation Program for Pesticide Residues (LAPPR). This is a voluntary program intended for laboratories that conduct pesticide residue analysis of food products. The program was developed by a working group on pesticide residues made up of representatives from government and the private sector, and operating under SCC's Testing Accreditation Sub-Committee.

Accredited laboratories must meet all the pertinent provisions of International Orga-

Sustainable development
(continued on page 32)

Fig. 4.5 Example of the high-complexity document.

properties of the texture. Maragos and Sun [Maragos and Sun, 1993] developed a theoretical approach for measuring the fractal dimension of arbitrary continuous-time signals by using morphological erosion and dilation function operations to create covers around a signal's graph at multiple scales. Matteo Baldoni [Baldoni et al., 1998] proposed fractal encodings based on Iterated Function Systems (IFS) to extract information, that is relevant in object recognition tasks.

This chapter will present a new approach based on *Modified Fractal Signature (MFS)* or *Modified Fractal Feature (MFF)* for document page segmentation. The MFS does not need iterative breaking or merging, and can divide a document into blocks in only one step. It is anticipated that this approach can be widely used to process various types of documents including even some with high geometrical complexity.

4.2 Document Analysis Based on Modified Fractal Signature (MFS)

This section consists of three subsections, namely, first, the basic idea of the modified fractal signature (MFS) will be presented. In this section, the preliminaries of δ parallel bodies will be presented in next subsection. To solve the problem of computing fractal dimension of a document image, a special form of the δ parallel body, blanket, is applied, and it will be stated in this section. Then a new method for document analysis using the modified fractal signature (MFS) will be described. Finally, a significant algorithm will be presented.

4.2.1 *Basic Idea of Modified Fractal Signature (MFS)*

A diagram of the process to obtain the modified fractal signature (MFS) is illustrated in Figure 4.6.

First, a document image is mapped onto a gray-level function as shown by the arrow form block 1 to block 2 in Figure 4.6.

Furthermore, this function can be mapped onto a gray-level surface, and from the area of such surface, the fractal dimension of the document image can be approximated. This stage is represented by the arrow from block 2 to block 3 in Figure 4.6.

However, directly calculating the area of the gray-level surface of the document image is an obscure task. To simplify this computation, a special equivalent technique of the Minkowski dimension technique, which was mentioned in Chapter 2, is applied in this study, referred as δ Parallel Bodies. This stage is represented by the arrow from block 3 to the most left block (block 4) in Figure 4.6.

Block 4 consists of three sub-blocks. We analyze the box counting methodology from the standpoint of computing fractal features. Using the δ parallel body of the gray-level surface of the document image, we first thicken the gray-level surface, so that it becomes a three-dimensional parallel body. Then, we calculate the volume of that body, since the calculation of a volume is much easier than that of a gray-level surface. Direct computation of fractal dimension of a document image is not used in this method. Alternately, the volume of a δ parallel body is estimated to approximate the fractal dimension. We focus on fractal dimension for computing the MFS.

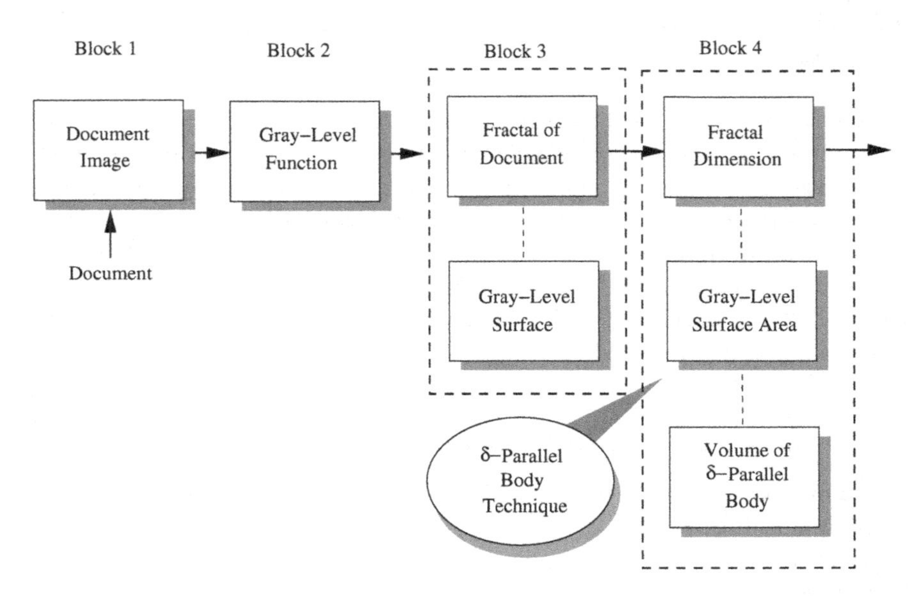

Fig. 4.6 Diagram of fractal feature approach to document analysis.

4.2.2 *δ-Parallel Bodies*

It is worthwhile discussing another equivalent definition of box computing dimension which is a rather different form.

Definition 4.1 δ-parallel body F_δ can be defined by:

$$F_\delta = \{x \in I\!R^n : |x - y| \le \delta, \quad \text{for} \quad y \in F\}. \tag{4.1}$$

It is obvious that F_δ is a set of points, where the distance between F and any element of F_δ is not greater than δ. We consider the rate at which the n-dimensional volume of F_δ shrinks as $\delta \to 0$. Specifically, in $\mathbb{R}^3$, we consider the following examples:

- F is a set containing only one point, i.e. F is a single point: F_δ is a ball (Figure 4.7a) with volume of
$$Vol(F_\delta) = \frac{4}{3}\pi\delta^3;$$
- F is a set containing a segment of straight line with length of L: F_δ is a "cylinder" (Figure 4.7b) with volume of
$$Vol(F_\delta) \sim \pi L \delta^2;$$
- F is a set containing a segment of curve with length of L: F_δ is "sausage-like" (Figure 4.7c) with volume of
$$Vol(F_\delta) \sim \pi L \delta^2;$$
- F is a set containing a plane with area of $\mathcal{A}$: F_δ is a "brick" (Figure 4.7d) with volume of
$$Vol(F_\delta) \sim 2\mathcal{A}\delta.$$

In each case, the following formula holds:

$$Vol(F_\delta) \sim \beta\delta^{3-D}. \tag{4.2}$$

where β is a constant, and the integer D is the *Fractal Dimension* of F.

Let $F = \{X_{i,j}\}$, $i = 0, 1, ..., K$, $j = 0, 1, ..., L$ be a document image with multi-gray level, and $X_{i,j}$ be the gray level of the (i, j)-th pixel. In a certain measure range, the gray level surface of F can be viewed as a fractal. The surface area can be used to approximate its fractal dimension.

Particularly in document processing, the gray level function F is a non-empty and bounded set in $\mathbb{R}^3$ for either text areas, graphics areas or background areas in a document. Thus, the δ-parallel body can be applied, and a special technique that is referred to as *Blanket Technique* is used to thicken the function F. It leads to a set F_δ, which is still a non-empty and bounded set in $\mathbb{R}^3$. According to the definition of *Minkowski Dimension*

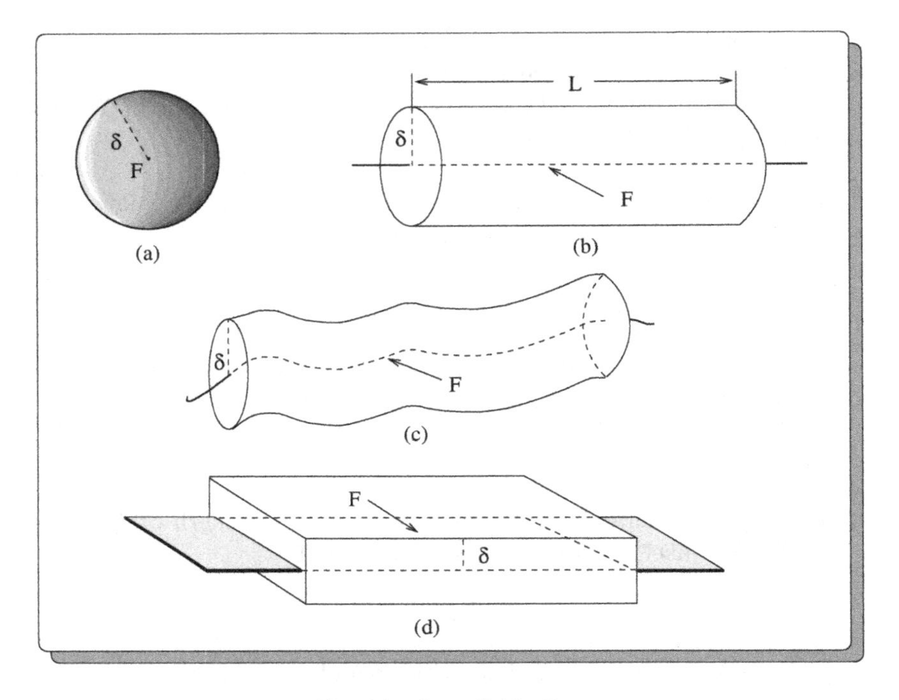

Fig. 4.7 δ-parallel bodies.

and Theorem 2.8, we can conclude that if

$$\lim_{\delta \to 0} \frac{Vol^3(F_\delta)}{\delta^{3-D}} = \beta > 0,$$

then

$$D = \dim_M F = \dim_B F,$$

where β denotes a constant, and $Vol^3(F_\delta)$ stands for the volume of the blanket F_δ. Therefore, when δ is sufficiently small, we have

$$Vol^3(F_\delta) \approx \beta \delta^{3-D}.$$

Let $A(\delta)$ be the area of the surface of the blanket, it can be represented as

$$\begin{aligned} A(\delta) &= \frac{Vol^3(F_\delta)}{2\delta} \\ &\approx \frac{\beta \delta^{2-D}}{2}. \end{aligned} \tag{4.3}$$

In fact, $A(\delta)$ is the area of the gray level surface of the image of a document.

To simplify the representation, we use the notion of β instead of $\frac{\beta}{2}$. Thus, Eq. (4.3) can be rewritten as

$$A(\delta) \approx \beta\delta^{2-D}, \qquad \text{if } \delta \text{ is sufficiently small,} \tag{4.4}$$

where β denotes a constant, D stands for the fractal dimension of a document image.

According to Eq. (4.4), the fractal dimension D can be computed from the area $A(\delta)$. More precisely, taking the logarithm of both sides in Eq. (4.4) yields:

$$\log_2 A(\delta) \approx \log_2 \beta + (2 - D) \log_2 \delta,$$

$$2 - D \approx \frac{\log_2 A(\delta)}{\log_2 \delta} - \frac{\log_2 \beta}{\log_2 \delta},$$

$$D \approx 2 - \frac{\log_2 A(\delta)}{\log_2 \delta} + \frac{\log_2 \beta}{\log_2 \delta}. \tag{4.5}$$

It should be noted that when δ is sufficiently small, the term $\frac{\log_2 \beta}{\log_2 \delta}$ in Eq. (4.5) approaches zero, namely,

$$\frac{\log_2 \beta}{\log_2 \delta} \approx 0.$$

Consequently, Eq. (4.5) becomes

$$D \approx 2 - \frac{\log_2 A(\delta)}{\log_2 \delta}. \tag{4.6}$$

This is an important formula to approximate the fractal dimension of a document image.

4.2.3 *Blanket Technique to Extract Fractal Feature*

To compute the fractal dimension, we need to measure the area of the gray level surface. In our study, a "blanket" approach is used for this purpose [Peleg et al., 1984]. The idea of the blanket technique is based on the equivalent definition of the BCD shown in Eq. (4.8), i.e. the δ-parallel body. The basic idea will be presented below:

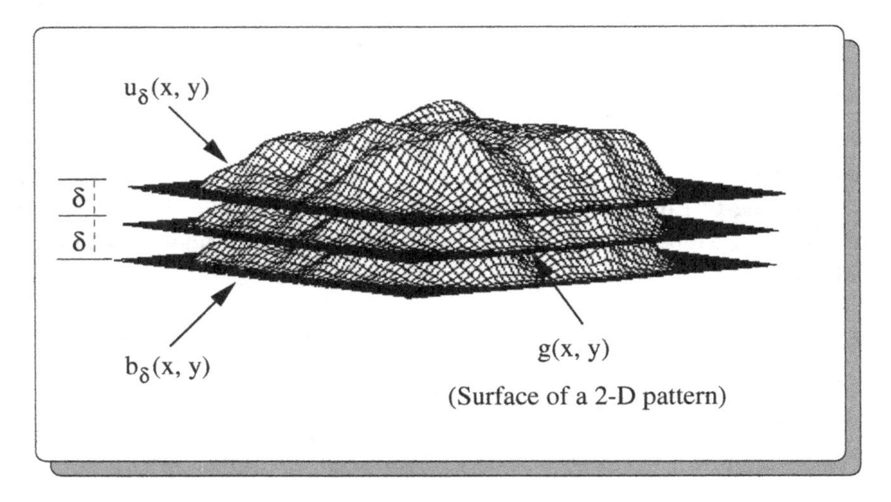

Fig. 4.8 Surface and its blankets.

In the blanket technique, all points of the three-dimensional space at distance δ from the gray level surface are considered. These points construct a "blanket" of thickness 2δ covering this surface. A graphical illustration is shown in Figure 4.8. The document image is represented by a gray-level function $g(i,j)$. The covering blanket is defined by its upper surface $u_\delta(i,j)$ and its lower surface $b_\delta(i,j)$. Initially, $\delta = 0$ and given the gray-level function equals the upper and lower surfaces, namely:

$$g(i,j) = u_0(i,j) = b_0(i,j).$$

For $\delta = 1, 2, ...,$ the blanket surfaces are defined iteratively as follows:

$$u_\delta(i,j) = \max\left\{u_{\delta-1}(i,j) + 1, \quad \max_{|(m,n)-(i,j)|\leq 1} u_{\delta-1}(m,n)\right\} \tag{4.7}$$

$$b_\delta(i,j) = \min\left\{b_{\delta-1}(i,j) - 1, \quad \min_{|(m,n)-(i,j)|\leq 1} b_{\delta-1}(m,n)\right\} \tag{4.8}$$

The image pixels (m,n) with distance less than one from (i,j) are taken to be the four immediate neighbors of (i,j). Similar expressions exist when the eight-neighborhood is desired. A point $f(x, y)$ will be included in the blanket for δ when $b_\delta(x, y) < f(x, y) < u_\delta(x, y)$. The blanket definition uses the fact that the blanket of the surface for radius δ includes all the points of the blanket for radius $\delta - 1$, together with all the points within radius 1 from the surfaces of that blanket (Figure 4.9). Eq. (4.7) ensures that the new upper surface u_δ is higher than $u_{\delta-1}$ by at least 1, and also at distance at least 1 from $u_{\delta-1}$ in the horizontal and vertical directions.

The volume Vol_δ of the blanket is computed from u_δ and b_δ:

$$Vol_\delta = \sum_{i,j}(u_\delta(i, j) - b_\delta(i, j)). \tag{4.9}$$

Definition 4.2 As the volume Vol_δ of the blanket is measured with radius δ, the area of a fractal surface can be deduced, which is called *fractal signature (FS)*

$$A_\delta = \frac{Vol_\delta}{2\delta}, \tag{4.10}$$

or

$$A_\delta = \frac{Vol_\delta - Vol_{\delta-1}}{2}. \tag{4.11}$$

In this study, the latter will be used.

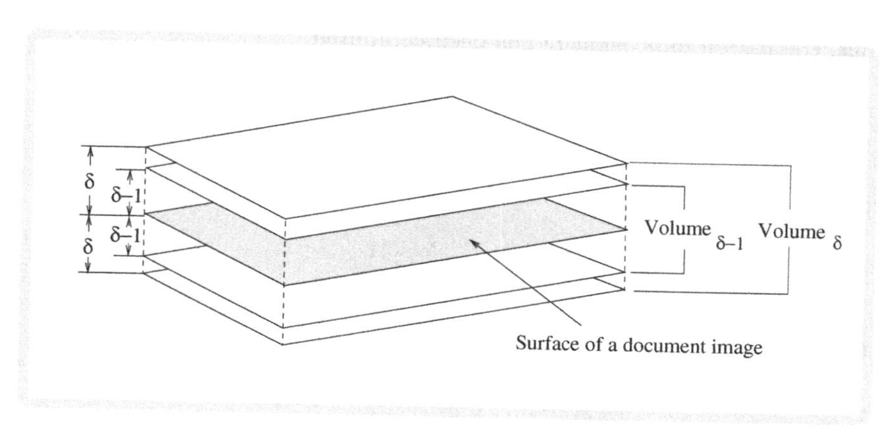

Fig. 4.9 Calculation of the area of a surface using volumes of two blankets δ and $\delta - 1$.

The area of a fractal surface behaves according to Eq. (4.4), namely:

$$A(\delta) \approx \beta\delta^{2-D}, \quad \delta = 1, 2, ...,$$

from which the fractal dimension D can be computed. Since the dimension can be regarded as a slope on a log-log scale, to get the dimension, only two points are needed. We use two values of δ to compute the fractal dimension, namely, we take δ_1 and δ_2, then

$$A_{\delta_1} \approx \beta\delta_1^{2-D}, \tag{4.12}$$

$$A_{\delta_2} \approx \beta\delta_2^{2-D}. \tag{4.13}$$

When Eq. (4.12) is divided by Eq. (4.13), we have

$$\frac{A_{\delta_1}}{A_{\delta_2}} \approx \frac{\delta_1^{2-D}}{\delta_2^{2-D}}.$$

Taking the logarithm of the both sides yields:

$$2 - D \approx \frac{\log_2 A_{\delta_1} - \log_2 A_{\delta_2}}{\log_2 \delta_1 - \log_2 \delta_2},$$

$$D \approx 2 - \frac{\log_2 A_{\delta_1} - \log_2 A_{\delta_2}}{\log_2 \delta_1 - \log_2 \delta_2}. \tag{4.14}$$

Thus, the fractal dimension D has been computed.

According to the property of proportions, the following is true:

$$\text{if } \frac{a_1}{b_1} = \frac{a_2}{b_2}, \quad \text{then } \frac{a_1}{b_1} = \frac{a_1 - a_2}{b_1 - b_2}.$$

Therefore, formula

$$\frac{\log_2 A_{\delta_1}}{\log_2 \delta_1} = \frac{\log_2 A_{\delta_2}}{\log_2 \delta_2}$$

yields

$$\frac{\log_2 A_{\delta_1}}{\log_2 \delta_1} = \frac{\log_2 A_{\delta_1} - \log_2 A_{\delta_2}}{\log_2 \delta_1 - \log_2 \delta_2}.$$

Consequently, Eqs. (4.6) and (4.14) are equivalent, namely:

$$\left(D \approx 2 - \frac{\log_2 A(\delta)}{\log_2 \delta}\right) \equiv \left(D \approx 2 - \frac{\log_2 A_{\delta_1} - \log_2 A_{\delta_2}}{\log_2 \delta_1 - \log_2 \delta_2}\right) \tag{4.15}$$

Recall the formula of Eq. (4.6), which is used to approximate the fractal dimension of the document images, i.e.

$$D \approx 2 - \frac{\log_2 A(\delta)}{\log_2 \delta} \qquad \delta \text{ is sufficiently small.}$$

Several points are worth noting:

- Why do the different document images have different fractal dimensions? The essential distinction of document images is their values of $A(\delta)$.
- The value of $A(\delta)$ depends on the volume $Vol^3(F_\delta)$ of the thickened blanket F_δ only.
- In summary, they can be represented as

$$D \Longleftrightarrow A(\delta) \Longleftrightarrow Vol^3(F_\delta).$$

Consequently, in this chapter, the volume $Vol^3(F_\delta)$ of the thickened blanket F_δ is applied to identify different blocks in a document, instead of using the fractal dimension.

4.3 Algorithm of Modified Fractal Signature (MFS)

In this section, a very significant characteristic of the fractal signature will be discussed first. It can be used to identify different kinds of blocks in the document. This is the basic idea of this approach for document analysis. In this section, the steps to compute the modified fractal signature (MFS) will also be presented in the second subsection.

4.3.1 *Identification of Different Blocks of Document by Fractal Signature*

From the definition of fractal signature (FS), i.e. equation Eq. (4.10), it is clear that the fractal signature is completely determined by the area of the surface which is a mapping of the gray-level function representing a document image. Consequently, the fractal signature reflects certain characteristics of the document image.

Consider a page of document F, which consists of many regions such as texts, graphics, and background areas, we have

$$F = \{\Im_T, \Im_G, \Im_B\}$$

where

- $\Im_T$ represents a set of *text areas*,
- $\Im_G$ stands for a set of *graphic areas*,
- $\Im_B$ denotes a set of *background areas.*

Different regions $\Im_T$, $\Im_G$ and $\Im_B$ have different gray-level functions $g_{\Im_T}$, $g_{\Im_G}$ and $g_{\Im_B}$. The different gray-level functions have different surfaces. Furthermore, the different surfaces have different areas $A(\Im_T)$, $A(\Im_G)$ and $A(\Im_B)$ from which different fractal signatures can be estimated.

$$\begin{aligned}
\Im_T &\Longrightarrow g_{\Im_T} \Longrightarrow A(\Im_T) \\
\Im_G &\Longrightarrow g_{\Im_G} \Longrightarrow A(\Im_G) \\
\Im_B &\Longrightarrow g_{\Im_B} \Longrightarrow A(\Im_B),
\end{aligned}$$

where $g_{\Im_T}$, $g_{\Im_G}$ and $g_{\Im_B}$ denote the gray-level functions of the text, graphics and background regions respectively; and $A(\Im_T)$, $A(\Im_G)$ and $A(\Im_B)$ denote the fractal signatures of the text, graphics and background regions respectively. For instance, the gray-level functions of the text region and background region are provided below:

(1) $\Im_T$: The gray-level function for a text block is

$$g_{\Im_T}(x,y) = \begin{bmatrix}
6 & 2 & 5 & 6 & 2 & 6 & 2 & 6 \\
2 & 2 & 2 & 2 & 2 & 2 & 2 & 2 \\
5 & 2 & 6 & 5 & 2 & 5 & 2 & 6 \\
2 & 2 & 2 & 2 & 2 & 2 & 2 & 2 \\
6 & 2 & 6 & 5 & 2 & 6 & 2 & 5 \\
2 & 2 & 2 & 2 & 2 & 2 & 2 & 2 \\
6 & 2 & 6 & 4 & 2 & 5 & 2 & 4 \\
2 & 2 & 2 & 2 & 2 & 2 & 2 & 2
\end{bmatrix},$$

where the elements of values 2's denote the background, i.e. the intervals between characters and text lines. The elements with values which are not 2's indicate the pixels of the texts. This function can be mapped onto a building-surface as shown in Figure 4.10(a). In the same region, the area of this type of surface is greater than that of a plane.

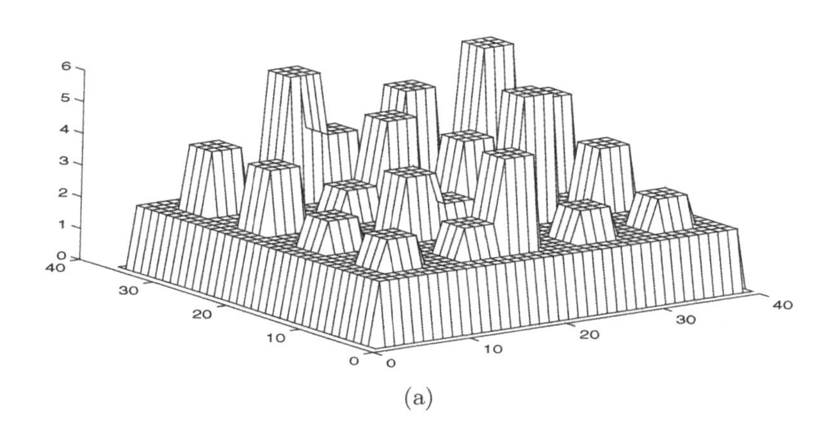

(a)

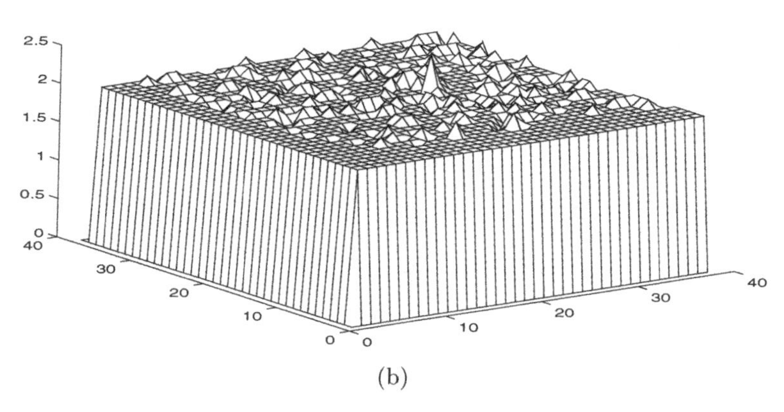

(b)

Fig. 4.10 Graphical example of representing gray-level functions.

(2) $\Im_B$: The gray-level function of a block of the background is

$$g_{\Im_B}(x,y) = \begin{bmatrix} 2 & 2.1 & 2 & 2.1 & 2 & 1.9 & 2 & 2 \\ 2 & 2 & 1.9 & 2 & 2.1 & 2 & 2.1 & 2 \\ 2 & 2.1 & 2 & 2.1 & 2 & 2 & 1.9 & 2 \\ 2 & 2.1 & 2 & 2 & 2.5 & 2.1 & 2 & 2 \\ 2 & 1.9 & 2 & 1.9 & 2 & 2.1 & 2 & 2 \\ 2 & 2.1 & 1.9 & 2 & 2 & 2 & 2.1 & 2 \\ 2 & 2 & 2 & 2.1 & 2.1 & 1.9 & 2 & 2 \\ 2 & 2 & 2 & 2 & 2 & 2 & 2 & 2 \end{bmatrix},$$

where most of the elements of this matrix have the same values of 2, that means only a few changes in gray level, since this is a block of the background. The elements with values, which are not 2's, represent noise. This function can be mapped onto a plane as shown in Figure 4.10(b). In the same region, the area of this surface is less than that of a building-surface.

Consequently, in the same region, the fractal signature of a plane is less than that of a building-surface, that means the fractal signature of a text block is greater than that of a background block, i.e., $A(\Im_T) > A(\Im_B)$. Thus we have the following theorem:

Theorem 4.1 *Let $\Im_T$ be a text block of a document, and $\Im_B$ be a background block of the document. If both blocks have the same geometrical size, then the fractal signature of the text block is greater than that of the background, namely,*

$$A(\Im_T) > A(\Im_B).$$

Proof (The proof of it will be excluded, because it is very simple). ∎

A graphical example can be illustrated in Figure 4.11(a) and (b). Figure 4.11(a) is a part of a document with texts while Figure 4.11(b) shows the result of using the fractal signature, the darken parts indicate the higher values of the fractal signatures, which represent the text regions.

This is a very significant characteristic of the fractal signature. It can be used to identify different kinds of blocks in the document. This is the basic idea of this approach for document analysis.

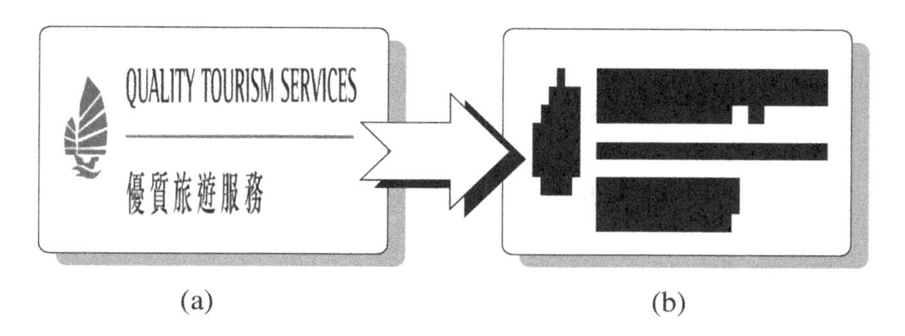

(a) (b)

Fig. 4.11 A portion of a poster is broken into three blocks.

4.3.2 *Modified Fractal Signature (MFS)*

An image $g(x, y)$ can be seen as a survey in $\mathbb{R}^3$. According to the Minkowski dimension algorithm, evaluating the $vol(g_\delta(x, y))$ in $\mathbb{R}^3$ under different δ is the key step for obtaining the dimension. In our algorithm, the value of $Vol^3(F_\delta)$ is evaluated by the upper surface $u_\delta(x, y)$ and the lower surface $b_\delta(x, y)$ of the $g(x, y)$. A graphical illustration is shown in Figure 4.8. Steps are listed below:

- The whole image F is divided into several non-overlapping sub-images $R_k(x, y)$, and each sub-image has size $N \times N$. For $k = 1$ to n do $R_k(x, y)$ is mapped onto a gray-level function $g_k(x, y)$, here, $k = 1, 2, ..., n$.
- Initially, taking $\delta = 0$, the upper layer $u_0^k(x, y)$ and lower layer $b_0^k(x, y)$ of the blanket are chosen as the same as the gray-level function $g_k(x, y)$, namely:

$$u_0^k(x, y) = b_0^k(x, y) = g_k(x, y). \tag{4.16}$$

- Taking $\delta = \delta_1$,

(a). $u_{\delta_1}(x, y)$ is computed according to Eq. (4.7), i.e.:

$$u_{\delta_1}(x, y) = max\left\{u_0(x, y) + 1, \quad \max_{|(i,j)-(x,y)|\leq 1} u_0(i, j)\right\};$$

(b). $b_{\delta_1}(x,y)$ is computed according to Eq. (4.8), i.e.:

$$b_{\delta_1}(x,y) = min\left\{b_0(x,y) - 1, \quad \min_{|(i,j)-(x,y)|\leq 1} b_0(i,j)\right\};$$

(c). The volume Vol_{δ_1} of the blanket is computed by Eq. (4.9), i.e.:

$$Vol_{\delta_1} = \sum_{x,y}(u_{\delta_1}(x,y) - b_{\delta_1}(x,y)).$$

- Taking $\delta = \delta_2 = \delta_1 + 1$,

(a). $u_{\delta_2}(x,y)$ is computed according to

$$u_{\delta_2}(x,y) = max\left\{u_{\delta_1}(x,y) + 1, \quad \max_{|(i,j)-(x,y)|\leq 1} u_{\delta_1}(i,j)\right\};$$

(b). $b_{\delta_2}(x,y)$ is computed according to

$$b_{\delta_2}(x,y) = min\left\{b_{\delta_1}(x,y) - 1, \quad \min_{|(i,j)-(x,y)|\leq 1} b_{\delta_1}(i,j)\right\};$$

(c). The volume Vol_{δ_1} of the blanket is computed by

$$Vol_{\delta_2} = \sum_{x,y}(u_{\delta_2}(x,y) - b_{\delta_2}(x,y)).$$

- The sub fractal signature A_δ^k is computed by Eq. (4.11), namely:

$$A_\delta^k = \frac{Vol_{\delta_2} - Vol_{\delta_1}}{2}.$$

- Combining sub fractal signatures A_δ^k, $k = 1, 2, ..., n$ into the whole fractal signature (FS) gives:

$$A_\delta = \bigcup_{k=1}^{n} A_\delta^k.$$

- For $\delta = 1, 2, ...,$

$$\begin{cases} u_\delta(x,y) = max\left\{u_{\delta-1}(x,y)+1, \max_{0\neq|(i,j)-(x,y)|\leq 1} u_{\delta-1}(i,j)\right\}; \\ b_\delta(x,y) = min\left\{b_{\delta-1}(x,y)-1, \min_{0\neq|(i,j)-(x,y)|\leq 1} b_{\delta-1}(i,j)\right\}. \end{cases} \tag{4.17}$$

The volume of $Vol_\delta^3(x,y)$ of the blanket is

$$Vol_\delta = \sum_{x=1}^{weight} \sum_{y=1}^{height} (u_\delta(x,y) - b_\delta(x,y)). \tag{4.18}$$

Based on the Minkowski dimension, we have

$$V_\delta \approx K\delta^{3-D},$$

where K can be looked as a constant.
Since

$$V_{\delta 1} \approx K\delta_1^{3-D}, \qquad V_{\delta 2} \approx K\delta_2^{3-D},$$

then

$$3 - D \approx \frac{\log_2 V_{\delta 1} - \log_2 V_{\delta 2}}{\log_2 \delta_1 - \log_2 \delta_2}.$$

- In order to obtain a better representation, a linear factor will be extracted by least square method (LSM). Fitting a straight line to a series of data points is one of the most commonly performed procedures in data analysis.
Let

$$x_i = \log_2 \delta i, \qquad y_i = \log_2 V_{\delta i},$$

for N points, the best fit value of m is calculated based on the least square method from summation of the x_i and y_i values as

$$m = \frac{N\sum x_i \cdot y_i - \sum x_i \sum y_i}{N\sum x_i^2 - (\sum x_i)^2}.$$

Then the modified fractal dimension (MFD) D can be calculated by

$$D = 3 - m.$$

This m isolates the feature that changes from scale $k-1$ to k. For image that are fractal only within a certain range of scale this is a necessary precaution. The final estimate of fractal feature is obtained by linear regression over the range of k as a curve. We call m as the modified fractal signature (MFS) of the set F.

4.4 Experiments

Experiments have been conducted to prove the effectiveness of the fractal-based approach for document processing. All experiments have been conducted in both personal computer system, and Sun SPARCstation computer system. An HP scanner with resolution of 100-600 DPI is employed to capture the image of the documents which are scanned with multi-gray levels.

In the experiments, the whole image F is divided into several non-overlapping sub-images $R_k(x,y)$, and each sub-image has size $N \times N$. The algorithm presented in the previous section is used. Here, we concentrate our study to finding the upper and lower layers of the blanket for a given surface, which represents a part of document.

Initially, taking $\delta = 0$, the upper layer $u_0^k(x,y)$ and lower layer $b_0^k(x,y)$ of the blanket are chosen as the same as the gray-level function $g_k(x,y)$. For example, a gray-level function $g_k(x,y)$ is

$$g_k(x,y) = \begin{bmatrix} 1 & 1 & 2 & 3 \\ 1 & 2 & 4 & 5 \\ 2 & 4 & 3 & 2 \\ 1 & 1 & 2 & 1 \end{bmatrix},$$

as illustrated in Figure 4.12(a).

Taking $\delta = 1$, the upper layer $u_1(x,y)$ is computed according to Eq. (4.7), i.e.

$$u_1(x,y) = \max\left\{ u_0(x,y)+1, \quad \max_{|(i,j)-(x,y)|\le 1} u_0(i,j) \right\}.$$

The result can be found in

$$u_1(x,y) = \begin{bmatrix} 2 & 2 & 4 & 5 \\ 2 & 4 & 5 & 6 \\ 2 & 5 & 4 & 5 \\ 2 & 4 & 3 & 2 \end{bmatrix},$$

which can be shown in Figure 4.12(b). The lower layer $b_1(x,y)$ is computed according to Eq. (4.8), i.e.:

$$b_1(x,y) = \min\left\{b_0(x,y) - 1, \quad \min_{|(i,j)-(x,y)|\leq 1} b_0(i,j)\right\}.$$

The result can be found in

$$b_1(x,y) = \begin{bmatrix} 0 & 0 & 1 & 2 \\ 0 & 1 & 2 & 2 \\ 2 & 1 & 2 & 1 \\ 0 & 0 & 1 & 0 \end{bmatrix},$$

which can be shown in Figure 4.12(c).

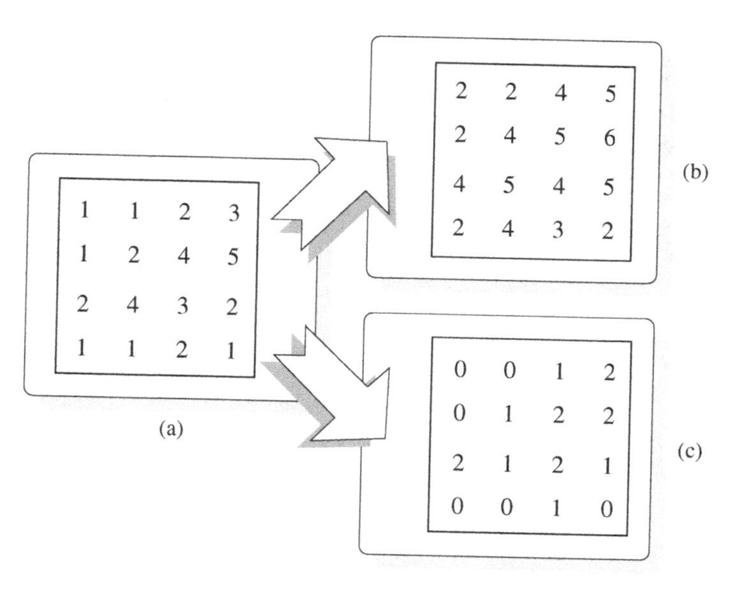

Fig. 4.12 An example of gray-level function and its upper and lower layers.

Two groups of documents have been tested in the experiments. The first group consists of two samples of documents where the document blocks have regular shapes. The geometric structures for these documents are not too complicated. Figure 4.13(a) shows a portion of a page of newspaper. Figure 4.14(b) illustrates the document blocks obtained from Figure 4.13(a) by document analysis using the fractal signature. Furthermore, from these document blocks, the document geometric structure can be extracted and illustrated in Figure 4.14. In this figure, $H1$, $H2$, and $H3$ stand for the "Headline Blocks" which represent titles of articles. $T1$, $T2$, ..., $T6$ indicate the "Text Line Blocks" corresponding to texts of different papers in the page. $G1$ means that this document contains only one graphic block.

Another example of the first group can be shown in Figure 4.15.

The second group of documents used in our experiments are illustrated in Figures 4.16 and 4.17, where the document blocks have irregular shapes. The geometric structures for these documents are complicated. We used the traditional methods, such as projection profile cuts, run-length smoothing (RLSA) to process them will fail because the document pages cannot be divided into correct blocks. The new approach has been applied to these complicated documents, and the positive results have been obtained shown in Figures 4.16 and 4.17.

The choice of the size of the sub-block plays an important role in this fractal-based approach. Some experiment results are shown in Figures 4.18 and 4.19. For each example, we apply the fractal signature with different sub-block sizes, i.e. 4×4 pixels, 8×8 pixels, and 16×16 pixels sub-block size.

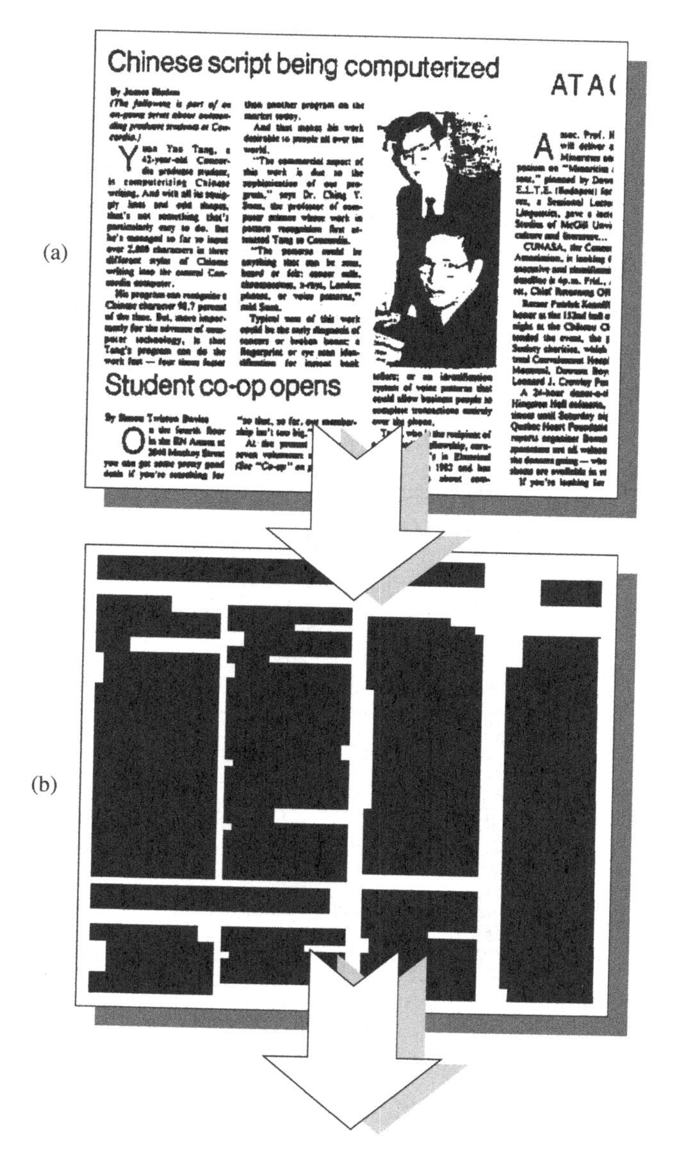

Fig. 4.13 A portion of a page of newspaper is broken into blocks by the fractal signature.

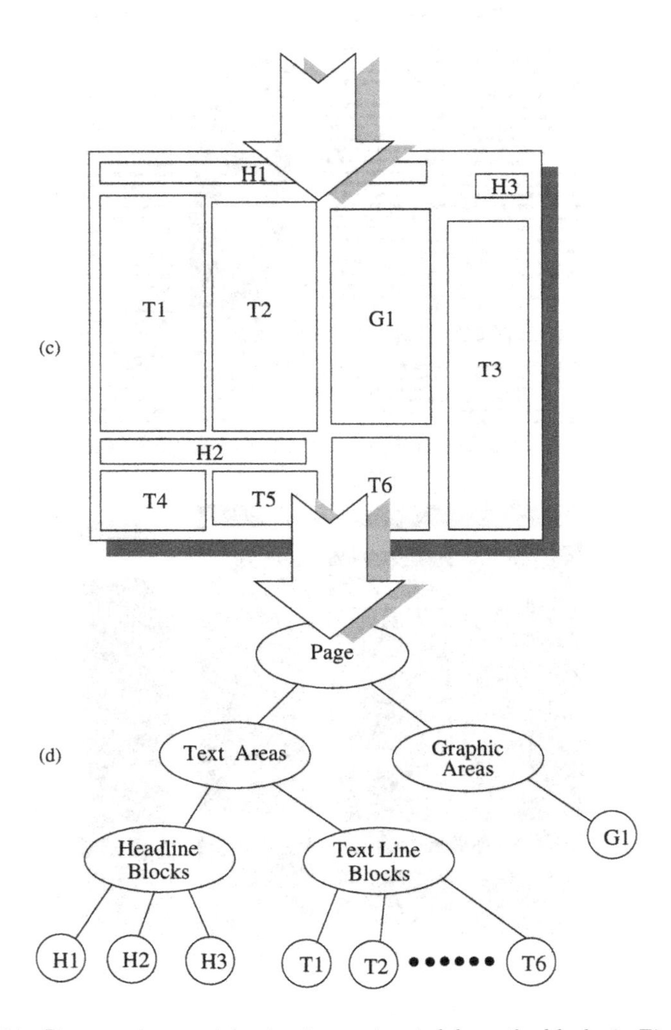

Fig. 4.14 Document geometric structure extracted from the blocks in Figure 12.

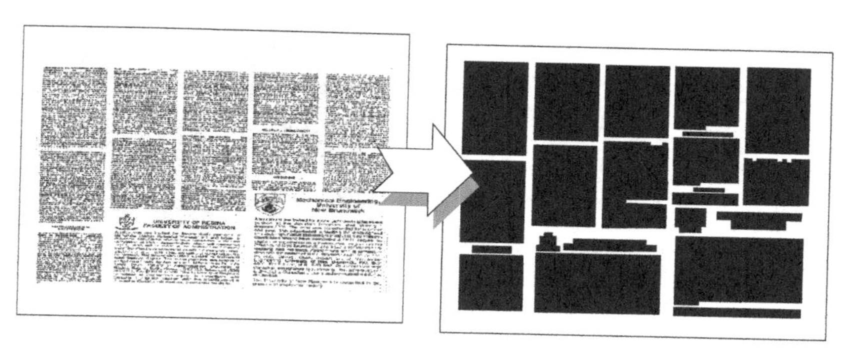

Fig. 4.15 An example of document analysis of the document image with regular blocks by the fractal signature.

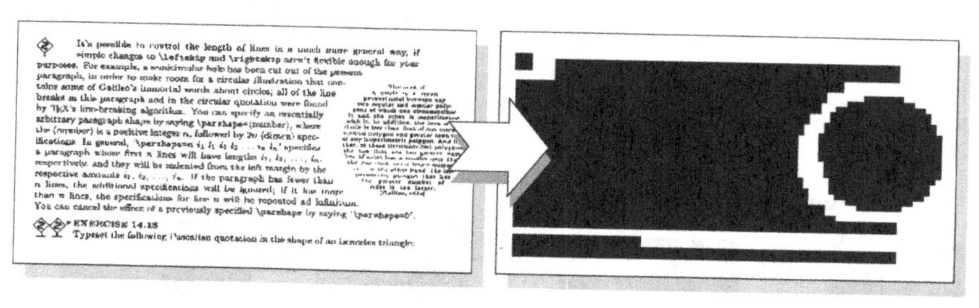

It's possible to control the length of lines in a much more general way, if simple changes to \leftskip and \rightskip aren't flexible enough for your purposes. For example, a semicircular hole has been cut out of the present paragraph, in order to make room for a circular illustration that contains some of Galileo's immortal words about circles; all of the line breaks in this paragraph and in the circular quotation were found by TeX's line-breaking algorithm. You can specify an essentially arbitrary paragraph shape by saying \parshape=⟨number⟩, where the ⟨number⟩ is a positive integer n, followed by $2n$ ⟨dimen⟩ specifications. In general, '\parshape= i_1 l_1 i_2 l_2 ... i_n l_n' specifies a paragraph whose first n lines will have lengths l_1, l_2, ..., l_n, respectively, and they will be indented from the left margin by the respective amounts i_1, i_2, ..., i_n. If the paragraph has fewer than n lines, the additional specifications will be ignored; if it has more than n lines, the specifications for line n will be repeated ad infinitum. You can cancel the effect of a previously specified \parshape by saying '\parshape=0'.

EXERCISE 14.18

Typeset the following Pascalian quotation in the shape of an isosceles triangle:

Fig. 4.16 An example of document analysis of the document image with irregular blocks by the fractal signature.

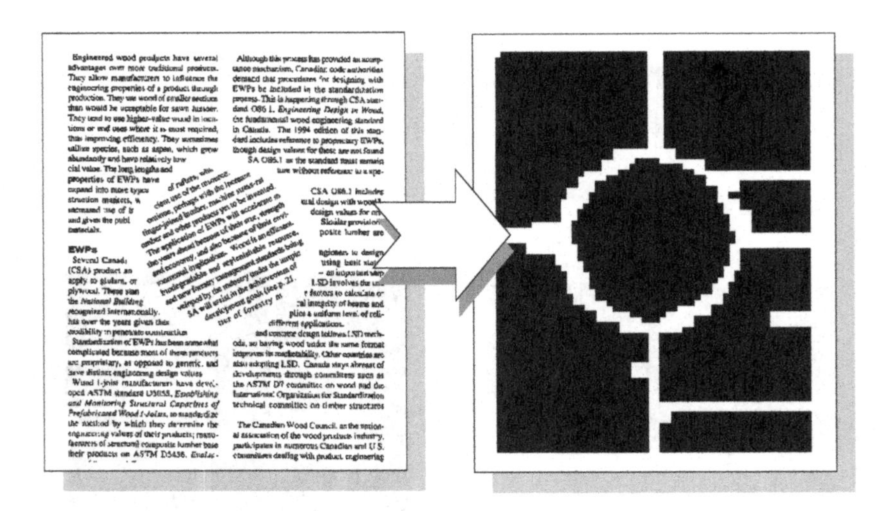

Engineered wood products have several advantages over more traditional products. They allow manufacturers to influence the engineering properties of a product through production. They use wood of smaller sections than would be acceptable for sawn lumber. They tend to use higher-value wood in locations or end uses where it is most required, thus improving efficiency. They sometimes utilize species, such as aspen, which grow abundantly and have relatively low
cial value. The long lengths and
properties of EWPs have
expand into more types
struction markets, w
increased use of li
and given the publ
materials.

EWPs

Several Canad
(CSA) product an
apply to glulam, or
plywood. These stan
the *National Building*
recognized internationally.
has over the years given the
credibility to penetrate construction

Standardization of EWPs has been somewhat complicated because most of these products are proprietary, as opposed to generic, and have distinct engineering design values.

Wood I-joist manufacturers have developed ASTM standard D5055, *Establishing and Monitoring Structural Capacities of Prefabricated Wood I-Joists*, to standardize the method by which they determine the engineering values of their products; manufacturers of structural composite lumber base their products on ASTM D5456. *Evalu-*

of rafters, wh-
cient use of the resource.
ontinue, perhaps with the increase
finger-jointed lumber, machine stress-rat
umber and other products yet to be invented.
The application of EWPs will accelerate in
the years ahead because of their size, strength
and economy, and also because of their environmental implications. Wood is an efficient,
biodegradable and replenishable resource.
and new forestry management standards being
veloped by the industry under the auspic
SA will assist in the achievement of
development goals (see p. 21.
tor of forestry m

Although this process has provided an acceptance mechanism, Canadian code authorities demand that procedures for designing with EWPs be included in the standardization process. This is happening through CSA standard O86.1, *Engineering Design in Wood*, the fundamental wood engineering standard in Canada. The 1994 edition of this standard includes reference to proprietary EWPs, though design values for these are not found
SA O86.1 as the standard must remain
ture without reference to a spe-

CSA O86.1 includes
ural design with wood
design values for
Similar provision
posite lumber are

ngineers to design
using limit sta
– an important step
LSD involves the
e factors to calculate
ral integrity of beams and
plies a uniform level of reli-
different applications.
and concrete design follows LSD methods, so having wood under the same format improves its marketability. Other countries are also adopting LSD. Canada stays abreast of developments through committees such as the ASTM D7 committee on wood and the International Organization for Standardization technical committee on timber structures

The Canadian Wood Council, as the national association of the wood products industry, participates in numerous Canadian and U.S. committees dealing with product engineering

Fig. 4.17 An example of document analysis of the document image with irregular blocks by the fractal signature.

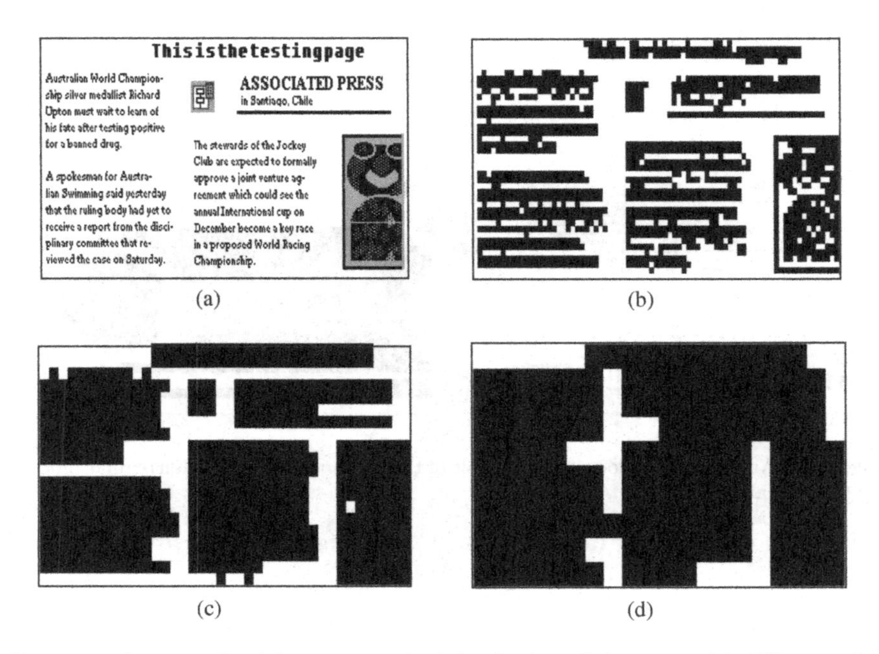

Thisisthetestingpage

Australian World Championship silver medallist Richard Upton must wait to learn of his fate after testing positive for a banned drug.

A spokesman for Australian Swimming said yesterday that the ruling body had yet to receive a report from the disciplinary committee that reviewed the case on Saturday.

ASSOCIATED PRESS
in Santiago, Chile

The stewards of the Jockey Club are expected to formally approve a joint venture agreement which could see the annual International cup on December become a key race in a proposed World Racing Championship.

Fig. 4.18 An example of document analysis by the fractal signature with different sub-block sizes.

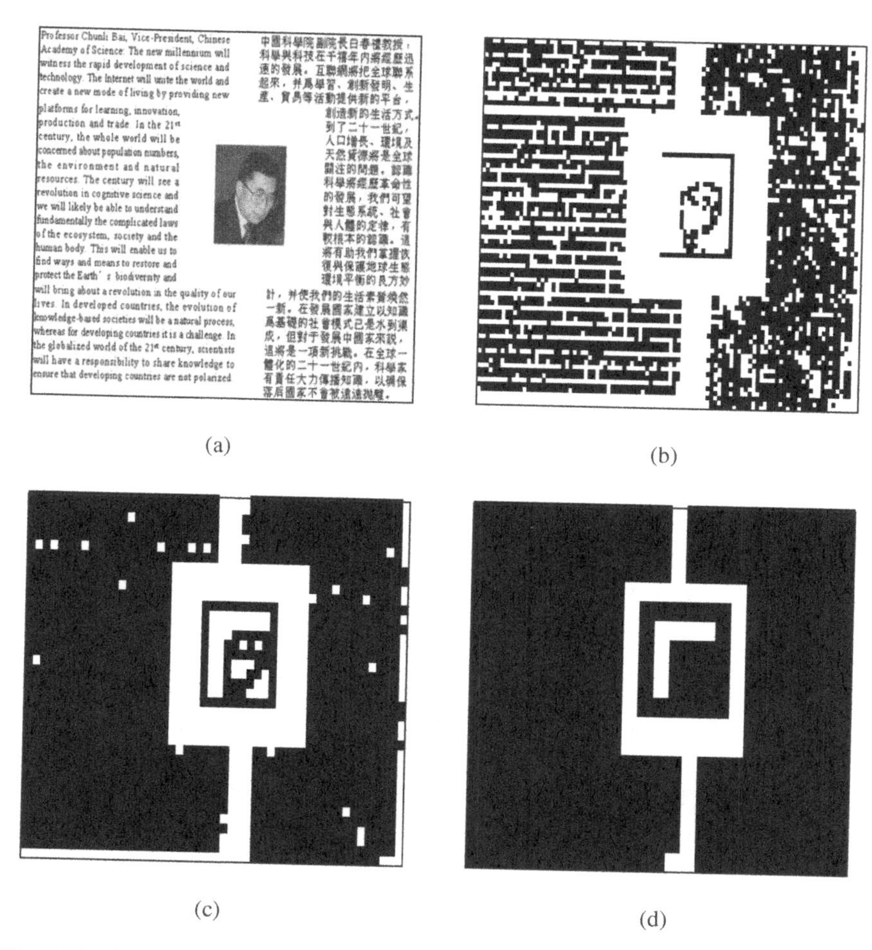

Fig. 4.19 An example of document analysis by the fractal signature with different sub-block sizes.

Chapter 5

Text Extraction by Wavelet Decomposition

5.1 Introduction

In this chapter, we propose a new method for segmentation using wavelet decomposition of *pseudo-motion image*. The aim of this approach is to detect the position of a special kind of objects in an image. In [Wiskott, 1999], L. Wiskott presented a method for segmentation from motion of objects using Gabor- and Mallat-wavelets. Wiskott's method works on an sequence of images, which is assumed that objects move fronto-parallel on a fixed background. In our method, we work on a single image using a motion of eyeshot: the translation of the function. The translation of one image provides a sequence of moving images. We call this sequence to be a *pseudo-motion image* and the translation to be a *pseudo-motion*. When a pseudo-motion of an image occurs, we study what will happen in its wavelet decomposition coefficients. The fact is: the coefficients will oscillate. Meantime, the amplitudes of the oscillation will give the information about the frequency of the function. Thus we can use this significant property to segment the areas of different frequency of the function.

In section 2, we establish the formulas of wavelet decomposition under translate, which formulas are essential in the calculation of pseudo-motion image. Section 3 proposes our algorithm, the segmentation algorithm based on wavelet decomposition with pseudo-motion, which is called WDPM Algorithm. In section 4, some experiments demonstrate the performance of the method.

5.2 Wavelet Decomposition of Pseudo-Motion Functions

In this section, the wavelet decomposition of a function with pseudo-motion will be considered. A *pseudo-motion* of a function means a translation of a function. The translation can provide a sequence of function from one single function. We will present how to calculate the wavelet decomposition coefficients of this sequence of functions.

A multiresolution analysis gives a family of orthonormal bases for the function space L^2. Therefore, each function in L^2 can be expressed as an (infinite) linear combination of the bases. Such expressions are called wavelet decomposition of the function. By studying the coefficients of the wavelet decomposition, one can detects some special properties of the function. Furthermore, when the function is translated, the coefficients of the wavelet decomposition will be changed, too. Such changes will indicate the change of the frequency of the function.

5.2.1 *One Variable Case*

At first, we consider the wavelet decomposition of one-variable functions with pseudo-motion. Suppose that $\phi(x)$ is a scaling function and $\psi(x)$ is the associated wavelet function. For any function g, set

$$g_{j,k}(x) = 2^{-j/2} g(2^{-j}x - k). \tag{5.1}$$

Let $V_j = \overline{\text{Span}\{\phi_{j,k}\ ;\ k \in \mathbb{Z}\}}$, $W_j = \overline{\text{Span}\{\psi_{j,k}\ ;\ k \in \mathbb{Z}\}}$. Then $\{V_j\}$ is a multiresolution analysis (MRA). The associated orthogonal wavelet decompositions of function spaces are

$$\begin{aligned} V_j &= W_{j+1} \oplus V_{j+1} \\ &= W_{j+1} \oplus \cdots \oplus W_{j+n} \oplus V_{j+n}, \\ L^2(\mathbb{R}) &= \oplus_{j \in \mathbb{Z}} W_j, \end{aligned}$$

where all these sums are orthogonal sums. Furthermore, $\{\phi_{j,k}; k \in \mathbb{Z}\}$ is an orthonormal basis of V_j and $\{\psi_{j,k}; k \in \mathbb{Z}\}$ is an orthonormal basis of W_j.

Let A_j and B_j be the orthogonal projective operator from $L^2(\mathbb{R})$ to V_j and W_j respectively. For any $f \in L^2(\mathbb{R})$, we have

$$\begin{aligned} A_j(f) &= A_{j+1}(f) + B_{j+1}(f) \\ &= A_{j+n}(f) + \textstyle\sum_{k=1}^{n} B_{j+k}(f). \end{aligned}$$

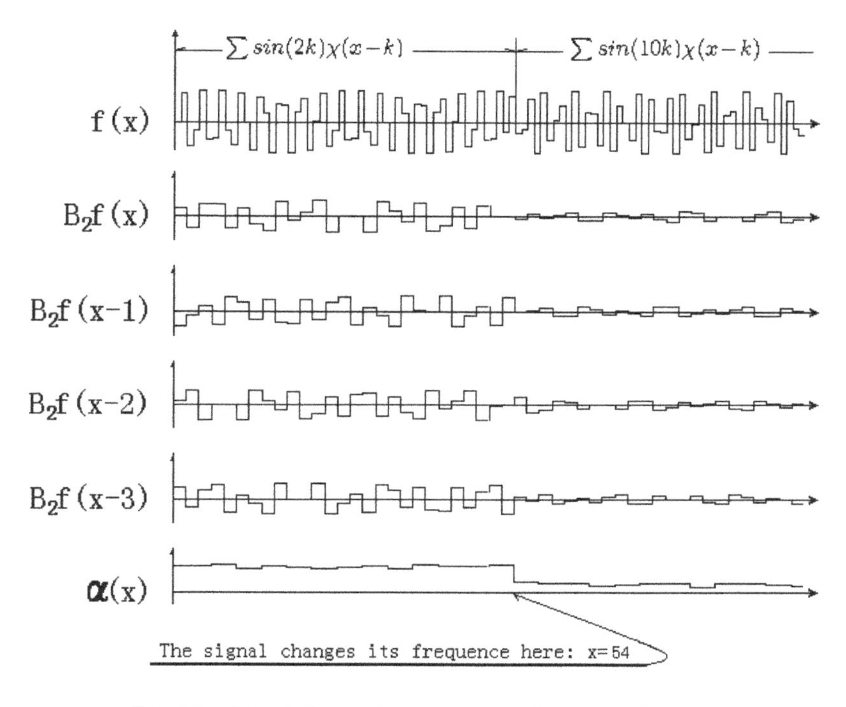

Fig. 5.1 Detect the position where the frequency changes.

This is referred to as *wavelet decomposition* of the function f. The coefficients of $A_j(f)$ and $B_j(f)$ are important to represent the properties of f. The coefficients of $A_{j+1}(f)$ and $B_{j+1}(f)$ can be calculated from the coefficients of $A_j(f)$ by the well known Mallat algorithm.

To create a pseudo-motion function, the original function $f(x)$ is translated. We define $\mathfrak{T}_m$ to be the translation by m: $(\mathfrak{T}_m f)(x) = f(x - m)$, $m \in \mathbb{R}$. For any single function f, we can obtain the pseudo-motion function, which is the sequence of functions $\mathfrak{T}_m f$ for m in some interval. For different m, the wavelet decomposition $A_j(\mathfrak{T}_m f)$ and $B_j(\mathfrak{T}_m f)$ will be different. Thus, $A_j(\mathfrak{T}_m f)$ and $B_j(\mathfrak{T}_m f)$ will change when m moves. This is a dynamic process. By studying the characteristic of this process, we can detect the position where the frequency of f changes. To explain this phenomenon more clearly, we look at the following example. Let $\chi(x)$ be the character function of the unit interval $[0,1)$. If we define $\eta(x) = \frac{1}{\sqrt{2}}(\chi_{-1,0} - \chi_{-1,1})$, then χ is the Haar scaling function and η is the associated wavelet function. We define a function

$$f(x) = \sum_{k<54} sin(2k)\chi(x-k) + \sum_{k\geq 54} sin(10k)\chi(x-k).$$

In this function, there exists a frequency change which occurs near the point $x = 54$. We use χ and η as the wavelet bases to perform wavelet decomposition for $\mathfrak{T}_m f(x)$ $(m = 0,1,2,3)$, we can obtain $B_2\mathfrak{T}_m f(x)$. Let

$$\alpha(x) = \max_{0\leq m\leq 3} B_2\mathfrak{T}_m f(x) - \min_{0\leq m\leq 3} B_2\mathfrak{T}_m f(x).$$

Figure 5.1 shows the graphic representation of functions f, $B_2\mathfrak{T}_m f$ and α. It is obviously that α indicates where the frequency of f is changed.

Now we present the calculation of the coefficients of wavelet decomposition with pseudo-motion. Suppose $f \in V_0$, and according to Mallat algorithm, for each m, we can compute $A_1(\mathfrak{T}_m f)$ and $B_1(\mathfrak{T}_m f)$ from $\mathfrak{T}_m f$, $A_2(\mathfrak{T}_m f)$ and $B_2(\mathfrak{T}_m f)$ from $A_1(\mathfrak{T}_m f)$, $A_3(\mathfrak{T}_m f)$ and $B_3(\mathfrak{T}_m f)$ from $A_2(\mathfrak{T}_m f)$, and so on. The above computation can be presented below:

For each m do: $\mathfrak{T}_m f \rightarrow A_1(\mathfrak{T}_m f)$ *and* $B_1(\mathfrak{T}_m f)$

  └→ $A_2(\mathfrak{T}_m f)$ *and* $B_2(\mathfrak{T}_m f)$

    └→ $A_3(\mathfrak{T}_m f)$ *and* $B_3(\mathfrak{T}_m f)$

      └→ ……

Hence, if one wants to obtain $A_K(\mathfrak{T}_m f)$ and $B_K(\mathfrak{T}_m f)$ for $m = M, M+1, \cdots, N$, all of the $A_k(\mathfrak{T}_m f)$ for $k = 1, 2, \cdots, K$ and $m = M, M+1, \cdots, N$ have to be calculated. This is too bother in the calculate. Fortunately, according to the following theorem, it can be concluded that we only have to calculate $A_k(\mathfrak{T}_m f)$ for $k = 1, 2, \cdots, K$ and $m = 0, 1, \cdots, 2^k - 1$. In fact, $f \in V_j$ if and only if $\mathfrak{T}_{2^j} f \in V_j$. $f \in V_j$ and $\mathfrak{T}_{2^j m} f \in V_j$ have the same sequence of coefficients in expression of wavelet basis but with a translation of subscript. Thus, we can have the following calculating process:

$f \to A_1(\mathfrak{T}_m f)$ and $B_1(\mathfrak{T}_m f)$ for $m = 0, 1$

↳ $A_2(\mathfrak{T}_m f)$ and $B_2(\mathfrak{T}_m f)$ for $m = 0, 1, 2, 3$

↳ $A_3(\mathfrak{T}_m f)$ *and* $B_3(\mathfrak{T}_m f)$ for $m = 0, 1, \cdots, 7$

↳ ......

Theorem 1 *Let $\{V_j\}$ be a MRA with scaling function $\phi(x)$ and wavelet function $\psi(x)$. Suppose $J, m \in \mathbb{Z}$, $p = 0$ or 1. Let*

$$f = \sum_{k\in\mathbb{Z}} s_k \phi_{J,k} \in V_J,$$

$$A_{J+1}\mathfrak{T}_{2^J(2m+p)} f = \sum_{k\in\mathbb{Z}} t_k \phi_{J+1,k} (\in V_{J+1}),$$

and

$$B_{J+1}\mathfrak{T}_{2^J(2m+p)} f = \sum_{k\in\mathbb{Z}} v_k \psi_{J+1,k} (\in W_{J+1}).$$

Then the coefficients t_k, v_k and s_k can be computed by

$$t_k = \sum_{n\in Z} \bar{h}_{n-2(k-m)} s_{n-p},$$

$$v_k = \sum_{n\in Z} \bar{g}_{n-2(k-m)} s_{n-p},$$

$$s_k = \sum_{n\in\mathbb{Z}} h_{k+p-2n} t_{n+m} + \sum_{n\in\mathbb{Z}} g_{k+p-2n} v_{n+m},$$

where $h_k = <\phi, \phi_{-1,k}> = \sqrt{2}\int_{\mathbb{R}} \phi(x)\bar{\phi}(2x-k)dx$ and $g_k = <\psi, \phi_{-1,k}> (= (-1)^{k-1}h_{-k-1})$.

Proof Since

$$\mathfrak{T}_{2^J(2m+p)} f = \sum_{k\in\mathbb{Z}} s_k \phi_{J,k+2m+p} = \sum_{k\in\mathbb{Z}} s_{k-2m-p} \phi_{J,k} \in V_J,$$

thus, the coefficients of $A_{J+1}\mathfrak{T}_{2^J(2m+p)} f \in V_{J+1}$ and $B_{J+1}\mathfrak{T}_{2^J(2m+p)} f \in V_{J+1}$ can be calculated by Mallat algorithm:

$$t_k = \sum_{n\in\mathbb{Z}} \bar{h}_{n-2k} s_{n-2m-p} = \sum_{n\in\mathbb{Z}} \bar{h}_{n-2k+2m} s_{n-p},$$

$$v_k = \sum_{n \in \mathbb{Z}} \bar{g}_{n-2k} s_{n-2m-p} = \sum_{n \in \mathbb{Z}} \bar{g}_{n-2k+2m} s_{n-p}.$$

According to Mallat reconstruction formula,

$$s_{k-2m-p} = \sum_{n \in \mathbb{Z}} h_{k-2n} t_n + \sum_{n \in \mathbb{Z}} g_{k-2n} v_n.$$

We can obtain

$$\begin{aligned} s_k &= \sum_{n \in \mathbb{Z}} h_{k+2m+p-2n} t_n + \sum_{n \in \mathbb{Z}} g_{k+2m+p-2n} v_n \\ &= \sum_{n \in \mathbb{Z}} h_{k+p-2n} t_{n+m} + \sum_{n \in \mathbb{Z}} g_{k+p-2n} v_{n+m} \end{aligned}$$

■

Based on Theorem 1, for $f \in V_0$, we can calculate the coefficients of $A_1(\mathfrak{T}_0 f)$ and $B_1(\mathfrak{T}_1 f)$ from the coefficients of f, calculate the coefficients of $A_2(\mathfrak{T}_{2p+q} f)$ and $B_2(\mathfrak{T}_{2p+q} f)$ from the coefficients of $A_1(\mathfrak{T}_q f)$ for $p, q = 0, 1$, calculate the coefficients of $A_3(\mathfrak{T}_{4m+2p+q} f)$ and $B_3(\mathfrak{T}_{4m+2p+q} f)$ from the coefficients of $A_2(\mathfrak{T}_{2p+q} f)$ for $m, p, q = 0, 1$, and so on.

5.2.2 *Two Variables Case*

Now we turn to the wavelet decomposition with pseudo-motion on plane $\mathbb{R}^2$. Suppose $\phi^{(0)}(x)$ and $\phi^{(1)}(x)$ are two scaling functions on $\mathbb{R}$. The associated wavelet functions are $\psi^{(0)}(x)$ and $\psi^{(1)}(x)$ respectively. Let

$$\begin{cases} \Phi^{(00)}_{i,j,k}(x,y) = \phi^{(0)}_{i,j}(x)\phi^{(1)}_{i,k}(y) \\ \Phi^{(01)}_{i,j,k}(x,y) = \phi^{(0)}_{i,j}(x)\psi^{(1)}_{i,k}(y) \\ \Phi^{(10)}_{i,j,k}(x,y) = \psi^{(0)}_{i,j}(x)\phi^{(1)}_{i,k}(y) \\ \Phi^{(11)}_{i,j,k}(x,y) = \psi^{(0)}_{i,j}(x)\psi^{(1)}_{i,k}(y) \end{cases} \tag{5.2}$$

We define closed subspaces $V_I^{(uv)} = \overline{Span\{\Phi^{(uv)}_{I,j,k} | j, k \in \mathbb{Z}\}}$ for $I \in \mathbb{Z}$, $u, v = 0, 1$. Then, we obtain the tensor product multiresolution analysis on $\mathbb{R}^2$ and we have the orthogonal decomposition of subspaces

$$V^{(00)}_{i-1} = V^{(00)}_i \oplus V^{(01)}_i \oplus V^{(10)}_i \oplus V^{(11)}_i.$$

Let $P_i^{(uv)}$ be the orthogonal projective operator from $L^2(\mathbb{R})$ to $V_i^{(uv)}$, $u, v = 0, 1$. Then for any $f \in V_J^{(00)}$, we have the two-dimensional wavelet decomposition

$$f = P^{(00)}_{J+1} f + P^{(01)}_{J+1} f + P^{(10)}_{J+1} f + P^{(11)}_{J+1} f.$$

For $m, n \in \mathbb{R}$, define $\mathfrak{T}_{m,n}$ to be the translation by (m, n):

$$(\mathfrak{T}_{m,n}f)(x, y) = f(x - m, y - n).$$

Similar to the one-dimensional case, we can prove the following wavelet decomposition algorithm of the translated functions.

Theorem 2 *Let $I, m, n \in \mathbb{Z}$, $p, q = 0$ or 1. Suppose*

$$F(x, y) = \sum_{j,k\in\mathbb{Z}} s_{j,k}\Phi^{(00)}_{I,j,k}(x, y) \ \in V_I^{(00)},$$

$$(P^{(uv)}_{I+1}\mathfrak{T}_{2^I(2m+p),2^I(2n+q)}F)(x, y) = \sum_{j,k\in\mathbb{Z}} t^{(uv)}_{j,k}\Phi^{(uv)}_{I+1,j,k} \ (\in V^{(uv)}_{I+1}).$$

Then the coefficients can be calculated by

$$\begin{array}{rcl} t^{(00)}_{\beta,\gamma} & = & \sum_{j,k\in\mathbb{Z}} s_{j-p,k-q}\bar{h}^{(0)}_{j-2(\beta-m)}\bar{h}^{(1)}_{k-2(\gamma-n)} \\ t^{(01)}_{\beta,\gamma} & = & \sum_{j,k\in\mathbb{Z}} s_{j-p,k-q}\bar{h}^{(0)}_{j-2(\beta-m)}\bar{g}^{(1)}_{k-2(\gamma-n)} \\ t^{(10)}_{\beta,\gamma} & = & \sum_{j,k\in\mathbb{Z}} s_{j-p,k-q}\bar{g}^{(0)}_{j-2(\beta-m)}\bar{h}^{(1)}_{k-2(\gamma-n)} \\ t^{(11)}_{\beta,\gamma} & = & \sum_{j,k\in\mathbb{Z}} s_{j-p,k-q}\bar{g}^{(0)}_{j-2(\beta-m)}\bar{g}^{(1)}_{k-2(\gamma-n)} \end{array} \quad (5.3)$$

where $h^{(u)}_k = \sqrt{2}\int_{\mathbb{R}}\phi^{(u)}(x)\bar{\phi}^{(u)}(2x-k)dx$, $g^{(u)}_k = (-1)^{k-1}h^{(u)}_{-k-1}$ and $u, v =$ $0, 1$.

Theorem 2 gives the similar convenience in calculation of the coefficients of the wavelet decomposition of the pseudo-motion function on two variable.

5.3 Segmentation of Different Areas of Document Image

In this section, we focus our major aim on segmenting the different areas in digital image including document image.

5.3.1 *Segmentation of Areas of Different Frequency*

A digital image G with $M \times N$ pixels can be considered to be a two-variable real function $F(x, y) \in L^2(\mathbb{R}^2)$, and $F(i, j)$ is the gray level of (i,j)-point of the image. In different area, the frequency of the function $F(x, y)$ will be different. Therefore, it is significant to detect the difference of the frequency of $F(x, y)$ in different area.

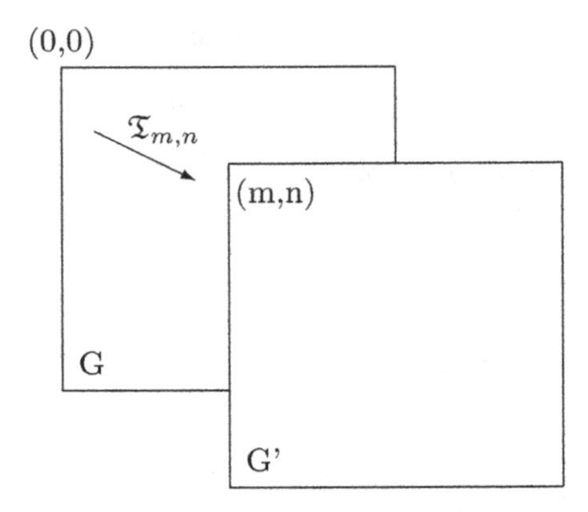

Fig. 5.2 The operator $\mathfrak{T}_{m,n}$ moves the image G to G'. If the gray level funcion of G is $F(x, y)$, then the gray level function of G' is $\mathfrak{T}_{m,n}F$.

Now we translate the function $F(x, y)$, or equivalently, we move the image. This translation provides a sequence of moving images. We call this sequence to be a *pseudo-motion image* and the translation to be a *pseudo-motion* of images. The graphic illustration of the pseudo-motion can be found in Figure 5.2. When a pseudo-motion of an image occurs, the coefficients of its wavelet decomposition will oscillate. Meantime, the amplitudes of the oscillation will give the information about the frequency of the function. Thus we can use this significant property to segment the areas of different frequency of the function.

We select a suitable pair of scaling functions, and thereafter, construct the tensor product miltiresolution analysis. In this work, the compact supported wavelet bases [Daubechies, 1992] are used. According to the selected wavelet basis, we can achieve the wavelet decompositions of $\mathfrak{T}_{m,n}F$ at levels $I = 0, 1, 2, \cdots$ for $m, n \in \mathbb{Z}$. The coefficients of these decompositions will be changed if m, n are changed. In other words, the coefficients will *oscillate* while m, n are moving. The amplitude of these oscillate will be used to segment the different areas of $F(x, y)$ with difference frequency.

Let $\phi^{(u)}$, u=0,1, be two scaling functions and $\psi^{(u)}$ be the associated wavelet functions. Suppose $\Phi^{(uv)}_{i,j,k}(x, y)$ are defined by Eq. (5.2), and $V^{(uv)}_I$ are defined by the closure subspace of $L^2(\mathbb{R}^2)$ spanned by $\{\Phi^{(uv)}_{I,j,k}|j, k \in \mathbb{Z}\}$.

Then $\{\Phi_{I,j,k}^{(uv)} | j, k \in \mathbb{Z}\}$ is an orthonormal basis of $V_I^{(uv)}$. Let G be an input $M \times N$ pixels digital image with gray level function $F(x, y)$. For any $m, n \in \mathbb{R}$, the coefficients of the projection of $\mathfrak{T}_{m,n}F$ on $V_I^{(uv)}$ by $C_{I,j,k}^{(uv)}(m, n)$ can be denoted by

$$P_I^{(uv)}(\mathfrak{T}_{m,n}F) = \sum_{j,k\in\mathbb{Z}} C_{I,j,k}^{(uv)}(m,n)\Phi_{I,j,k}^{(uv)},$$

where

$$\begin{aligned} C_{I,j,k}^{(uv)}(m,n) &= < \mathfrak{T}_{m,n}F, \Phi_{I,j,k}^{(uv)} > \\ &= \int\int \mathfrak{T}_{m,n}F(x,y)\overline{\Phi_{I,j,k}^{(uv)}(x,y)}dxdy. \end{aligned} \tag{5.4}$$

For the computation of the coefficients $C_{I,j,k}^{(uv)}(m, n)$, we use the integral formula Eq. (5.4) only when the scale level $I = 0$. For $I = 1, 2, \cdots$, we can calculate the values of $C_{I,j,k}^{(uv)}(m, n)$ from the coefficients $C_{0,j,k}^{(00)}(0, 0)$, $j, k \in \mathbb{Z}$, by Theorem 2:

- $C_{I,j,k}^{(uv)}(m + 2^I p, n + 2^I q) = C_{I,j-p,k-q}^{(uv)}(m, 0)$
  for $u, v = 0, 1,\ I, j, k, m, n, p, q \in \mathbb{Z}$.
- $C_{I+1,J,K}^{(uv)}(m + 2^I p, n + 2^I q)$ are linear combinations of
  $\{C_{I,j,k}^{(00)}(m, n) | j, k \in \mathbb{Z}\}$
  for $I, J, K \in \mathbb{Z}$, $u, v, p, q = 0, 1$ and $m, n = 0, 1, \cdots, 2^I - 1$.

Thus, we can calculate
$\{C_{1,j,k}^{(uv)}(0, 0)\}$ and $\{C_{1,j,k}^{(uv)}(1, 0)\}$ from $\{C_{0,j,k}^{(00)}(0, 0)\}$,
$\{C_{2,j,k}^{(uv)}(0, 0)\}$ and $\{C_{2,j,k}^{(uv)}(2, 0)\}$ from $\{C_{1,j,k}^{(00)}(0, 0)\}$,
$\{C_{2,j,k}^{(uv)}(1, 0)\}$ and $\{C_{2,j,k}^{(uv)}(3, 0)\}$ from $\{C_{1,j,k}^{(00)}(1, 0)\}$, etc.
This processing can be illustrated in Figure 5.3.

When u, v, I, j, k are fixed, $C_{I,j,k}^{(uv)}(m, n)$ is a two variable function on $\mathbb{R}^2$. When (m, n) is running around a fixed point (m_0, n_0), $C_{I,j,k}^{(uv)}(m, n)$ will give the information of the frequency of $F(x, y)$ at the Ith level wavelet decomposition near $(m_0 + 2^I j, n_0 + 2^I k)$. If we define

$$M_I^{(u,v)}(j,k) = \max_{0\le m,n<2^I}\{C_{I,j,k}^{(uv)}(m,n)\} - \min_{0\le m,n<2^I}\{C_{I,j,k}^{(uv)}(m,n)\}, \tag{5.5}$$

then $M_I^{(u,v)}(j, k)$ is the amplitude of the oscillate of the coefficients $C_{I,j,k}^{(uv)}(m, n)$ for m and n running from 0 to $2^I - 1$. $M_I^{(u,v)}(j, k)$ can be used to detect the different frequency areas of $F(x, y)$.

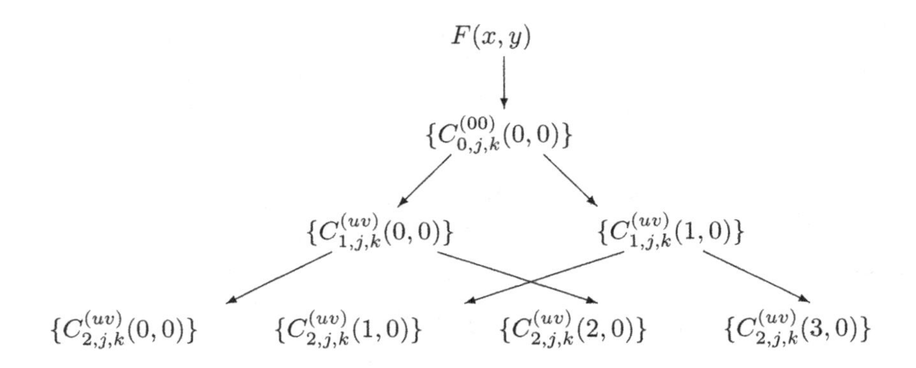

Fig. 5.3 The calculate of $\{C^{(uv)}_{I,j,k}(m,0)\}$.

5.3.2 *WDPM Algorithm*

This approach can be represented by the following algorithm. Since this algorithm is based on the wavelet decomposition with pseudo-motion, it is referred as *WDPM Algorithm*.

WDPM Algorithm

(1) Calculate the coefficients $C^{(00)}_{0,j,k}(0,0)$ from $F(x,y)$ by formula Eq. (5.4).
(2) $I = 1$.
(3) For $m, n = 0, \cdots, 2^{I-1} - 1$, $p, q = 0, 1$, calculate the coefficients $C^{(uv)}_{I,j,k}(m + 2^I p, n + 2^I q)$ $(u, v = 0, 1)$ from $C^{(00)}_{I-1,j,k}(m,n)$ by Theorem 2.
(4) Calculate $M^{(u,v)}_I(j,k)$ $(u, v = 0, 1)$ from $C^{(uv)}_{I,j,k}(m,n)$ follow formula Eq. (5.5).
(5) Try to segment the different frequency areas of the two-variable function $F(x,y)$ form $M^{(u,v)}_I(j,k)$. If it is not successful, let $I = I + 1$ and go to Step 3. If it is successful, end the algorithm.

Figure 5.4 shows the steps of the total detection processing. The coefficients $C^{(uv)}_{I,j,k}(m,n)$ obtained in Step 3 are just the wavelet decomposition coefficients of $F(x,y)$ under the pseudo-motion from $m, n = 0$ to $m, n = 2^I - 1$. $M^{(u,v)}_I(j,k)$ in Step 4 is the amplitude of the oscillate.

The analysis of the time complexity of the algorithm is presented below:

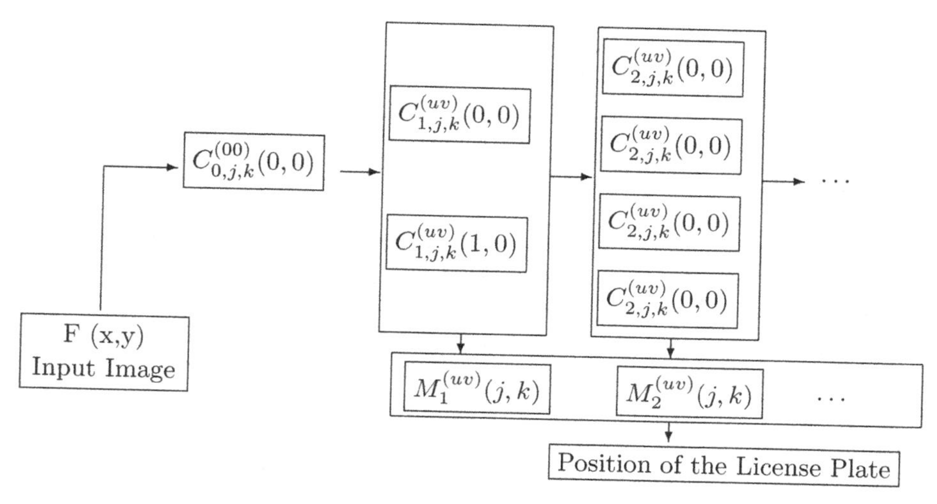

Fig. 5.4 Processing steps.

Step 1 is calculated according to formula Eq. (5.4). Now, the wavelet bases we used here are compact supported. Suppose $\text{Supp}\phi^{(u)} = [0, \beta^{(u)})$, $u = 0, 1$, then $\Phi^{(00)}_{0,j,k}(x, y) = 0$ if $x \notin [j, \beta^{(0)} + j)$ or $y \notin [k, \beta^{(1)} + k)$. So, we have

$$\begin{aligned} & C^{(00)}_{0,j,k}(0,0) \\ = \ & \textstyle\sum_{m=0}^{\beta^{(0)}-1} \sum_{n=0}^{\beta^{(1)}-1} F(m+j, n+k) \int_{m+j}^{m+1+j} dx \int_{n+k}^{n+1+k} \overline{\Phi^{(00)}_{0,j,k}(x,y)} dy \\ = \ & \textstyle\sum_{m=0}^{\beta^{(0)}-1} \sum_{n=0}^{\beta^{(1)}-1} F(m+j, n+k) \int_{m}^{m+1} dx \int_{n}^{n+1} \overline{\Phi^{(00)}_{0,0,0}(x,y)} dy. \end{aligned}$$

We denote $\int_m^{m+1} dx \int_n^{n+1} \overline{\Phi^{(00)}_{0,0,0}(x,y)} dy$ in terms of $P(m, n)$, then

$$C^{(00)}_{0,j,k}(0,0) = \sum_{m=0}^{\beta^{(0)}-1} \sum_{n=0}^{\beta^{(1)}-1} F(m+j, n+k) P(m,n) \tag{5.6}$$

On the other hand, the image is of size $M \times N$, thus $F(x, y)$ is vanished if $x \notin [0, M)$ or $y \notin [0, N)$. Consequently $C^{(00)}_{0,j,k}(0,0) = 0$ if $j \notin [1-\beta^{(0)}, M]$ or $k \notin [1-\beta^{(1)}, N]$. Hence we only need to calculate $C^{(00)}_{0,j,k}(0,0)$ by formula Eq. (5.6) for $j = 1-\beta^{(0)}, \cdots, M$ and $k = 1-\beta^{(1)}, \cdots, N$. Therefore, the time expenses in step 1 is $O(MN)$.

Step 3 is calculated according to formula Eq. (5.3). Here the number of nonzero elements of $h^{(u)}_k$ (and $g^{(u)}_k$) is $2\beta^{(u)} + 2$. So, the time expense to compute each single coefficient is a fixed value (depending on $\beta^{(u)}$). For each fixed I, u, v, m and n, the number of nonzero coefficients $C_{I,j,k}^{(uv)}(m,n)$ will be about $O(\frac{M}{2^I} \times \frac{N}{2^I})$. Notice that m, n run from 0 to 2^I. Hence, for any I, the number of nonzero coefficients $C_{I,j,k}^{(uv)}(m,n)$ which need to be calculated is about $O(M \times N)$. Therefore, the time expense in step 3 is $O(MN)$.

Furthermore, because $M_I^{(uv)}(j,k)$ are determined by $C_{I,j,k}^{(uv)}(m,n)$ following formula Eq. (5.5), it is obvious that the time expenses in step 4 and step 5 are also $O(MN)$.

Finally, at each time when the level I grows one, the size of $\{C_{I,j,k}^{(uv)}\}$ will be half in both horizontal and vertical. So, the values of level I will take $0, 1, 2, \cdots, \min(\log_2 M, log_2 N)$. Hence, the total time complexity of the process is $O(MN \log(MN))$ if all the $M_I(j,k)$ are calculated for all useful $I = 1, 2, \cdots$. But in general only the small I's are needed. In most cases, the levels $I = 1$ and 2 are enough. Consequently, the total time expense of WDPM algorithm is $O(MN)$.

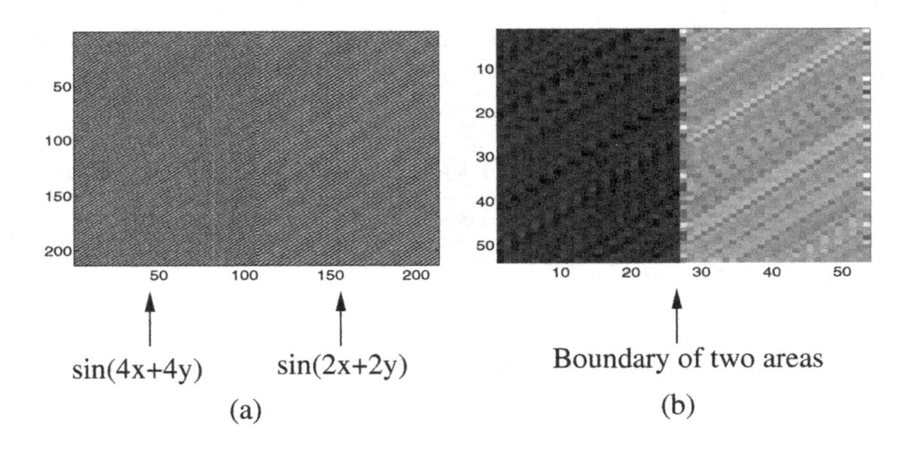

Fig. 5.5 Segment of different frequency area.

Figure 5.5 shows an example of the use of the WDPM algorithm. Here, we choose the Haar basis to generate the tensor product multiresolution analysis on $\mathbb{R}^2$. Figure 5.5(a) is an image of size 213×213 pixels with gray level function

$$F(x, y) = \begin{cases} sin(4x + 4y), & x < 107, \\ sin(2x + 2y), & x \geq 107. \end{cases}$$

This image has two areas of different frequency divided by the vertical line $x = 107$. Figure 5.5(b) is the gray level image of $M_2^{(11)}(j, k)$ which points out the boundary of the two areas of the different frequency.

5.4 Experiments

In this section, we use two experiments to demonstrate the performance of WDPM algorithm. The first experiment is to find the position of license plates of vehicles in digital images. The second one is to detect the text areas of document images with gray-level background.

5.4.1 *Position of License Plate*

The problem studied in this experiment is how to determine the position of license plate in a digital image of vehicles. The license plate can be regarded as a rectangle region on which the gray level function has a distinguished frequency.

5.4.1.1 *Choose of the Bases*

The first step is to find a suitable pair of scaling functions $\phi^{(0)}(x)$ and $\phi^{(1)}(y)$ to construct the tensor product wavelet basis $\{\Phi_{i,j,k}^{(uv)}\}$. The different choice will make many differences in the procession, such as the information about the position of license plate, the number of the non-zero coefficients, the amount of the calculate, the accuracy of the result, etc.

At first, we will determine $\phi^{(0)}(x)$. We require $\phi^{(0)}(x)$ has the following properties:

(1) $\phi^{(0)}(x)$ is continuous. In fact, we assume that $F(x, y)$ is a continuous function of x for any fixed y.

(2) $\phi^{(0)}(x)$ has a small support set. Equivalently, the number of non-zero elements of the frequency response $\{h_j^{(0)}\}$ is small. Consequently, the amount of the calculation in the wavelet decomposition procession will be less.

Fig. 5.6 A sample of plate.

(3) If the region $a \leq x \leq b$, $c \leq y \leq d$ in the image is the location of the license plate, then the following numbers will be small:

$$\delta_y = \int_a^b [F(x,y) - \sum_{k \in \mathbb{Z}} < F(x,y), \phi^{(0)}(x-k) > \phi^{(0)}(x-k)]^2 dx.$$

That means, if we project $F(x,y)$ into the space $V_0^{(00)}$ of the multiresolution analysis associated with $\phi^{(0)}(x)$, then the error will be small in the region of the plate. Thus, most of the information about the license plate will be kept.

According to the above first two requirements, we will select $\phi^{(0)}(x)$ from the Daubechies bases ${}_N\phi$ with $N = 2, 3, \cdots$, which are continuous and compact supported [Daubechies, 1992]. To determine which one is our choice, the third condition has to be considered.

Figure 5.6 is a typical sample of the image of a license plate. The values of δ_y for $\phi^{(0)}(x) =_N \phi$, $N = 2, \cdots, 10$, $y =$ 25, 27, 29, 31, 33, 35, 37, are calculated and listed below:

N\y	25	27	29	31	33	35	37
2	10.0416	10.3002	9.8862	10.8877	10.9222	10.9263	6.9926
3	10.5506	10.8719	10.5204	11.4175	11.5507	11.2533	7.2893
4	8.9239	9.0215	8.6362	9.4998	9.5038	9.7823	6.4870
5	8.9546	9.0980	8.6443	9.6806	9.6095	9.9829	6.3920
6	9.6827	9.9475	9.5746	10.5105	10.5561	10.5116	6.8045
7	9.8116	9.9608	9.5784	10.4354	10.5108	10.5800	6.9563
8	9.4578	9.5519	9.1110	10.0443	10.0662	10.3515	6.7264
9	9.2338	9.4706	9.0642	10.0549	10.0513	10.1665	6.5361
10	9.4297	9.6909	9.3250	10.2681	10.2782	10.2871	6.6701

From these results we can claim that ${}_4\phi$ is a good choice. In fact, more samples show that ${}_5\phi$ is also suitable for keeping δ_y small. But the numbers of non-zero elements of the frequency response of them are 8 and 10 respectively. We select the less one. Thus, the choice is $\phi^{(0)}(x) ={}_4 \phi(x)$.

Now, we turn to determine $\phi^{(1)}(y)$. At this time, the totally influence of $\phi^{(0)}(x)$ and $\phi^{(1)}(y)$ will be studied together. Notice that there are many vertical lines in the words of the plate. Hence, we select $\phi^{(1)}(y)$ to be a generalization of *the Haar wavelet* with support $[0, T)$. We define ${}_T\chi$ by ${}_T\chi(y) = 1/\sqrt{T}$ if $y \in [0, T)$ and vanish elsewhere. Set ${}_T\xi(y) ={}_T \chi(2y) -{}_T \chi(2y - T)$, then, $\{{}_T\chi(y - kT)|k \in \mathbb{Z}\}$ and $\{{}_T\xi(y - kT)|k \in \mathbb{Z}\}$ are orthonormal systems and we can obtain ${}_T\chi(y - kT) \perp_T \xi(y - k'T)$. Therefore, we can use ${}_T\chi(y)$ and ${}_T\xi(y)$ as the scaling function and wavelet function respectively. In this case, we set ${}_T\chi_{j,k}(y) = 2^{-j/2}{}_T\chi(2^{-j}x - kT)$ and ${}_T\xi_{j,k}(y) = 2^{-j/2}{}_T\xi(2^{-j}x - kT)$. All the results and formulas established before still hold under these assumptions.

Figures 5.7(a) and (b) show the difference between ${}_4\phi(x)_4\phi(y)$ and ${}_4\phi(x)_T\chi(y)$ for $T = 10$. Figure 5.7(c) is a sample image of a part of a plate. There are three letters on it. Figure 5.7(d) is the map of the associated gray level function $F(x, y)$ of Figure 5.7(c). It is obvious that ${}_4\phi(x)_T\chi(y)$ is better for $F(x, y)$ than ${}_4\phi(x)_4\phi(y)$.

Suppose $F(x, y)$ is the gray level function of digital image G. Let $V_{i,j,k}^{(uv)}$ be the multiresolution analysis on $\mathbb{R}^2$ determined by ${}_4\phi(x)_T\chi(y)$. Then we

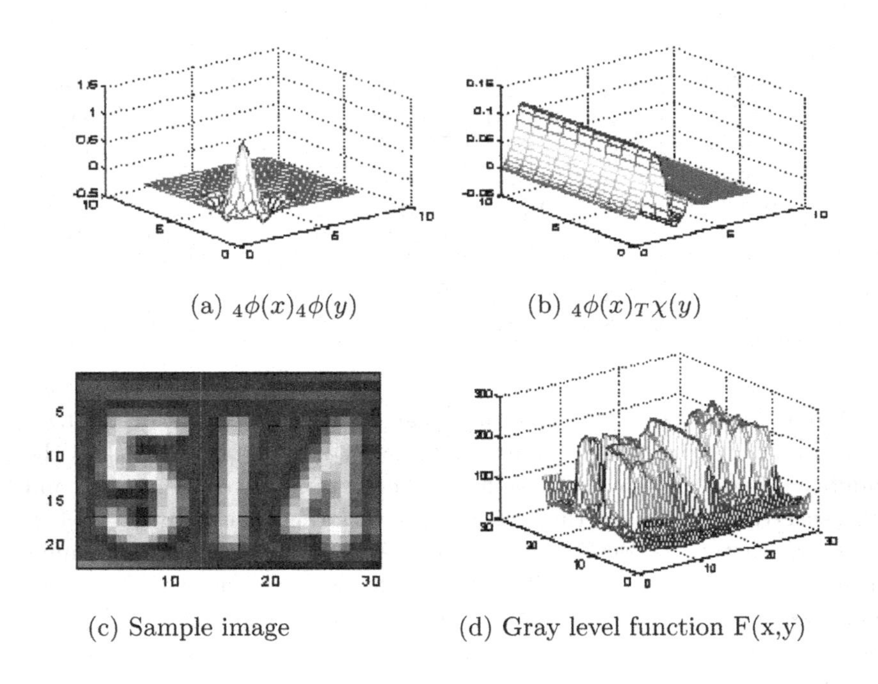

(a) ${}_4\phi(x){}_4\phi(y)$ (b) ${}_4\phi(x){}_T\chi(y)$

(c) Sample image (d) Gray level function F(x,y)

Fig. 5.7 Difference between ${}_4\phi(x){}_4\phi(y)$ and ${}_4\phi(x){}_10\chi(y)$.

can project $F(x,y)$ into the subspace $V_0^{(00)}$:

$$P_0^{(00)}F = \sum_{j,k\in\mathbb{Z}} C_{j,k}\Phi_{0,j,k}^{(00)},$$

where $C_{j,k} =< F, \Phi_{0,j,k}^{(00)} >$. When T is different, the coefficients $C_{j,k}$ of the projection are different. In fact, if image G is of size $m \times n$ pixels, there will be only $(m+12)\lceil n/T \rceil$ coefficients. The larger is the T, the less is the number of the coefficients will be considered. But on the other hand, when T becomes large, the projection will become unclear. Figure 5.8 shows the different results by choosing different values of T. Figure 5.8(a) is the original sample image, it contains two plates of size 20×86 and 10×43. Figure 5.8(b)-(f) is its projection in $V_0^{(00)}$ with $T = 2, 4, 6, 8, 10$ respectively. It seems that $T = 4$ is a good choice.

Hence, $\phi^{(0)}(x) =_4 \phi(x)$ and $\phi^{(1)}(y) =_4 \chi(y)$ are chosen as the bases in our experiments.

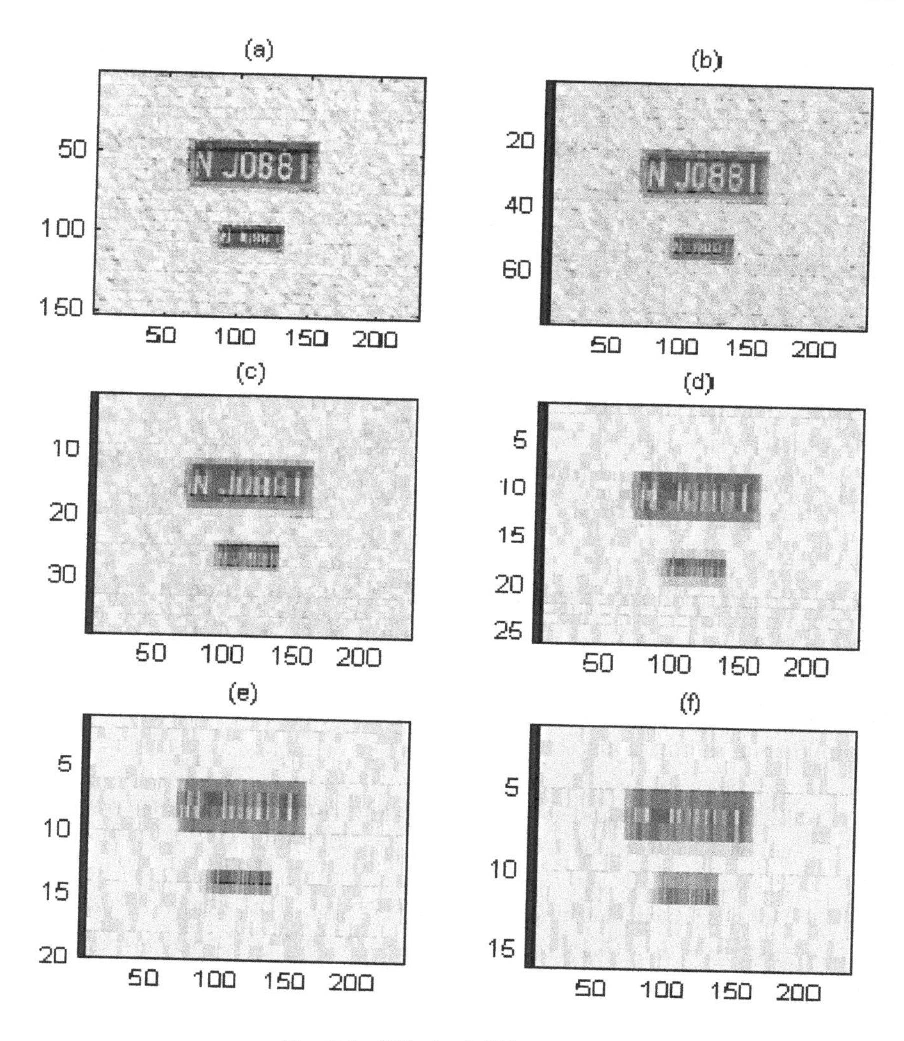

Fig. 5.8 Effect of different T.

5.4.1.2 *Experimental Results*

Some digital images of 640×480 pixels are used as samples to illustrate the effect of our method. The images used here are captured by a digital camera. The program was designed using Visual Basic 5.0 and have been running on a Pentium II MMX (TM) system. The time expense of each sample, from inputting the image to outputting the position of the plate, is $0.4 \sim 0.6$ seconds. The results of four samples are shown in Figures 5.9~5.12. In all of these figures, (a) is the original input image G, i.e. $F(x, y)$; (b) is the projection of F in $V_0^{(00)}$, i.e. $C_{0,j,k}^{(00)}(0,0)$; (c), (d) and (e) are $M_I^{(1,0)}(j,k)$ for $I = 1, 2$, and 3 respectively; (f) is the detection result. According to the property of the license plate, only the oscillate of $F(x, y)$ along x-axis are considered here. Thus, instead of Eq. (5.5), we set

$$M_I^{(u,v)}(j,k) = \max_{0 \le m < 2^I} \{C_{I,j,k}^{(uv)}(m,0)\} - \min_{0 \le m < 2^I} \{C_{I,j,k}^{(uv)}(m,0)\}. \tag{5.7}$$

In Samples 1 and 3, we can detect all plates in the image, including the plate of small size. But in Sample 2, we failed to get that small one. All the plates are determined from $M_1^{(10)}(j,k)$, except the big plate in sample 3, which is detected from $M_2^{(10)}(j,k)$.

5.4.2 *Localization of Text Areas of Document Images*

Now we present how to use our algorithm to detect the text areas of document images. As a typical document image is a binary image, we use the Haar wavelet basis here. For the convenience of the computation, we set the following assumptions:

- In both x-axis and y-axis, we use the same scaling function $\phi^{(0)}$, which is defined by $\phi^{(0)}(x) = 1$ if $x \in [0, 1)$ and $\phi^{(0)}(x) = 0$ otherwise.
- Define the wavelet function $\phi^{(1)}$ by $\phi^{(1)}(x) = \phi^{(0)}(2x) - \phi^{(0)}(2x-1)$.
- Instead of Eq. (5.1), define $\phi_{i,j}^{(u)}(x) = 2^{-i}\phi^{(u)}(2^{-i}x - j)$ for $i, j \in \mathbb{Z}$, $u = 0, 1$.
- Similar to the Eq. (5.2), we construct the tensor product two-dimensional wavelet basis by

$$\Phi_{I,j,k}^{(uv)}(x, y) = \phi_{I,j}^{(u)}(x)\phi_{I,k}^{(v)}(y)$$

  for $I, j, k \in \mathbb{Z}$, $u, v = 0, 1$.
- Still use Eq. (5.4) and Eq. (5.5) to define $C_{I,j,k}^{(uv)}(m, n)$ and $M_I^{(uv)}(j,k)$ respectively.

Under these settings, the coefficients in the formula Eq. (5.3) become very simple. Instead of the irrational number $\frac{1}{\sqrt{2}}$, we have $h_0^{(u)} = h_1^{(u)} = g_0^{(u)} = -g_1^{(u)} = 1$ and $h_k^{(u)} = g_k^{(u)} = 0$ if $k \neq 0, 1$, where $u = 0$ and 1. In this case, the formula Eq. (5.3) becomes

$$\begin{aligned} C_{I+1,j,k}^{(uv)}(m + 2^I p, \quad & n + 2^I q) \\ = \quad & [C_{I,2j-p,2k-q}^{(00)}(m,n) + (-1)^u C_{I,2j-p+1,2k-q}^{(00)}(m,n) \\ & +(-1)^v C_{I,2j-p,2k-q+1}^{(00)}(m,n) \\ & +(-1)^{u+v} C_{I,2j-p+1,2k-q+1}^{(00)}(m,n)], \end{aligned} \tag{5.8}$$

where $I, j, k, m, n \in \mathbb{Z}$, $u, v, p, q = 0, 1$. Besides, $C_{0,j,k}^{00} = F(j,k)$. Therefore, the step 1 in the algorithm can be omitted here.

The sample images were obtained from some books by a scanner. The texts on these samples have several different types: big letters or small letters, black letters or white letters, solid letters or empty letters, horizontal letters or slope letters, simple background or complex background.

The text areas of all these different types can be detected from $M_2^{(10)}(j,k)$. Figures 5.13 and 5.14 show the results. (a) is the original image. (b) is the image of $M_2^{(10)}(j,k)$. (c) is the result, where the highlight areas are the detedted text areas.

In the lower part of the sample image Figure 5.13(a), there are some texts located over the complex background. We detected them successfully. The sample image shown in Figure 5.14(a) contains a histogram and a piegraph. In the result (c), these two graphs are moved away.

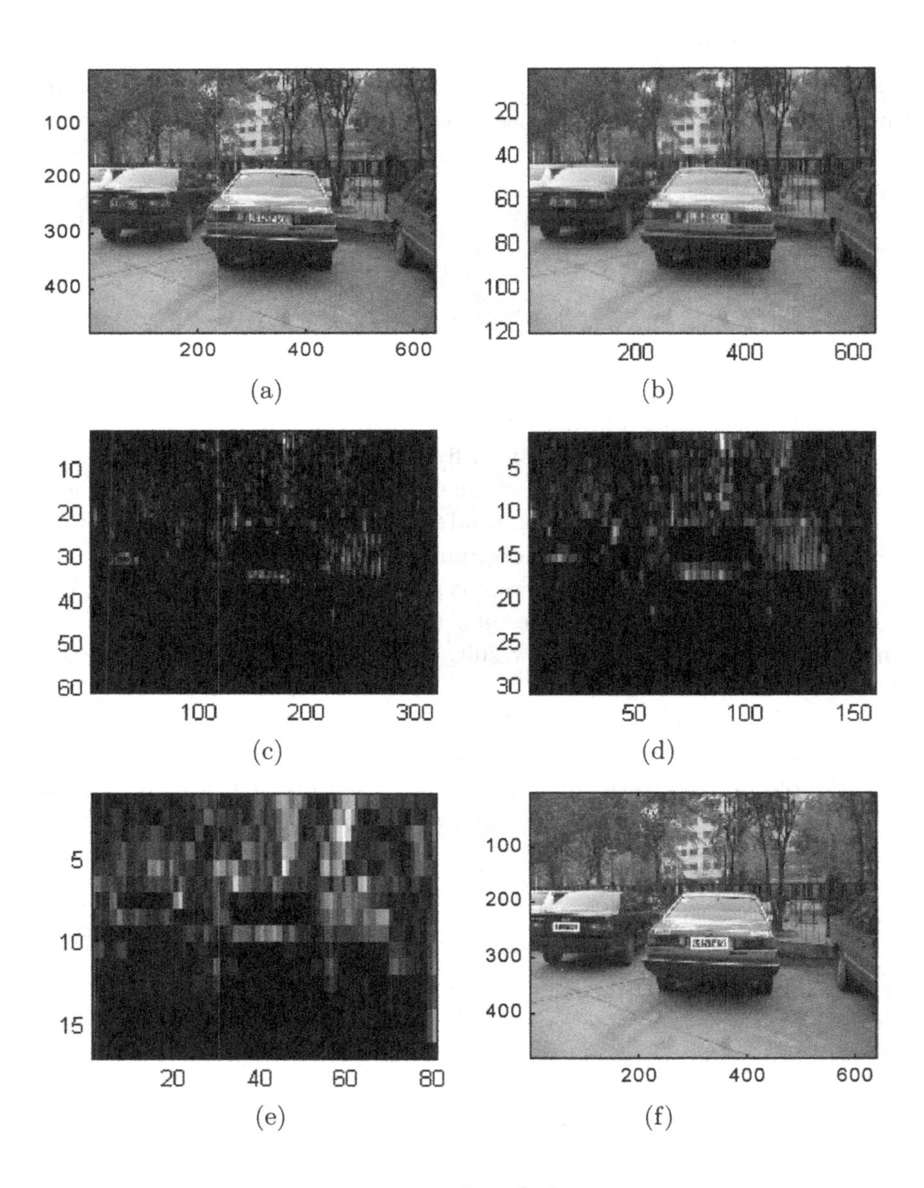

Fig. 5.9 Sample 1.

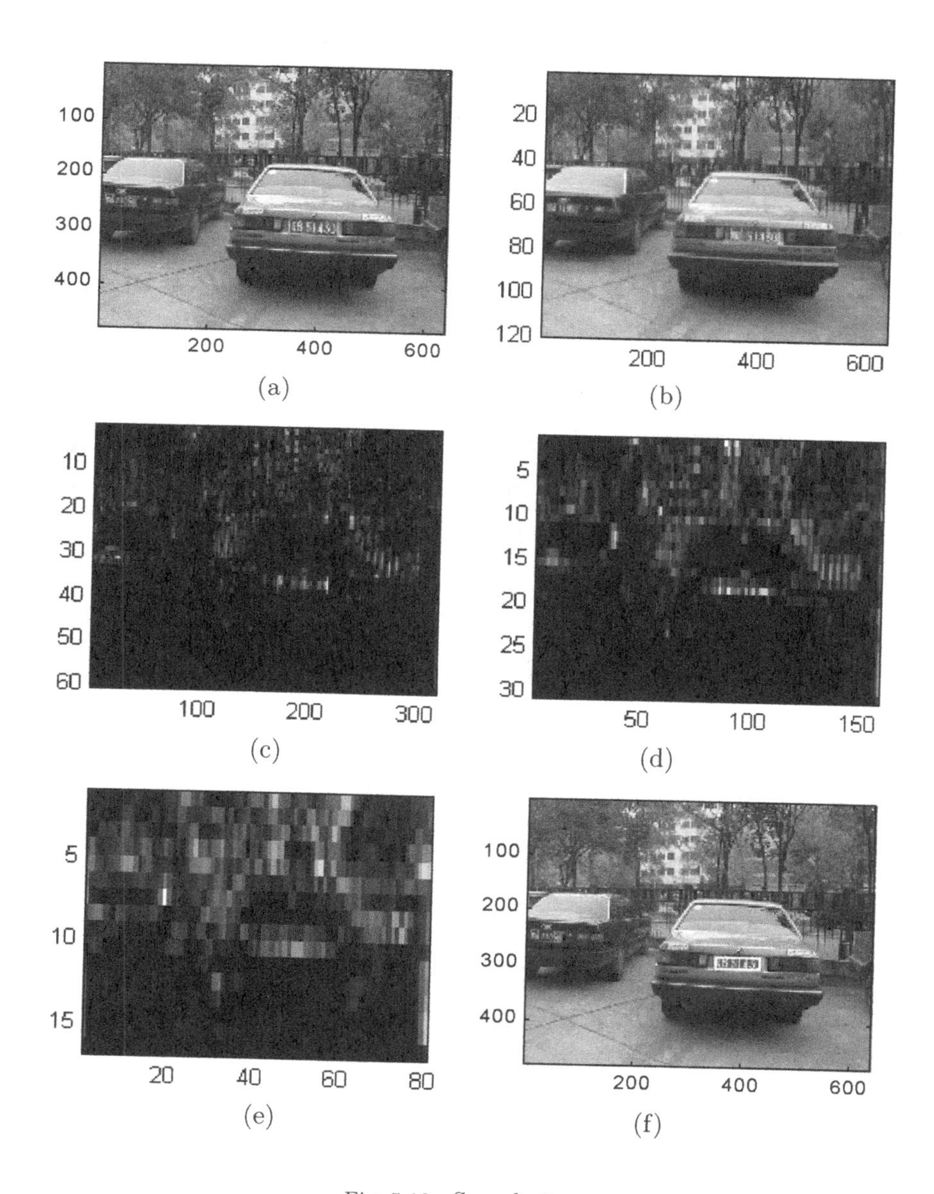

Fig. 5.10 Sample 2.

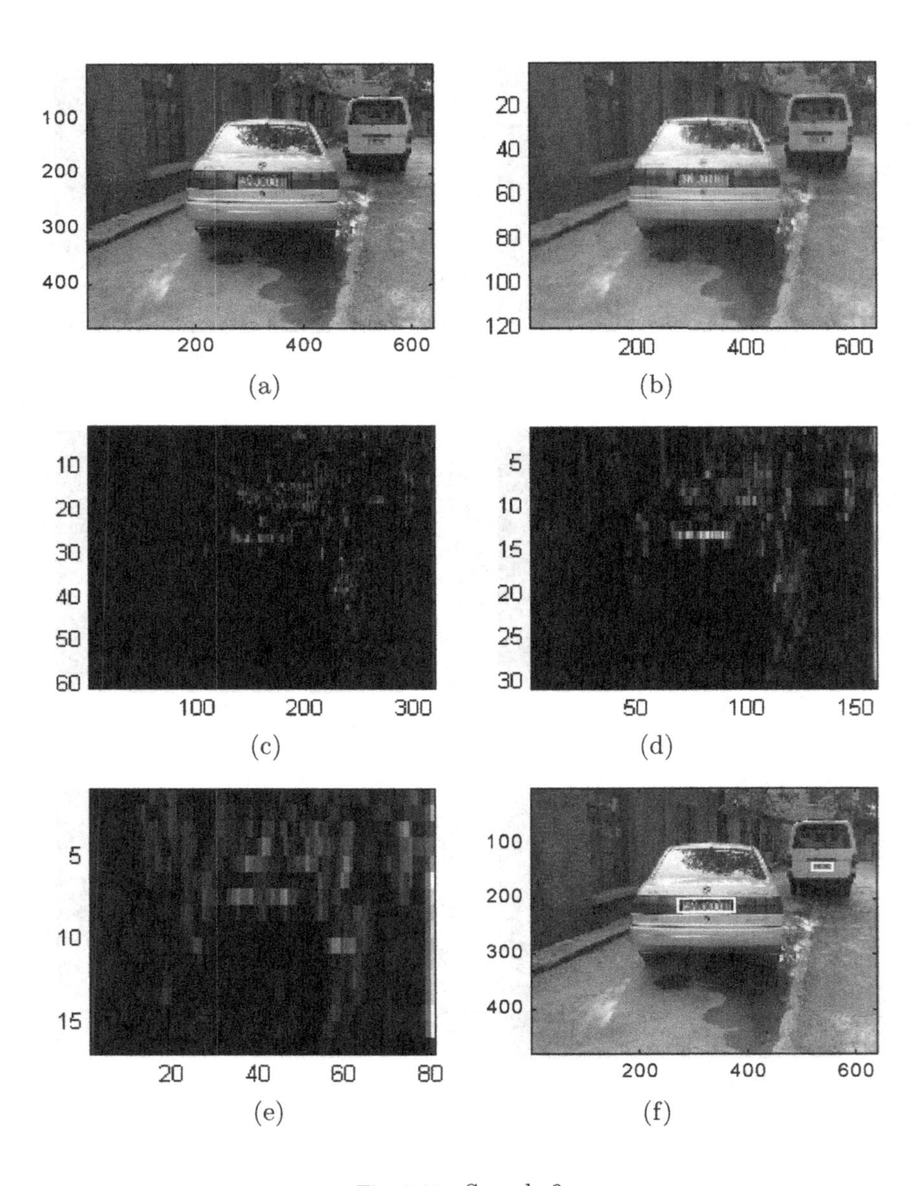

Fig. 5.11 Sample 3.

(a) (b) (c) (d) (e) (f)

Fig. 5.12 Sample 4.

Fig. 5.13 Sample 5 - English Text extraction from document.

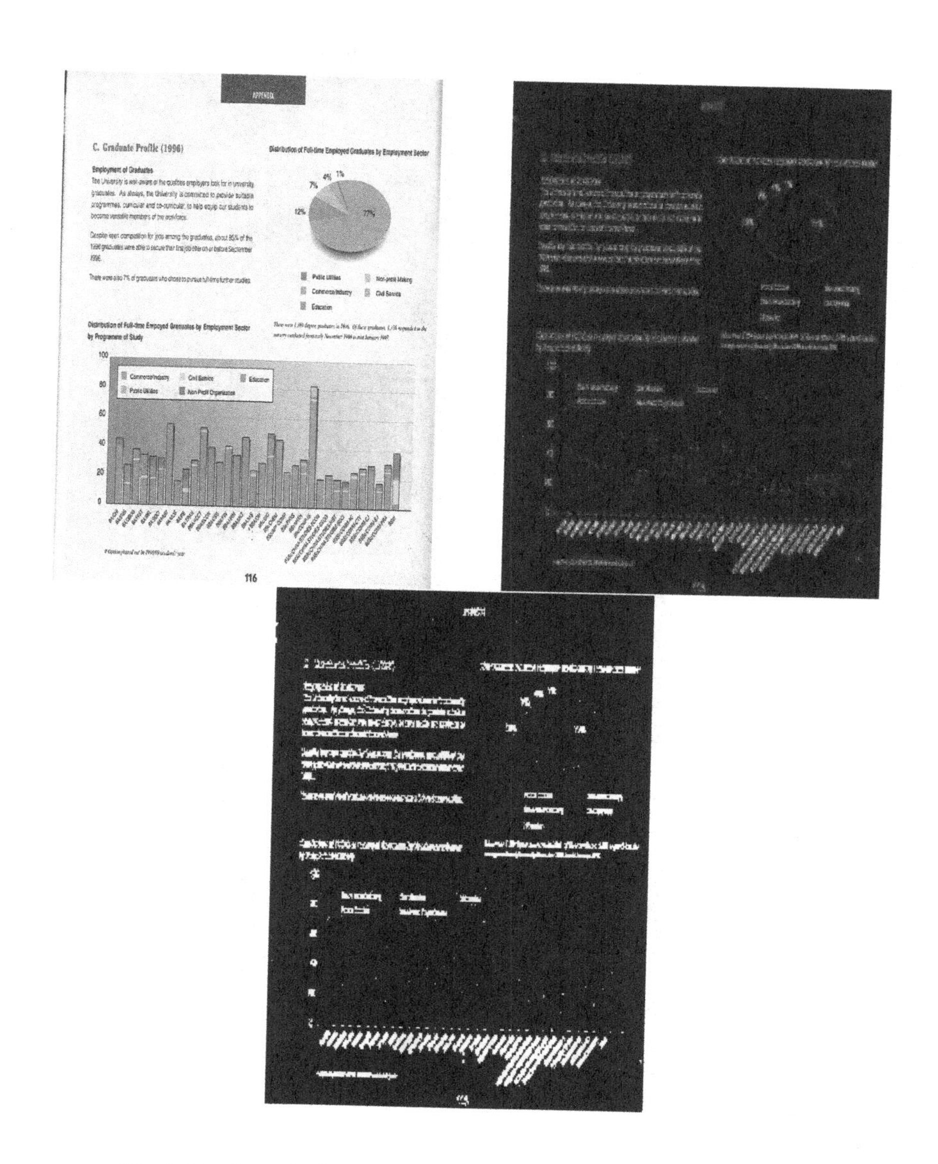
APPENDIX

C. Graduate Profile (1996)

Employment of Graduates

The University is well aware of the qualities employers look for in university graduates. As always, the University is committed to provide suitable programmes, curricular and co-curricular, to help equip our students to become versatile members of the workforce.

Despite keen competition for jobs among the graduates, about 85% of the 1996 graduates were able to secure their first job offer on or before September 1996.

There were also 7% of graduates who chose to pursue full-time further studies.

Distribution of Full-time Employed Graduates by Employment Sector

Distribution of Full-time Employed Graduates by Employment Sector by Programme of Study

116

Fig. 5.14 Sample 6 - English Text extraction from document.

Fig. 5.15 Sample 7 - Chinese Text extraction from document.

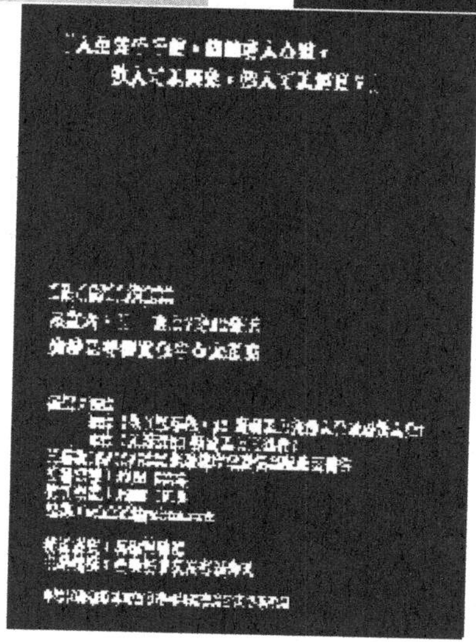

Fig. 5.16 Sample 8 - Chinese Text extraction from document.

Chapter 6

Rotation Invariant by Fractal Theory with Central Projection Transform (CPT)

The problem of invariant pattern recognition is considered to be a highly complex and difficult one. The technique of rotation invariance is one of the most important branches of invariant pattern recognition. The document image, such as printed or handwritten character, is regarded as a two-dimensional pattern. In this chapter, we are concerned in particular with the feature extraction of two-dimensional patterns related to rotation invariance in the plane.

6.1 Introduction

An invariant pattern recognition technique is to obtain a representation of the pattern in a form, which is invariant under some transformation of the original pattern. It can classify the different samples of a pattern as belonging to a particular class. Once we process an object with different rotation angles, a pattern many produce several samples which have characteristics different from the original one. In such a case, a unique representation of different samples for each possible input same pattern can actually be a disadvantage. All of these require an invariant representation which retains enough information for distinct classes to be distinguished. Indeed, if all members of a class are identical, or nearly so, in the invariant representation this can greatly reduce the size of the training set required by many recognition approaches. This is our interest to develop the stable and rotation invariant representations for the pattern with different rotated angles. In this chapter, the proposed feature extraction technique has rotation invariant properties. This method is based on the fractal theory combining the central projection transform. The details of central projection transform will be presented in the later section of this chapter.

As fractal geometry is able to describe the world in which we live it would seem like an ideal tool for image processing and pattern recognition (including document image recognition). The tendency of generalized pattern recognition techniques to be computationally expensive is the main motivation behind using fractal geometry for pattern recognition. As has already been stated, one of the most powerful aspects of fractal geometry is its ability to express complex sets as a few parameters. If these parameters can be extracted for a general shape and comparisons made between shapes using these parameters then this could form the basis of a completely new recognition technique.

The document image, such as printed or handwritten character, is regarded as a two-dimensional pattern. In this study, we are concerned with the invariant perception of two-dimensional patterns under rotation in the plane. This corresponds to the ability of humans to recognize patterns such as typed or handwritten characters independently of their orientation, which they do unthinkingly when reading a document such as an architectural drawing. Since images and their constituent features are usually transformed by projective transformations in many applications, it is necessary to determine characteristics invariant to such transformations. In particular, much attention has been paid to the identification of complete series of invariants in order to describe an object by it invariants in a one-to-one manner. These characteristic measures should exhibit invariance or tolerance with respect to transformations. This is why invariants have proved to be useful tools.

6.1.1 *Rotations*

A document image in the Cartesian (x, y) domain may be defined as a function $f : \mathbb{R}^2 \rightarrow \mathbb{R}$, $f(x, y) = z$, where z is the intensity of the image at coordinates (x, y). The invariant pattern recognition problem is to recognize images as being some sense "the same" even though they have undergone a variety of allowed transformations. Now rotate this coordinate system about the (perpendicular) z-axis. Since lengths don't depend on orientation, the scale does not change and the grid still appears as composed of parallel, equally spaced straight lines. Lines of constant $x^{'}$ and $y^{'}$ in the new coordinate system appear in the old one as in Figure 6.1.

The transformation of coordinates can be readily obtained by considering

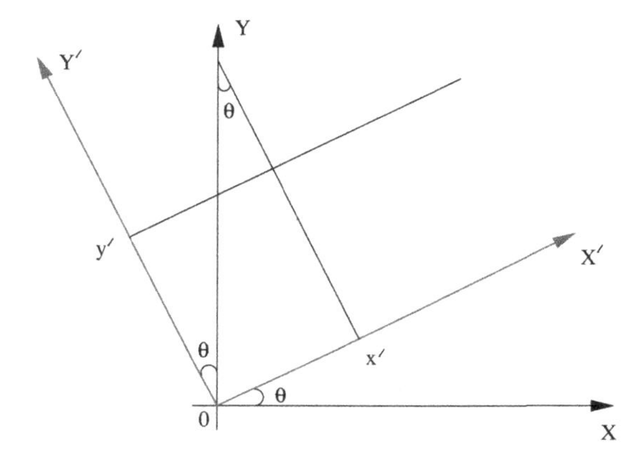

Fig. 6.1 The coordinate system is rotated by an angle θ about an axis perpendicular to the (x, y) domain.

the equations for these particular straight lines in the original coordinate system. The equation of a straight line always takes the form

$$y = mx + C,$$

where m is the slope and C the intercept on the y-axis. Consideration of Figure 6.1 shows that the line of constant $y^{'}$ is given by

$$y = x \tan\theta + y^{'} / \cos\theta,$$

while that of constant $x^{'}$ is given by

$$y = -x \cos\theta + x^{'} / \sin\theta.$$

Inverting these, one immediately obtains

$$\begin{aligned} x^{'} &= x \cos\theta + y \sin\theta \ ; \\ y^{'} &= -x \sin\theta + y \cos\theta \ ; \end{aligned}$$

and of course we also have $z^{'} = z$.

The corresponding active rotation is again just the inverse and is obtained by changing θ to $-\theta$. Since $\cos(-\theta) = \cos\theta$ and $\sin(-\theta) = -\sin\theta$, this results in the sine terms in the above expression changing sign.

In general, a rotation can be described by a rotation angle and a rotation axis. The specification of these is equivalent to specifying three rotation angles for an arbitrary rotation. There is more than one way of defining these three angles. We need not bother here with details though.

6.1.2 *Rotation Invariants*

Rotation of an image is corresponding to coordinate transformations of the form

$$\begin{bmatrix} x' \\ y' \end{bmatrix} = \begin{bmatrix} \cos\theta & -\sin\theta \\ \sin\theta & \cos\theta \end{bmatrix} \begin{bmatrix} x \\ y \end{bmatrix}.$$

Group theory tells us that the only one-dimensional invariants for two-dimensional rotations are given by the magnitude of $h_k r$ [Lenz, 1990], the circular Fourier transform of the image $f(r, \theta)$ represented in polar coordinates:

$$h_k r = \int_0^{2\pi} f(r,\theta) f(r, \theta + \alpha) d\theta, \qquad k \text{ integer.}$$

Below we will first see how this relates to correlation invariants and then how it is related to moment invariants.

- **Correlations**

  The first, second and third order correlations $g_1, g_2(a), g_3(a_1, a_2)$ of the image $f(x, y)$ are defined [Giannakis and Tsatsanis, 1990], followed by the general definition of the kth order correlation. $A_i^T = [\alpha_i \ \beta_i]$ is a two-dimensional vector; a point (x, y) on the image is represented by $f(x)$, where $x^T = [x \ y]$; the integrals are over the range $[-\infty, +\infty]$:

$$\begin{aligned} g_1 &= \int f(x)dx; \\ g_2(a) &= \int f(x)f(x+a)dx; \\ g_3(a_1, a_2) &= \int f(x)f(x+a_1)f(x+a_2)dx. \end{aligned}$$

The kth order correlation is given by

$$g_k(a_1, a_2, ..., a_{k-1}) = \int f(x)f(x+a_1)...f(x+a_{k-1})dx. \quad (6.1)$$

The second order angular correlation of the image $f(r, \theta)$ in polar coordinates is

$$g_2(\alpha, r) = \int_0^{2\pi} f(r, \theta)f(r, \theta + \alpha)d\theta;$$

the extension to higher order correlations is obvious from Eq. (6.1). It can easily be shown that the circular Fourier transform of $g_2(\alpha, r)$ equals the squared magnitude of $h_k(r)$. As before one can use higher order correlations to avoid losing information. Angular correlations of discrete images are not a natural extension of the continuous case because of the varying number of pixels at a given radius.

- **Moments**

  Moment invariants were first introduced by Hu [Hu, 1962; Hu, 1961] based on his fundamental theorem of moment invariants. It turns out that this theorem, which has recently been corrected by Resis [Reiss, 1991], only needs to derive invariants to affine transformations. The regular moment m_{pq} of an image $f(x, y)$ is defined as

$$m_{pq} = \int_{-\infty}^{+\infty} x^p y^q f(x, y)dxdy, \qquad p,\ q = 0, 1, 2, \ldots, \quad (6.2)$$

where (x, y) are Cartesian coordinates, f is a non-negative continuous image function with bounded support so that integration within the available image plane is sufficient to gather all the signal information. In this chapter, only binary image functions are considered, so $f(x, y)$ can only be 0 or 1. This means that the integral Eq. (6.2) can be seen as a one to one mapping of the continuous, finite area image $f(x, y)$ onto the infinite discrete moment matrix $M \times M$ with entries m_{pq}.

The central moments μ_{pq} are defined as

$$\mu_{pq} = \int_{-\infty}^{+\infty} \int_{-\infty}^{+\infty} (x - \overline{x})^p \quad (6.3)$$

where $\overline{x}$ and $\overline{y}$ are the coordinates of the image centroid given by

$$\overline{x} = \frac{m_{10}}{m_{00}}, \qquad \overline{y} = \frac{m_{01}}{m_{00}}.$$

For binary images, m_{00} represents the total number of pixels in the image.

The first four of Hu's translation and rotation invariants are given by [Hu, 1962]:

$$\begin{aligned}
\phi_1 &= \mu_{20} + \mu_{02}, \\
\phi_2 &= (\mu_{20} - \mu_{02})^2 + 4\mu_{11}^2, \\
\phi_3 &= (\mu_{30} - 3\mu_{12})^2 + (3\mu_{21} - \mu_{03})^2, \\
\phi_4 &= (\mu_{30} + \mu_{12})^2 + (\mu_{21} + \mu_{03})^2.
\end{aligned}$$

These can most easily be derived using complex moments c_{pq}, which were introduced by Davis [Davis, 1977] and re-introduced by Abu-Mostafa and Psaltis [Abu-Mostafa and Psaltis, 1984b]; they are defined as:

$$c_{pq} = \int_{-\infty}^{+\infty}\int_{-\infty}^{+\infty} (x + jy)^p (x - jy)^q dxdy, \tag{6.4}$$

where $j^2 \equiv -1$. If we map to polar coordinates, so that $x + jy = re^{j\theta}$, Eq. (6.4) becomes

$$\begin{aligned}
c_{pq} &= \int_{-\infty}^{+\infty}\int_{-\infty}^{+\infty} r^{p+q+1} e^{j(p-q)\theta} f(r,\theta) dr d\theta \\
&= \int_{0}^{+\infty} r^{p+q+1} \{\int_{0}^{2\pi} f(r,\theta) e^{jk\theta} d\theta\} dr, \qquad k = p - q, \\
&= \int_{0}^{+\infty} r^{p+q+1} h_k(r) dr.
\end{aligned} \tag{6.5}$$

From this we see that $|c_{pq}|$ is invariant to rotations of $f(r,\theta)$, and that c_{pq} is the projection of the image Fourier transform $h_{p-q}(r)$ onto the function r^{p+q+1}. In other words the continuous one dimensional function $h_{p-q}(r)$ is mapped onto an infinite vector with elements c_{pq}. If an image has $n-$fold rotational symmetry, all c_{pq}'s for which $p-q$ is not divisible by n are identically zero [Abu-Mostafa and Psaltis, 1984a] — for example, 'I' has twofold rotational symmetry, an equilateral triangle threefold, a square fourfold etc.

It is clear from Eq. (6.4) and the definition of moments that, for a centre image,

$$\begin{aligned} c_{11} &= \mu_{20} + j\mu_{02}, \\ c_{21} &= (\mu_{30} + \mu_{12})^2 + j(\mu_{21} + \mu_{02})^2, \qquad \textit{etc.} \end{aligned}$$

It is also apparent from Eq. (6.5) that using regular or central moments limits one to using radial moments of integer order $p + q + 1$. If one is interested in features invariant to rotation but not to translations, which would be the case if the images were normalized to be invariant to translations, the restriction on the powers of r can be lifted; indeed, one can use any radial weighting functions since the magnitude of $h_k(r)$ is invariant.

6.1.3 *Rotation Invariant of Discrete Images*

In this subsection, we will consider under what circumstances one can form functions of a sampled image that are absolutely invariant under rotations of the continuous image. It addresses the following two questions:

(1) Given I_0, an $N \times N$ sampled version of a continuous image, and I_θ, the $N \times N$ sampled version of the same image rotated by an angle θ, what conditions must be fulfilled for us to be able to compute I_θ given only I_0 and θ?
(2) Under what conditions can we compute the function of the I_θ that remain absolutely invariant over all values of $\theta \in [0, 2\pi]$?

We are used to thinking of a sampled version of a bandlimited image as containing all the information in the image, as long as the sampling rate is high enough. However, this is misleading because a bandlimited image has infinite spatial extent and hence requires an infinite number of samples to fully describe it. In the case of an $N \times N$ discrete image, we only have N^2 points with which to fully describe the continuous image; hence, the latter must be completely described by some known set of N^2 functions for the first question to be satisfied. The most straightforward case is when the continuous image is exactly representable by a linear combination of N^2 functions. If we let $f(x, y)$ represents, the continuous image and $v_{pq}(x, y), 1 \leq p, q \leq N$, represents the N^2 basis functions, then we want

$$f(x, y) = \sum_{p=1}^{N} \sum_{q=1}^{N} d_{pq} v_{pq}(x, y), \tag{6.6}$$

for some set of coefficients $\{d_{pq}\}$. Let the sampled image be g(k,l), then

$$g(k, l) = f(k_{x_o}, l_{y_0}), \qquad k, l = -\frac{N}{2}, -\frac{N}{2} + 1, \ldots, \frac{N}{2}, \tag{6.7}$$

and N is assumed to be even for the sake of simplicity. Combining Eq. (6.6) and Eq. (6.7) gives

$$\begin{aligned} g(k, l) &= \sum_{p=1}^{N} \sum_{q=1}^{N} d_{pq} v_{pq}(k_{x_0}, l_{y_0}), \qquad (6.8) \\ or \qquad g_{kl} &= V_{pqkl} d_{pq} \qquad \text{in tensor notation.} \end{aligned}$$

Rather than using tensor notation, we convert the matrix of values g_{kl} into an N^2-dimensional vector g with entries $g_i = g(k, l),\ i = k + N/2 + 1 + N * (l + N/2)$. This allows us to write Eq. (6.8) as

$$g = Vd; \qquad \text{V is an } N^2 \times N^2 \text{ matrix.}$$

We can now see that the condition of question 1 will be satisfied if the continuous image I_0 is exactly represented by a linear combination of N^2 basis functions $v_{pq}(x, y)$ that result in a non-singular matrix V. We can determine the coefficients using $d = V^{-1}g$.

Now we are in a position to address the second question. Before doing this though, it should be noted that the previous result only assumes that I_0 can be exactly represented by the N^2 basis functions; no assumption is made about I_θ being represented by the $v_{pq}(x, y)$. In fact, if V is invertible, every $I_\theta,\ \theta \in [0, 2\pi]$, will in general produce a different vector of coefficients d for each different value of θ, but combining these coefficients with the basis set as in Eq. (6.6) will generally result in a different continuous form that of the rotated continuous image. If we are to compute invariants we would like I_θ to be exactly represented by the N^2 basis functions for all values of θ, i.e., letting $f(r, \theta)$ be the image in polar coordinates and $\{V_{pq}(r, \theta)\}$ the basis in polar coordinates, we want the following to be satisfied: for all

$\{d_{pq}\}$ and ϕ there exists $\{d'_{pq}\}$ such that

$$f(r,\theta+\phi) = \sum_p \sum_q d_{pq} v_{pq}(r,\theta+\phi) = \sum_p \sum_q d'_{pq} v_{pq}(r,\theta). \qquad (6.9)$$

If the above is satisfied we can hope to find how $\{d_{pq}\}$ is related to $\{d'_{pq}\}$ and hence functions of the d_{pq} that are invariant to rotations.

Equation (6.9) will hold if the set of functions $v_{pq}(r,\theta+\phi)$ spans the same space as $\{V_{pq}(r,\theta)\}$ for all ϕ. To make the notation easier, let $v_k(r,\theta) = V_{pq}(r,\theta)$, where $1 \leq q \leq Q$ and $k = p + Q \times (q-1)$. $\{v_k(r,\theta)\}$ spans the same space as $\{v_k(r,\theta+\phi)\}$ for all ϕ if

$$v_k(r,\theta+\phi) = \sum_{n=1}^{N^2} G_{kn}(\phi) v_n(r,\theta) \qquad \text{for all } \phi,\ k = 1,\ldots,N^2, \quad (6.10)$$

where $G_{kn}(\phi)$ is some periodic function with period 2π. Eq. (6.10) can be written using vectors and matrices by defining $v(r,\theta)^T = [V_1\ (r,\theta), \ldots, v_{N^2}(r,\theta)]$ and $G(\phi)$ as the matrix with entry $G_{kn}(\phi)$ in row k and column n:

$$v(r,\theta+\phi) = G(\phi) v(r,\theta), \qquad (6.11)$$

and $G(\phi)$ is periodic with period 2π. Eq. (6.11) already tells us that $v(r,\theta)$ must be a separable function: $v_k(r,\theta) = h_k(r) s_k(\theta)$, so Eq. (6.11) becomes

$$s(\theta+\phi) = G(\phi) s(\theta). \qquad (6.12)$$

Considering the case $\phi = 0$, and using the fact that $s(\theta) = G^{-1}(\phi) s(\theta+\phi)$, gives us

$$\begin{aligned} G(0) &= I, \quad \text{the } N^2 \times N^2 \text{ identity matrix;} \\ G(-\phi) &= G^{-1}(\phi). \end{aligned} \qquad (6.13)$$

An example of matrices $G(\phi)$ that satisfy Eq. (6.13) are the rotation matrices that define a rotation by an amount ϕ about some axis in N^2 dimensions.

In order to find invariants given a matrix $G(\phi)$, we need first to find the set of functions $s(\theta)$ that satisfy Eq. (6.12) before proceeding to see how the coefficients d_k in Eq. (6.9) change with ϕ. In general finding the form of $s(\theta)$ is not easy; however, two special cases allow us to generate invariants easily. The first is when $G(\phi)$ is diagonal, and hence $s_k(\theta+\phi) = G_{kk}(\phi) s_k \theta$. It is easy to show that in this case the only suitable function $G_{kk}(\phi)$ is $G_{kk}(\phi) =$

$\exp(jl_k\phi)$, l_k an integer. Typically $l_k = k$. It is also straightforward to show that this value of $G_{kk}(\phi)$ results in $s_k(\theta) = \exp(jk\theta)$:

$$G(\phi)\ diagonal\ \Rightarrow G_{kk}(\phi) = e^{jl_k\phi} \Rightarrow s_k(\theta) = e^{jl_k\theta}.$$

Putting the above into Eq. (6.9) gives (remembering that $k = p+Q*(q-1)$)

$$\sum_k d_k h_k(r)e^{jl_k(\theta+\phi)} = \sum_k d_k^{'} h_k(r)e^{jl_k\theta}$$

from which we see that $d_k^{'}$ is related to d_k by

$$d_k^{'} = d_k e^{jl_k\phi}.$$

This in turn tells us that the magnitude of the coefficients, $|d_k|$, is invariant to rotations of the continuous image.

Before describing the second special case, a quick word about using complex coefficients d_k. Since the image is real, it is described by N^2 values; however, a complex d has $2N^2$ parameters, so we expect a certain amount of redundancy among the $\{d_k\}$. The redundancy expresses itself in the fact that the l_k's form pairs (l_{k1}, l_{k2}) where $l_{k1} = -l_{k2}$. This means that only half the coefficients d_k are independent.

The second special case is when $G(\phi)$ is of the form

$$G(\phi) =$$

$$\begin{bmatrix} & \vdots & \vdots & \vdots & \vdots & \vdots & \\ \cdots & 0 & 0 & 0 & 0 & 0 & \cdots \\ \cdots & \cos l\phi & -\sin l\phi & 0 & 0 & 0 & \cdots \\ \cdots & \sin l\phi & \cos l\phi & 0 & 0 & 0 & \cdots \\ \cdots & 0 & 0 & \cos(l+1)\phi & -\sin(l+1)\phi & 0 & \cdots \\ \cdots & 0 & 0 & \sin(l+1)\phi & \cos(l+1)\phi & 0 & \cdots \\ \cdots & 0 & 0 & 0 & 0 & \cos(l+2)\phi & \cdots \\ & \vdots & \vdots & \vdots & \vdots & \vdots & \end{bmatrix}$$

where l is an integer, which gives us (from Eq. (6.12)).

$$\begin{aligned} s_k(\theta+\phi) &= \cos l\phi s_k(\theta) - \sin l\phi s_{k+1}(\theta); \\ s_{k+1}(\theta+\phi) &= \sin l\phi s_k(\theta) + \cos l\phi s_{k+1}(\theta). \end{aligned}$$

It is easy to show that these equations are satisfied by

$$s_k(\theta) = \cos l\theta, \quad s_{k+1}(\theta) = sinl\theta.$$

Invariants can be found by considering the coefficients of $v_1(r, \theta)$ and $v_2(r, \theta)$ with $l = 1$:

$$d_1 \cos(\theta + \phi) + d_2 \sin(\theta + \phi) = d_1' \cos\theta + d_2' \sin\theta.$$

Expanding $\cos(\theta + \phi)$ and $\sin(\theta + \phi)$ gives

$$d_1[\cos\phi \cos\theta - \sin\phi \sin\theta] + d_2[\sin\phi \cos\theta + \cos\phi \sin\theta] = d_1' cos\theta + d_2' \sin\theta.$$

Comparing coefficents of $\cos\theta$ and $\sin\theta$ gives

$$\begin{aligned} d_1' &= d_1 \cos\phi + d_2 \sin\phi; \\ d_2' &= -d_1 \sin\phi + d_2 \cos\phi; \end{aligned}$$

from which we see that $d_1'^2 + d_2'^2 = d_1^2 + d_2^2$. This extends to the cases where $l \neq 1$, so $d_k^2 + d_{k+1}^2$, k even, is invariant to image rotations.

Both special cases correspond to modeling the image with an angular Fourier series; in the second case we do not have the burden of using complex functions.

In conclusion we have seen in this section that the continuous image must be exactly representable by N^2 known basis functions to obtain invariants of the sampled version. We have also seen two equivalent cases where the invariants are easy to compute.

6.1.4 *Rotation Invariants in Pattern Recognition*

Invariants may be used as descriptors to characterize patterns, independently of the coordinate system, and independently of transformations of the pattern. The significance of invariants is the important geometric relationship. Most often, invariance is considered with respect to sets of transformations that are groups of transformations under the composition operation.

The idea of invariance arises from our own ability to recognize objects irrespective of such movement — if we look at a book from a number of different orientations, we have no difficulty in recognizing it as a book each time: we can say that a book has properties which are invariant to its size, position and orientation. Finding mathematical functions of a pattern that are invariant to the above transformations would thus provide us with a technique for recognizing objects using computers, as well as providing us with a possible model for part of human vision. If we wish to recognize an object using a computer, and we assume the computer has stored the models (or example views) of the objects to be recognized in its memory, and the object to be recognized corresponds to one of these models, the straightforward approach searches sequentially through the computer's memory, trying out each model and seeing whether it can be positioned in such a way as to produce a pattern that matches that of the object to be recognized, until a good match is found. Clearly, this is computationally intensive; ideally, we would like to be able to extract the correct model directly from the information contained in the pattern — this is precisely what the so-called pattern invariants allow us to do.

A number of different methods to extract the rotation invariant feature were investigated and as the examples, some of them are briefly described below. Besides being rotation invariant, some of the feature extraction methods presented are also translation or scale invariant.

6.1.4.1 *Boundary Curvature*

The outer boundary of two-dimensional pattern contains almost all of the shape information present in the original pattern [Otterloo, 1991]. It is therefore convenient to use the outer boundaries of character shapes as the basis for the extraction of shape features. An advantage of using the boundary rather than the complete pattern, is that the boundary can be represented in a very compact form (often as a one-dimensional sequence of numbers) and this allows the feature extraction process to be relatively fast. A problem with this approach is that it means that the extracted features are mirror invariant i.e. the feature values produced by shape which are reflected copies of one another will be identical. These multi-contour patterns will therefore not allow the classification stage of the OCR process to distinguish between these patterns.

6.1.4.2 *Fourier Descriptors*

The Fourier descriptor method, as introduced by Granlund [Granlund, 1972] and Arbter [arbter et al., 1990], also makes use of the outer boundary of the pattern as the basis for the features which it extracts. The boundary, in this case, is treated as a contour in the complex plane: each boundary pixel is represented by a complex number whose real part is derived from the pixel's x co-ordinate and whose imaginary part is derived from the pixel's y co-ordinate. By tracking along the character boundary and recording these complex numbers, the algorithm builds up a one-dimensional sequence of complex numbers which specify the pattern boundary. Though the Fourier descriptors can be easily constructed, they are also mirror invariant, similar to the above method

6.1.4.3 *Zernik Moments*

Zernike moments are the mapping of a pattern onto a set of complex basis functions which have two distinctive properties, orthogonality and rotation invariance of their magnitude [Abu-Mostafa and Psaltis, 1984b; Khotanzad and Hong, 1990]. The first property ensures that the contribution of each moment to the pattern is unique and independent. The second property allows the magnitude of Zernike moments extracted from a pattern to be the same at any orientation. Its advantage is that the reconstruction of the pattern from its Zernike moments is easily performed. This makes it possible to evaluate the importance of a particular moment, or set of moments, by reconstructing the pattern without these moments and comparing the resulting pattern to the original. By doing this it is possible to choose the most significant moments and use only those moments as features. This has the desirable effect of reducing the dimensionality of the feature vector.

However, one really needs to perform an experimental evaluation of a few of the most promising methods to decide which feature extraction method is the best in practice for each application. The evaluation should be performed on large data sets that are representative for the particular application. The number of maximum order of the Zernike moments has been yielded with the recognition rate, as well as with the computational complexity.

6.1.4.4 *Neural Networks*

Fukumi et al. [Fukumi et al., 1992] have proposed the following rotation-invariant first order network for rotation invariance in two-dimensional patterns. Each input pattern is represented by a discrete polar-coordinate grid. Invariance is achieved under a finite, though arbitrarily large, rotation group. This works by constructing a first hidden layer of nodes, each of which is connected to the input with weights that are shared among pixels with the same radius from the pattern center. The network structure from the first layer upwards has no weight sharing, because the output of the first hidden layer is an invariant feature vector. Fukumi et al. used their network in a simple experiment on coin recognition, in which it performed very well. A disadvantage of this network is that it discriminates imperfectly. This is because the invariant features calculated by the first hidden layer are function of the total activation in each radial circle of pixels. An arbitrary rearrangement of the pixel intensities within any given circle would have no effect on the network output.

A new method of feature extraction with rotation invariant property has been developed by our research team, which is presented in this chapter. One of the main contributions of this study is that a rotation invariant signature of two-dimensional contours is selected based on the fractal theory is proposed. The rotation invariant signature is a measure of the fractal dimensions, which is rotation invariant based on a series of central projection transform (CPT) groups. As the CPT is applied to a two-dimensional object, a unique contour is obtained. In the unfolding processing, this contour is further spread into a central projection unfolded curve, which can be viewed as a periodic function due to the different orientations of the pattern. We consider the unfolded curves to be non-empty and bounded sets in $\mathbb{R}^n$, and the box computing dimension is applied to them. It has been proved that the box computing dimensions of the central projection unfolded curves of a rotated pattern are equal to each other, i.e. the central projection unfolded curves with respect to the box computing dimension are rotation invariant. Some experiments with positive results are conducted. This approach is applicable to a wide range of areas such as image analysis, pattern recognition (including document image recognition), etc.

6.2 Preprocessing and Central Projection Transform (CPT)

6.2.1 *Preprocessing*

First, we convert a scanned image into digitized (binary) image, i.e. an image with pixels 0 (white) and 1 (black). After converting the image, we remove unnecessary pixels (0) from the original image. For this purpose, we use the following preprocessing.

- *Find the center of gravity and translate to the center of the image plan.*
  Removing the deformation of the position of the input pattern: When the pattern is input, the system will first find its center of gravity, then translate the input pattern to the center $(0, 0)$ of the pattern plan.
  For a binary image of the pattern in the two-dimensional Cartesian system:

$$f(x,y) = \begin{cases} 1 & \text{if } (x,y) \in \mathcal{D}, \\ 0 & \text{otherwise}, \end{cases} \tag{6.14}$$

$$(1 \leq x, y \leq M).$$

  The center of gravity $(\overline{X}, \overline{Y})$ for the pattern is given by

$$\begin{cases} \overline{X} = \frac{m_{10}}{m_{00}} \\ \overline{Y} = \frac{m_{01}}{m_{00}} \end{cases} \tag{6.15}$$

  where

$$m_{pq} = \sum_{x=1}^{M} \sum_{y=1}^{M} x^p y^q f(x,y)$$

  denotes geometrical moments of the pattern.
- *Scale the input pattern by* $\frac{D}{d}$
  If the present standard distance is D, then the input pattern is first scaled by $\frac{D}{d}$,

$$\begin{bmatrix} X \\ Y \end{bmatrix} = \begin{bmatrix} \frac{D}{d_j} & 0 \\ 0 & \frac{D}{d_j} \end{bmatrix} \begin{bmatrix} x_j \\ y_j \end{bmatrix}. \tag{6.16}$$

  where (X, Y) is the new coordinate of a point for a pattern sample with standard size, (x_j, y_j) is the coordinate of a point for a pattern sample with the jth kind of size, D is the standard size, d_j stands for the jth kind of size.

6.2.2 *Central Projection Transform (CPT)*

After preprocessing, the system will convert the Cartesian coordinate system into the polar coordinate system, then transform the pattern from the image domain into the area domain using central projection.

Central projection transform (CPT) is an improvement of the regional projection contour transform (RPCT) [Tang et al., 1993a]. The fundamental difference between the RPCT and CPT is the projection base. For the former, the projection bases of the RPCT are the horizontal, vertical and diagonal lines, while for the latter, the projection base is a point. The basic principle of the CPT is that all pixels of a pattern are projected onto the center of the image. With this transformation plus the contour chain operation, an image becomes a unique contour. Figures 6.2(b-d) describe these operations graphically. Two very significant properties can be achieved, namely,

- Contour unitization, and
- Shape invariance with respect to rotation.

Before describing these important characteristics, the basic definitions of CPT are described below, followed by a parallel algorithm.

6.2.2.1 *Basic Definitions of CPT*

The basic definitions of the central projection transform (CPT) will discussed in the subsection:

1. Central Projection

Definition 6.1 Let $f(x, y)$ be a pattern, and R the area of the pattern. Assume that $f(x, y) = 0$ lies outside the pattern. $\delta[...]$ denotes a delta function, which is zero everywhere except where its argument is zero, and whose integral from -∞ to ∞ equals one. If the projection angle from the x-axis is θ, $t = xsin\theta - ycos\theta$ gives the Euclidean distance of a line from the origin. The projection can be defined as follows [Pavlidis, 1982]:

$$p(\theta, t) = \sum_R f(x, y)\delta[xsin\theta - ycos\theta - t]. \tag{6.17}$$

Definition 6.2 Suppose the entire area R has been broken into several smaller regions R_i, $i = 1, 2, ..., k$, the projection can be divided into several

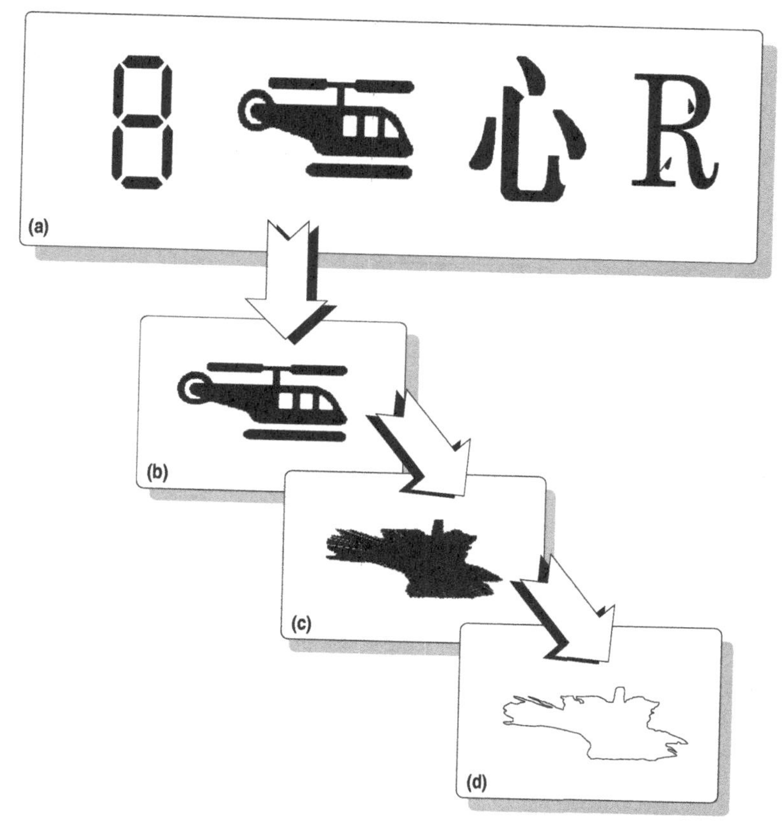

Fig. 6.2 A multi-contour pattern "helicopter" is transformed into a unique-contour pattern.

sub-projections which will take place in these regions such that

$$\begin{aligned} p(\theta, t) &= \sum_{i=1}^{k} p_i(\theta_i, t_i) \\ &= \sum_{i=1}^{k} \sum_{R_i} f_i(x, y) \delta_i [x sin\theta_i - y cos\theta_i - t_i]. \end{aligned} \tag{6.18}$$

Each sub-projection $p_i(\theta_i, t_i)$ is called a *regional projection.*

Central projection refers to mapping a multi-contour pattern into a unique contour pattern and producing a unique contour whose values are the sums of the values of the image points along the different radial directions. The formal definition of the central projection is presented below:

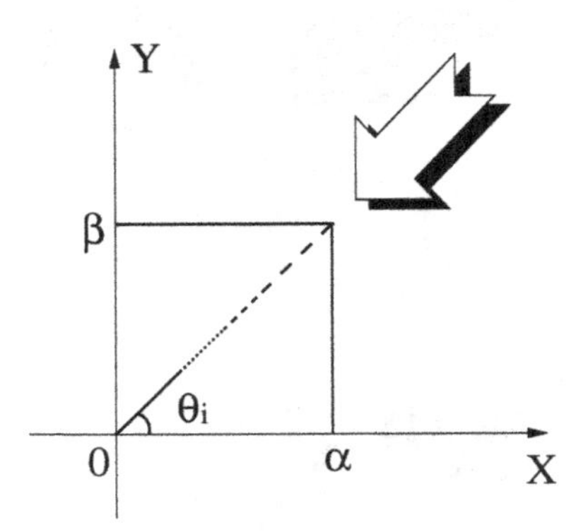

Fig. 6.3 Central projection.

Definition 6.3 Let $f(x, y)$ be a pattern, and θ_i be the angle between the projection direction and the x-axis as shown in Figure 6.3. $y = xtg\theta_i$ or $x = y\frac{1}{tg\theta_i}$ gives the relationship between coordinates x and y. The central projection with angle θ_i can be defined by

$$CP_x(\theta_i, x) = \sqrt{1 + tg^2\theta_i} \sum_{x=0}^{\alpha} f(x, xtg\theta_i), \tag{6.19}$$

or

$$CP_y(\theta_i, y) = \sqrt{1 + (\frac{1}{tg\theta_i})^2} \sum_{y=0}^{\beta} f(y\frac{1}{tg\theta_i}, y), \tag{6.20}$$

where, $CP_x(\theta_i, x)$ is called *x-based central projection*, while $CP_y(\theta_i, y)$ is referred to as *y-based central projection.*

2. Regional Central Projection (RCP)

The definition of regional central projection (RCP) is presented below:

Definition 6.4 Let the area of a pattern be divided into 8 sub-regions, R_i, $i = 1, 2, ..., 8$ as shown in Figure 6.4. The eight sub-regions are symbolized by A, B, C, D, E, F, G, and H such that

$$\begin{aligned} CP_R &= \bigcup_{i=1}^{8} P_R^i \\ &= A_R \cup B_R \cup C_R \cup D_R \cup \\ &\quad E_R \cup F_R \cup G_R \cup H_R. \end{aligned} \tag{6.21}$$

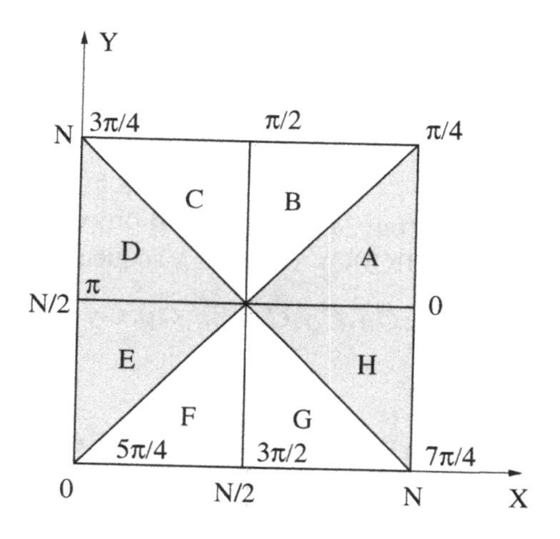

Fig. 6.4 Regional central projection (RCP).

Note that the symbol $\bigcup_{i=1}^{8}$ has been used instead of $\sum_{i=1}^{8}$ in the above definition, because overlaps may occur in it.

The projection of the whole pattern, P_R, is called *regional central projection (RCP)*, and the sub-projection P_R^i is called *sub-RCP 'i'*. In the concrete, A_R - H_R denote sub-RCPs 'A' - 'H' in sub-regions A - H respectively. Since sub-RCPs 'A', 'D', 'E' and 'H' can be established by x-based central projection, according to Eq. 6.19 we can obtain

$$A_R = CP_x(\theta_i, x) = \sqrt{1 + tg^2\theta_i} \sum_{x=0}^{\alpha} f(x, xtg\theta_i), \quad \theta_i \in (0, \frac{\pi}{4}) \tag{6.22}$$

Similarly, for D_R, E_R and H_R we have $\theta_i \in (\frac{\pi}{4}, \frac{\pi}{2})$, $\theta_i \in (\frac{3\pi}{4}, \pi)$ and $\theta_i \in$

$(\pi, \frac{5\pi}{4})$ respectively. Combining them together produces the following:

$$\left.\begin{array}{l} A_R \\ D_R \\ E_R \\ H_R \end{array}\right\} = \sqrt{1 + tg^2\theta_i} \sum_{x=0}^{\alpha} f(x, xtg\theta_i) \left\{\begin{array}{l} \theta_i \in (0, \frac{\pi}{4}) \\ \theta_i \in (\frac{3\pi}{4}, \pi) \\ \theta_i \in (\pi, \frac{5\pi}{4}) \\ \theta_i \in (\frac{7\pi}{4}, 2\pi) \end{array}\right. \tag{6.23}$$

Since sub-RCPs 'B', 'C', 'F' and 'G' can be established by the y-based central projection described by Eq. 6.20, the following description can be achieved:

$$\left.\begin{array}{l} B_R \\ C_R \\ F_R \\ G_R \end{array}\right\} = \sqrt{1 + (\frac{1}{tg\theta_i})^2} \sum_{y=0}^{\beta} f(y\frac{1}{tg\theta_i}, y) \left\{\begin{array}{l} \theta_i \in (\frac{\pi}{4}, \frac{\pi}{2}) \\ \theta_i \in (\frac{\pi}{2}, \frac{3\pi}{4}) \\ \theta_i \in (\frac{5\pi}{4}, \frac{3\pi}{2}) \\ \theta_i \in (\frac{3\pi}{2}, \frac{7\pi}{4}) \end{array}\right. \tag{6.24}$$

Definition 6.5 Let C_A - C_H be sub-contours for sub-RCP 'A' - 'H' respectively. The entire contour of the projected object is called a regional central projection contour (RCPC), which is a sequence of C_A - C_H, i.e.

$$\{C_R\} = C_A, C_B, C_C, C_D, C_E, C_F, C_G, C_H.$$

6.2.2.2 *Properties of CPT*

This subsection presents the important properties of the central projection transform (CPT).

1. Contour Unitization

Contour analysis has found wide applications in pattern recognition, computer vision, image processing and other areas [Ayache and Faugeras, 1986; Persoon and Fu, 1977; Dudani et al., 1972; Turney et al., 1985; Zahn and Roskies, 1972]. Many parallel algorithms have been developed for such an operation [Cass, 1988; Dinstein et al., 1991; Li et al., 1989; Tucker et al., 1988; Wu et al., 1989]. However multi-contour patterns are often met, such as

- compound patterns containing unconnected patterns and patterns with isolated noises,
- patterns comprising several holes in them.

Several examples are illustrated in Figure 6.2. For these patterns, contour tracing may encounter great difficulty due to

- more contours to deal with,
- the relations among the isolated parts have to be carefully established,
- more complicated feature extraction,
- more time required for processing, etc.

CPT can concentrate the compound pattern into an integral object with only one outer contour and no internal contours. An example can be found in Figure 6.2(d). It can be employed to recognize multi-contour patterns.

2. Shape Invariance

Another property of the CPT is shape invariance with respect to rotation. In practice, the object to be recognized may be rotated. Once an object is rotated in different angles, the different oriented images are produced. For instance, an image of the helicopter produces several oriented images due to the different fly angles. When the RPCT [Tang et al., 1993a] is applied to these images, various shapes are obtained as shown in Figure 6.5. It means that the RPCT is a rotation sensitive transformation.

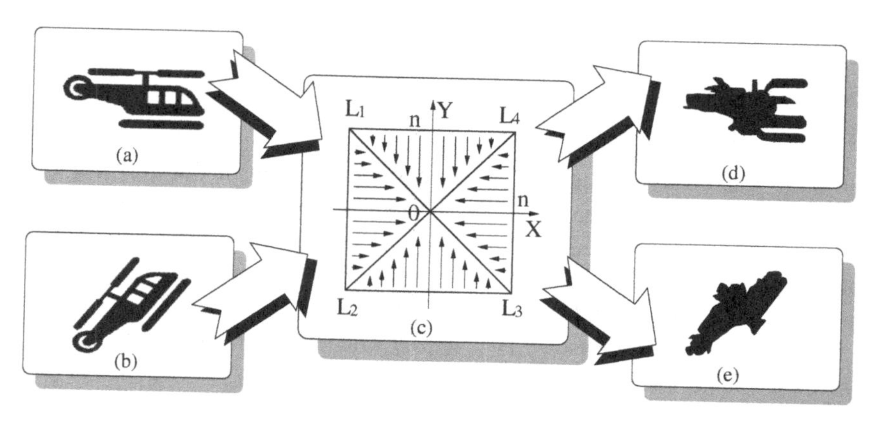

Fig. 6.5 Different oriented images of an object produce various shapes due to RPCT.

The main contribution of the CPT is an improvement over the RPCT. The proposed CPT can overcome the weakness of the rotation sensibility created by the RPCT. An example of CPT applying to the different oriented images produced from the same image, helicopter, can be found in

Figure 6.6. It is clear that the shapes are the same except the orientations. This is the property of *shape invariance* with respect to the rotation. We will use this property to achieve the rotation invariance features.

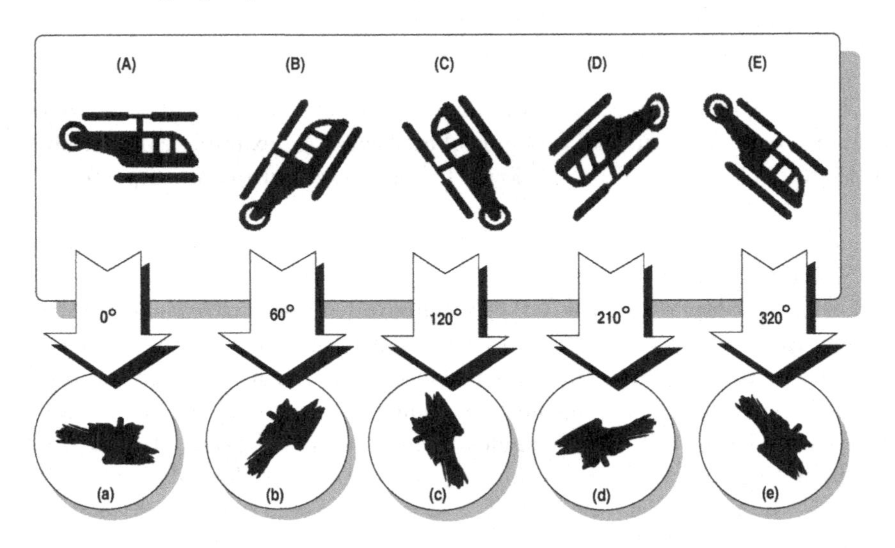

Fig. 6.6 Shape invariance with respect to the rotation by CPT.

6.2.2.3 *Parallel Algorithm for CPT*

The CPT has clear parallelism, which occurs not only in different regions but also in different operations. In the algorithm, all sub-regions can be processed at the same time. Meanwhile two major operations, the projections and contour tracing can also be executed concurrently. In this section, a parallel algorithm for CPT is developed.

It is evident from Eqs. (6.19), (6.20) and (6.21) that the projection base used in the CPT is the central point of the image, and all pixels in a pattern are to be concentrated onto it along different projecting directions. The whole pattern has been divided into 8 sub-areas and its projection can be expressed by 8 sub-projections. On examining Eq. (6.21), one can find that the computations of these 8 sub-projections are independent of each other, and each computation only needs the values $f(x, xtg\theta)$ or $f(\frac{y}{tg\theta}, y)$ of pixels (x, y) within their own sub-areas. Therefore, these computations can be done simultaneously. Moreover, because the computation of each sub-projection is the summation of values $f(x, xtg\theta)$ or $f(\frac{y}{tg\theta}, y)$ of pixels (x, y) in the different angles θ, the summations within the same sub-area can also be done in a parallel fashion. These parallelisms can be described in the following algorithm in which the endpoint for each projection direction is a point of the outer contour for the projected object. Therefore, its coordinates can be output as the coordinates of the contour.

Algorithm 6.1

parfor all pixels in the area $N \times N$ do

parbegin

parfor $x = 0$ to $\frac{N}{2}$ do

$D_R = \sqrt{1 + tg^2\theta_i}\sum_x f(x, xtg\theta_i), \quad \theta_i \in (\frac{3\pi}{4}, \pi)$

$E_R = \sqrt{1 + tg^2\theta_i}\sum_x f(x, xtg\theta_i), \quad \theta_i \in (\pi, \frac{5\pi}{4})$

and output the coordinates of their endpoints;

parfor $x = \frac{N}{2}$ to N do

$A_R = \sqrt{1 + tg^2\theta_i}\sum_x f(x, xtg\theta_i), \quad \theta_i \in (0, \frac{\pi}{4})$

$H_R = \sqrt{1 + tg^2\theta_i}\sum_x f(x, xtg\theta_i), \quad \theta_i \in (\frac{7\pi}{4}, 2\pi)$

and output the coordinates of their endpoints;

parfor $y = 0$ to $\frac{N}{2}$ do

$F_R = \sqrt{1 + (\frac{1}{tg\theta_i})^2}\sum_y f(y\frac{1}{tg\theta_i}, y), \quad \theta_i \in (\frac{5\pi}{4}, \frac{3\pi}{2})$

$G_R = \sqrt{1 + (\frac{1}{tg\theta_i})^2}\sum_y f(y\frac{1}{tg\theta_i}, y), \quad \theta_i \in (\frac{3\pi}{2}, \frac{7\pi}{4})$

and output the coordinates of their endpoints;

parfor $y = \frac{N}{2}$ to N do

$B_R = \sqrt{1 + (\frac{1}{tg\theta_i})^2}\sum_y f(y\frac{1}{tg\theta_i}, y), \quad \theta_i \in (\frac{1\pi}{4}, \frac{\pi}{2})$

$C_R = \sqrt{1 + (\frac{1}{tg\theta_i})^2}\sum_y f(y\frac{1}{tg\theta_i}, y), \quad \theta_i \in (\frac{\pi}{2}, \frac{3\pi}{4})$

and output the coordinates of their endpoints;

parend

6.2.2.4 *Contour Unfolding*

According to the above analysis (refers to Figure 6.2), as the central projection transform (CPT) is operated to a two-dimensional object, the unique contour can be produced. In order to facilitate further processing and achieving the rotation invariance property by the fractal dimension, the contour (closed curve) will be unfolded into a one-dimensional opened curve. Look at an example in Figure 6.7, where the original pattern is a two-

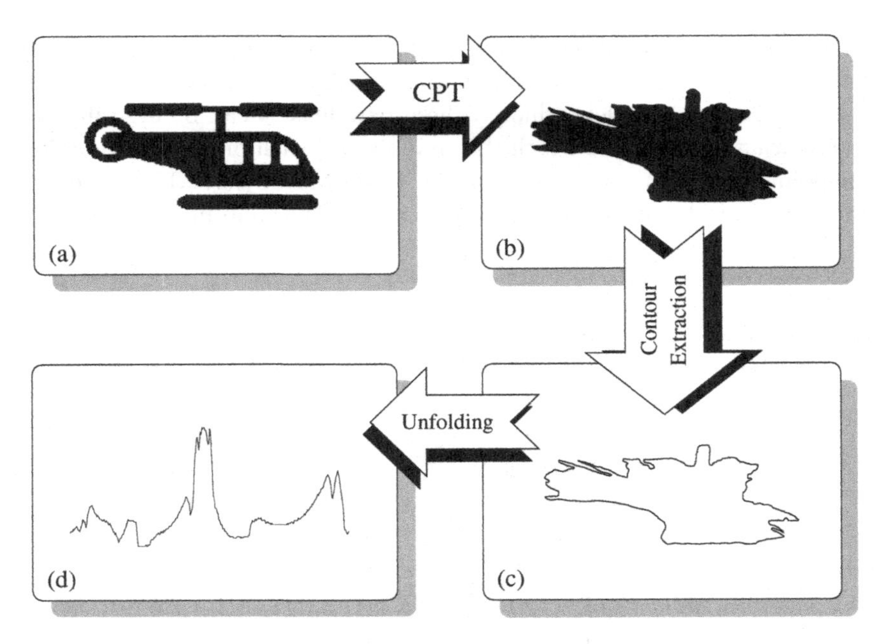

Fig. 6.7 A two-dimensional object shown in (a) is transformed into a unique contour pattern illustrated in (b) by the CPT. From this pattern, a unique contour is extracted (c), Thereafter, the unfolding operation is performed, and an one-dimensional opened curve is obtained in (d).

dimensional object, helicopter, shown in Figure 6.7(a). In terms of the CPT, it is transformed into a unique contour pattern presented in Figure 6.7(b), and thereafter, a unique contour can be found in Figure 6.7(c). Thus, this closed curve is unfolded, and an one-dimensional opened curve can be obtained, which is displayed in Figure 6.7(d).

We will apply a mathematical tool called *box computing dimension (BCD)* to such a periodic function to achieve the property of rotation invariance.

The CPT has an very significant property, shape invariance. Once an object is rotated in different angles, the different oriented images are produced. When the CPT is applied to these images, the shapes of the resulted images are same expect the orientations. Again, we use the sample "helicopter" as an example shown in Figure 6.14, where a helicopter is rotated in 0°, 60° and 320° respectively. We use the CPT to these rotated objects, the unique-contour images are obtained and displayed in the circles of bottom-left in the figure. It is clear that these images are shape-invariant. Next, we extract the contours and unfold them, so that three curves are produced and presented in the right sides of Figure 6.14(a), (b) and (c) respectively. We can consider these three curves to be a cyclic function $f(\theta)$ according to the different starting point from which we unfold a contour. The above analysis can be summarized by Eq. (6.25). From this equation

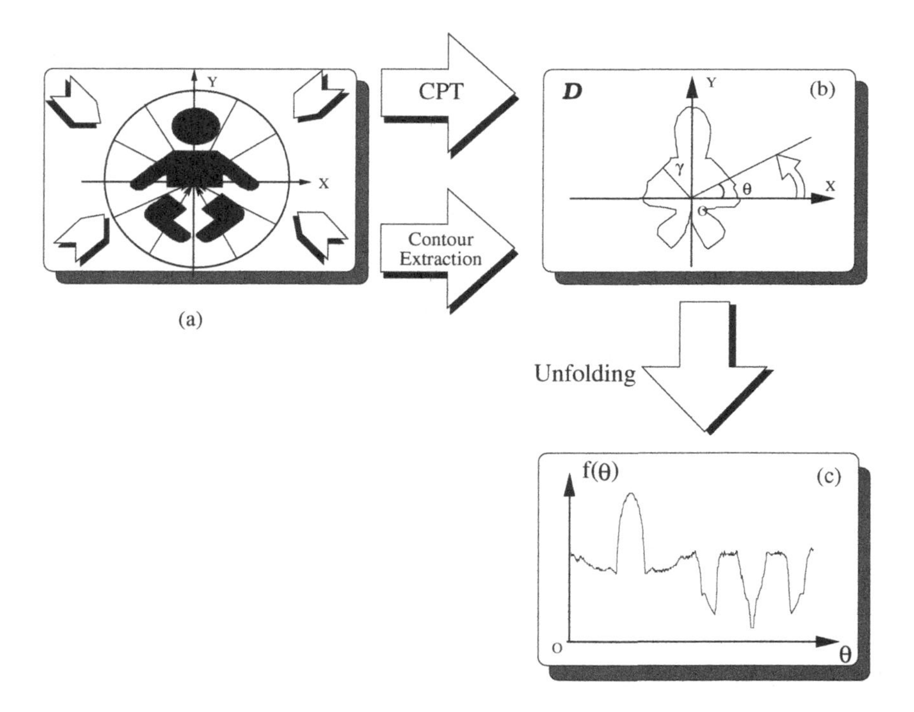

Fig. 6.8 An illustration of the central projection for the 2-D pattern.

and Figure 6.8, we can know that a rotated object can be represented by some rotated unique-contours in terms of the CPT. Each contour is spread by the unfolding professing, and a curve is obtained, which is called *central projection unfolded curve* of the pattern. All the central projection unfolded curves can be viewed as a periodic function.

$f(x,y)$	$\Longrightarrow f(x,y)*\alpha$	$\Longrightarrow CPT(f(x,y)*\alpha)$	$\Longrightarrow f(\theta)$
$\downarrow$	$\downarrow$	$\downarrow$	$\downarrow$
Original image	Rotated image	Central projection transformed image: it is a unique-contour pattern with the property of shape-invariance	Unfolded image: the rotated contour becomes a periodic function

(6.25)

6.3 Rotation Invariance Based on Box Computing Dimension

One of the basic characteristics of a fractal is its dimension. The basic concepts of fractal dimension is presented in previous chapter of this book. Estimates of the dimension tend to be very inaccurate as data size is reduced. There is a measurement of the complexity of a geometric object, which we will commonly call the "fractal dimension", although there are several varying concepts and terms, the big brother of which is the Hausdorff-Besicovitch dimension. The fractal dimension is always a nonnegative real number. There is a more familiar notion of dimension called the topological dimension, which is always a nonnegative integer. Continuous non-self-intersecting curves always have topological dimension *1*, and smooth surfaces always have dimension *2*, for instance. The fractal dimension is always greater than or equal to the topological dimension, and Mandelbrot defines a fractal set as one in which the fractal dimension is strictly greater than the topological dimension.

6.3.1 *Estimation of the 1-D Fractal Dimension*

There are a wide variety of computer algorithms for estimating the fractal dimension of a structure, such as the box computing algorithm, the pixel dilation method, the radial power spectrum method, and others. The box method [Barton and Larsen, 1985] consists of measuring the length of a curve of either points or lines density over a given area using boxes of various sizes. The idea is to count the number of boxes of size n necessary to cover the points or a curve in a certain area.

Here, we will demonstrate the *box computing algorithm*, because of its simplicity. This algorithm estimates how much of the available space is taken on with the fractal structure. First, an arbitrary grid is placed over the structure to be measured. Then, one counts how many boxes in the grid are filled by the fractal structure. For the grid size in the figure, the structure fills *6* boxes in Figure 6.9.

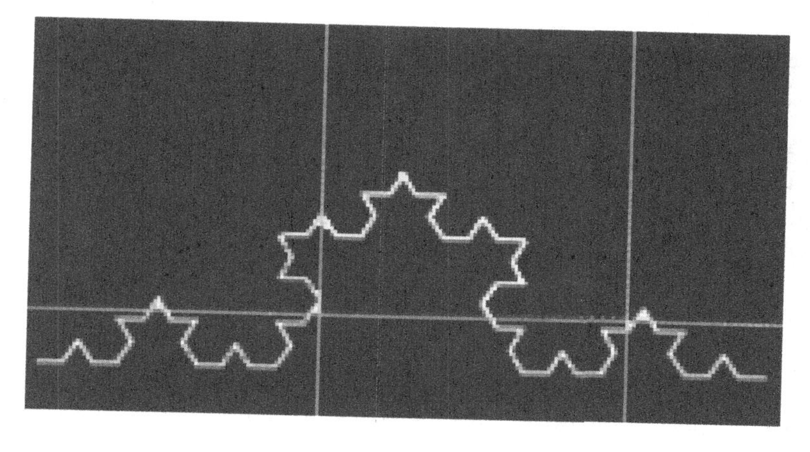

Fig. 6.9 The structure fiils 6 boxes.

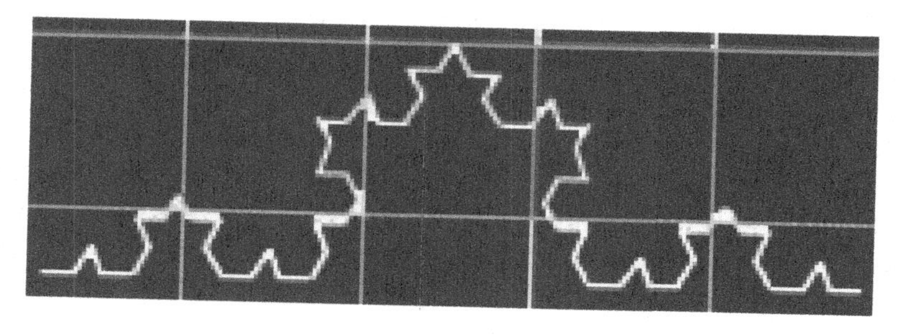

Fig. 6.10 The structure fiils 9 boxes.

The process is then repeated with a grid half the size of the previous one. With the grid below, the structure fills *9* boxes in Figure 6.10. The

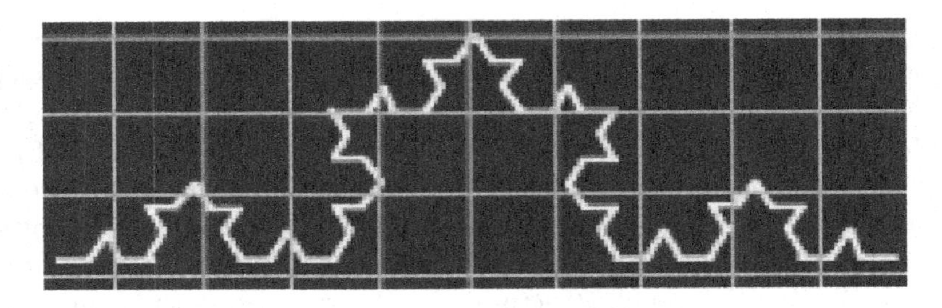

Fig. 6.11 The structure fiils 18 boxes.

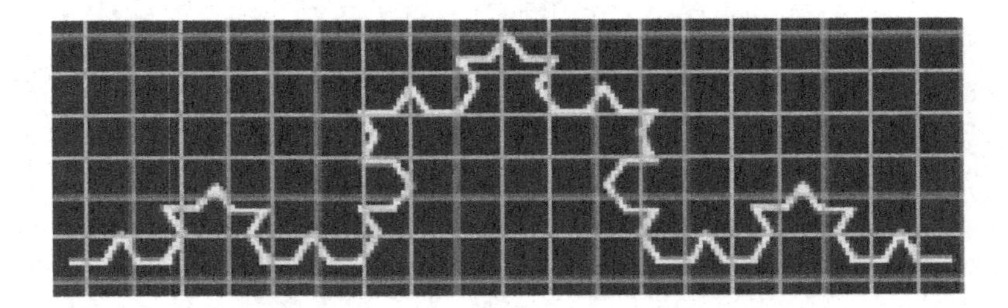

Fig. 6.12 The structure fiils 59 boxes.

process is again repeated with a grid half the size of the previous one. With the grid, the structure fills *18* boxes in Figure 6.11. This process can be carried on indefinitely, using smaller and smaller grids. For this demonstration, we will only count the boxes one more time. With the final grid below, the structure fills *59* boxes in Figure 6.12. So, our box counting data is summarized in Table 6.1. The data from these four countings are tabulated and plotted on a log-log graph plot as shown Figure 6.13. A linear regression is done to find the best fit. The slope of this line is used to calculate the fractal dimension. The best fit equation is $y = -0.8X + 7.8$. This fractal dimension is equal to *1* minus the slope of this line. Thus, for this fractal object, the fractal dimension $D = 1 - (-0.84) = 1.84$. As can be seen from these definitions of the previous chapter, we can compute the fractal dimension of the pattern as the rotation invariant signature (RIS).

6.3.2 *Rotation Invariant Signature (RIS)*

As the central projection transform (CPT) is applied to a two-dimensional object, a unique contour is obtained. In the unfolding processing, this con-

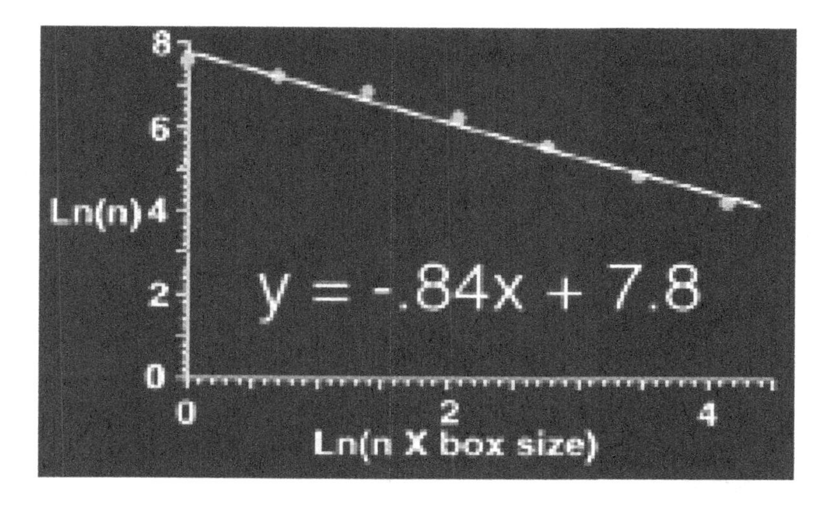

Fig. 6.13 The slope of this line is used to calculate the fractal dimension.

Table 6.1 Summary of box computing data.

Box Size	Numbers of Box Computing
1	6
1/4	9
1/16	18
1/256	59

tour is further spread into a central projection unfolded curve. In practice, the pattern can be arbitrarily rotated, and the angle of the rotation can be changed from 0 to 2π. Therefore, the unfolding processing can also be arbitrarily started at any point on the contour, and many functions are produced, say $f(t)$, $f(t+t_1)$, $f(t+t_2)$, ..., $f(t+t_n)$, which can be viewed as a periodic function of period 2π at $[0, 2\pi]$ with different initial phase t_i based on different starting unfolding pont. This can be found in Figure 6.14, where a cyclic function $f(\theta)$ is produced by the central projection transform (CPT) and unfolding for the helicopter image with different orientations.

Now, we analyze the central projection unfolded curve, which is represented by a periodic function, by the box computing dimension (BCD):

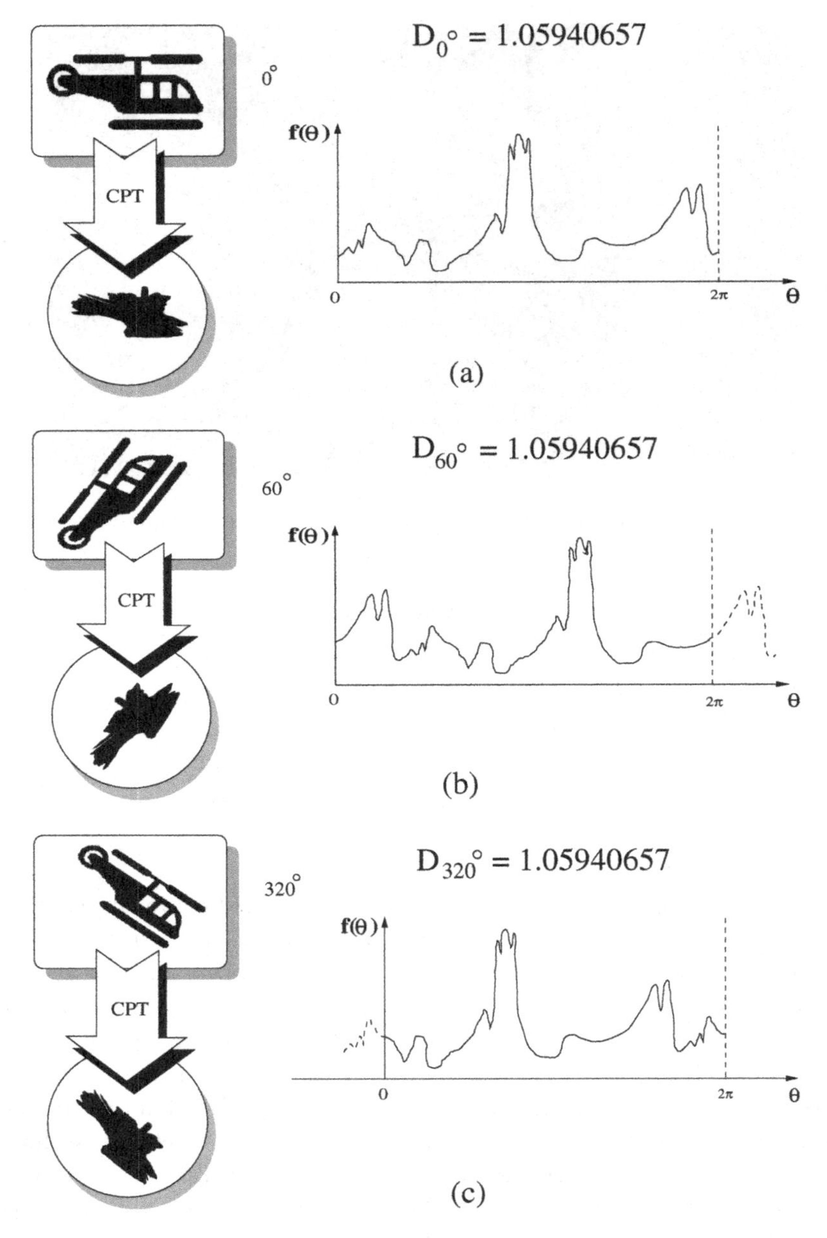

Fig. 6.14 A cyclic function $f(\theta)$ is produced by the central projection transform (CPT) and unfolding for the helicopter image with different orientations.

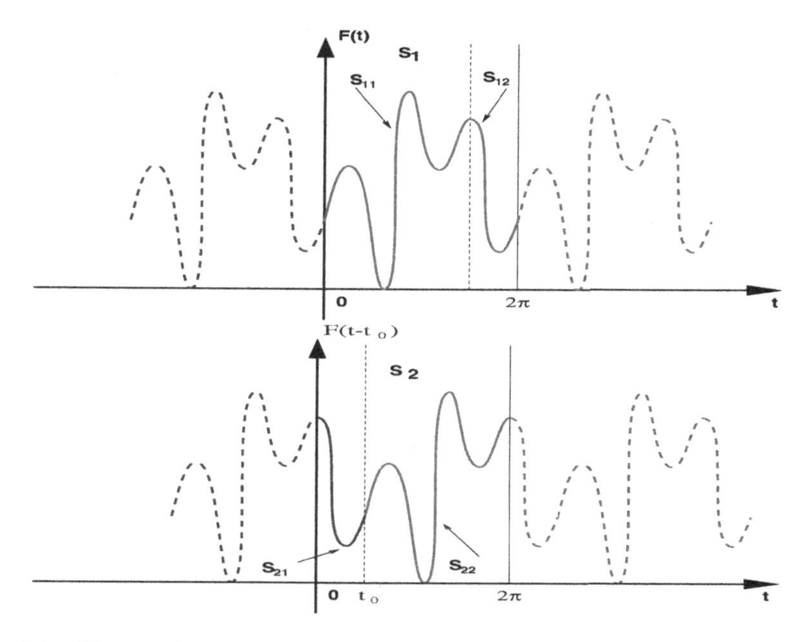

Fig. 6.15 Upper: The central projection unfolded curve of the original pattern that can be viewed as a periodic function $f(t)$ of period 2π. Bottom: The $f(t)$ is translated into the function $f(t+t_0)$ due to the rotation.

Suppose $f(t)$ represents the central projection unfolded curve for the original pattern (i.e. the rotation angle is zero) that can be referred to the upper of Figure 6.15. Let $f(t+t_0)$ be the central projection unfolded curve for the rotated object with a certain angle. It is illustrated in the bottom of Figure 6.15.

We have the following important theorem:

Theorem 6.1 *The box computing dimensions of the central projection unfolded curves of a rotated pattern are equal to each other, i.e. the central projection unfolded curves with respect to the box computing dimension are rotation invariant.*

The proof of Theorem 6.1 is very complicated. To do so, we have the following lemmas:

Lemma 6.1 *Let E and F be non-empty and bounded sets in $\mathbb{R}^n$. If $E \subset F$ then $dim_B F \geq dim_B E$.*

Proof We suppose $\delta < 1$, and let $N_\delta(E)$ and $N_\delta(F)$ denote the smallest number of sets of diameter at most δ, which can cover E and F, respectively.

Since $E \subset F$, we have $N_\delta(E) \leq N_\delta(F)$, thus,

$$\frac{log_2 V_\delta(E)}{-log_2\delta} \leq \frac{log_2 N_\delta(F)}{-log_2\delta}.$$

Taking limits as $\delta \to 0$ obtains

$$\lim_{\delta\to 0} \frac{log_2 N_\delta(E)}{-log_2\delta} \leq \lim_{\delta\to 0} \frac{log_2 N_\delta(F)}{-log_2\delta}.$$

According to Eqs. (2.15), we have

$$dim_B E = \lim_{\delta\to 0} \frac{log_2 N_\delta(E)}{-log_2\delta},$$

$$dim_B F = \lim_{\delta\to 0} \frac{log_2 N_\delta(F)}{-log_2\delta}.$$

Consequently, we have

$$\begin{aligned} dim_B E &= \lim_{\delta\to 0} \frac{log_2 N_\delta(E)}{-log_2\delta} \\ &\leq \lim_{\delta\to 0} \frac{log_2 N_\delta(F)}{-log_2\delta}. \end{aligned}$$

We simplify the equations by writing

$$dim_B F = \lim_{\delta\to 0} \frac{log_2 N_\delta(F)}{-log_2\delta},$$

such that

$$dim_B E \leq dim_B F. \tag{6.26}$$

The proof of the lemma is complete. ■

Lemma 6.2 *For E and F, which are the non-empty and bounded sets in $\mathbb{R}^n$, we have*

$$\lim_{\delta\to 0} \frac{\log_2 \max(N_\delta(E), N_\delta(F))}{log_2\delta} = \max(dim_B E, dim_B F)$$

Proof This proof consists of two steps:

(1) If $dim_B(E) \neq dim_B F$,
first we suppose

$$dim_B E \geq dim_B F,$$

and take

$$M = \frac{dim_B E - dim_B F}{2}.$$

There exists $\delta_0 > 0$, such that the following inequalities hold when $0 < \delta < \delta_0$

$$\left|\frac{\log_2 N_\delta(E)}{-\log_2 \delta} - dim_B E\right| < M,$$

$$\left|\frac{\log_2 N_\delta(F)}{-\log_2 \delta} - dim_B F\right| < M.$$

Therefore, we have

$$\frac{\log_2 N_\delta(E)}{-\log_2 \delta} \geq \frac{\log_2 N_\delta(F)}{-\log_2 \delta},$$

which indicates that we always have, when $0 < \delta < \delta_0$,

$$N_\delta(E) \geq N_\delta(F),$$

and

$$\max(N_\delta(E), N_\delta(F)) = N_\delta(E).$$

Consequently, we obtain

$$\begin{aligned} \lim_{\delta \to 0} \frac{\log_2 \max(N_\delta(E), N_\delta(F))}{-\log_2 \delta} &= \lim_{\delta \to 0} \frac{\log_2 N_\delta(E)}{-\log_2 \delta} \\ &= \max(dim_B(E), dim_B(F)) \end{aligned} \tag{6.27}$$

(2) If $dim_B E = dim_B F$
For $\forall \frac{\varepsilon}{2} > 0$, there exists $\delta_0 > 0$, when $\delta < \delta_0$, we have

$$\begin{aligned} &\left| \frac{\log_2 N_\delta(E)}{-\log_2 \delta} - dim_B E \right| < \frac{\varepsilon}{2}, \\ &\left| \frac{\log_2 N_\delta(F)}{-\log_2 \delta} - dim_B F \right| < \frac{\varepsilon}{2}, \end{aligned} \tag{6.28}$$

such that

$$\mid \frac{\log_2 N_\delta(E) - \log_2 N_\delta(F)}{-\log_2 \delta} \mid < \varepsilon.$$

That means

$$\mid \frac{\log_2 \frac{N_\delta(E)}{N_\delta(F)}}{-\log_2 \delta} \mid < \varepsilon. \tag{6.29}$$

Similarly, we can obtain

$$\mid \frac{\log_2 \frac{N_\delta(F)}{N_\delta(E)}}{-\log_2 \delta} \mid < \varepsilon. \tag{6.30}$$

If $0 < \delta < \delta_0$, the following holds

$$\begin{aligned} & \mid \frac{\log_2 \max(N_\delta(E), N_\delta(F))}{-\log_2 \delta} - dim_B E \mid \\ = & \mid \frac{\log_2 \frac{\max(N_\delta(E), N_\delta(F))}{N_\delta(E)} \times N_\delta(E)}{-\log_2 \delta} - dim_B F \mid \\ \leq & \mid \frac{\log_2(\frac{\max(N_\delta(E), N_\delta(F))}{N_\delta(E)})}{-\log_2 \delta} \mid + \mid \frac{\log_2 N_\delta(E)}{-\log_2 \delta} - dim_B F \mid \\ = & \Psi(1) + \Psi(2), \end{aligned}$$

where

$$\Psi(1) = \mid \frac{\log_2(\frac{\max(N_\delta(E), N_\delta(F))}{N_\delta(E)})}{-\log_2 \delta} \mid,$$

$$\Psi(2) = \mid \frac{\log_2 N_\delta(E)}{-\log_2 \delta} - dim_B F \mid.$$

Please note that, if $\max(N_\delta(E), N_\delta(F)) = N_\delta(E)$, we know

$$\Psi(1) = \mid \frac{log_2(\frac{N_\delta(E)}{N_\delta(E)})}{-\log_2 \delta} \mid = 0.$$

If $\max(N_\delta(E), N_\delta(F)) = N_\delta(F)$, according to Eq. (6.30), we have

$$\Psi(1) = \mid \frac{\log_2(\frac{N_\delta(F)}{N_\delta(E)})}{-\log_2 \delta} \mid < \varepsilon.$$

We have, from Eq. (6.28), that

$$\Psi(2) \leq \frac{\varepsilon}{2}.$$

Consequently, if $0 < \delta < \delta_0$, we have

$$\left| \frac{\log_2 \max(N_\delta(E), N_\delta(F))}{-\log_2 \delta} - dim_B E \right| \leq \frac{3}{2}\varepsilon, \tag{6.31}$$

which implies

$$\begin{aligned} \lim_{\delta \to 0} \frac{\log_2 \max(N_\delta(E), N_\delta(F))}{-\log_2 \delta} &= dim_B E \\ &= \max(dim_B E, dim_B F) \end{aligned} \tag{6.32}$$

Synthesizing 1 and 2 above, the proof is complete. ■

Lemma 6.3 *Let E and F be non-empty and bounded set in $\mathbb{R}^n$, we have*

$$dim_B(E \cup F) = \max(dim_B(E), dim_B(F)). \tag{6.33}$$

Proof To show

$$dim_B(E \cup F) = \max(dim_B(E), dim_B(F)),$$

we have to prove two cases, namely,

$$dim_B(E \cup F) \geq \max(dim_B(E), dim_B(F)). \tag{6.34}$$

and

$$dim_B(E \cup F) \leq \max(dim_B(E), dim_B(F)). \tag{6.35}$$

(1) According to Lemma 6.1, we can mathematically infer that

$$dim_B(E \cup F) \geq \max(dim_B E, dim_B F)$$

(2) It is easy to know that

$$N_\delta(E \cup F) \leq 2\max(N_\delta(E), N_\delta(F)).$$

From Eq. (2.15), we can easily find

$$dim_B(E \cup F) = \lim_{\delta \to 0} \frac{\log_2 N_\delta(E \cup F)}{-\log_2 \delta},$$

and we already verified, from Lemma 6.2, that

$$\lim_{\delta \to 0} \frac{log_2 \max(N_\delta(E), N_\delta(F))}{-log_2 \delta} = \max(dim_B E, dim_B F),$$

which implies that

$$\begin{aligned} dim_B(E \cup F) &= \lim_{\delta \to 0} \frac{\log_2 N_\delta(E \cup F)}{-\log_2 \delta} \\ &\leq \lim_{\delta \to 0} \frac{\log_2 2 \max(N_\delta(E), N_\delta(F))}{-\log_2 \delta} \\ &= \lim_{\delta \to 0} \frac{\log_2 \max(N_\delta(E), N_\delta(F))}{-\log_2 \delta} \\ &= \max(dim_B E, dim_B F). \end{aligned}$$

Therefore,

$$dim_B(E \cup F) \leq \max(dim_B E, dim_B F).$$

From the results in the above two cases, we can conclude that

$$dim_B(E \cup F) = \max(dim_B E, dim_B F).$$

The proof of the lemma is complete. ■

Based on the result from Lemma 6.3, we can now prove our objective theorem.

Proof of Theorem 6.1 The image of function $f(t)$ and $f(t = t_0)$ are respectively denoted as S_1 and S_2, where $t \in [0, 2\pi]$. It is easy to know, from Figure 6.15, that $S_1 = S_{11} \cup S_{12}$, $S_2 = S_{21} \cup S_{22}$. We can regard S_1 and S_2 as non-empty and bounded sets in $\mathbb{R}^n$. The box computing dimensions of S_1 and S_2 are $dim_B S_1$ and $dim_B S_2$ respectively.

According to Lemma 6.3, i.e. Eq. 6.33, we have

$$dim_B S_1 = max(dim_B S_{11}, dim_B S_{12}),$$

$$dim_B S_2 = max(dim_B S_{21}, dim_B S_{22}),$$

where

$$dim_B S_{11} = dim_B S_{22},$$

$$dim_B S_{12} = dim_B S_{21}.$$

Therefore, $dim_B S_1 = dim_B S_2$. ■

Thus the images S_1 and S_2 have the same box computing dimension. The box computing dimension is called the *rotation invariant signature* (RIS).

6.4 Experiments

In ths research, the rotation invariant signature (RIS) algorithm has been employed in our experiments. Therefore, the RIS algorithm will be presented firstly.

6.4.1 *Rotation Invariant Signature (RIS) Algorithm*

The novel RIS algorithm consists of four main processes:

- (1) preprocessing,
- (2) the central projection transform (CPT),
- (3) estimating the box computing dimension, and
- (4) extracting the feature with rotation invariant property.

The details of first two processes, i.e. the preprocessing and central projection transform (CPT) were already presented in previous subsections in this chapter. Here, we discuss the last two processes, namely, the estimation of the box computing dimension, and extraction of the feature with rotation invariant property.

6.4.1.1 *Estimation of the BCD*

The estimated BCD can be used to measure the fractal dimension of the non-self-intersecting curve. The procedure is analogous to moving a set of dividers of fixed length δ along this curve. Formally, the method finds the 'δ-cover' of the object, i.e. the number of pixels of length δ (or circles

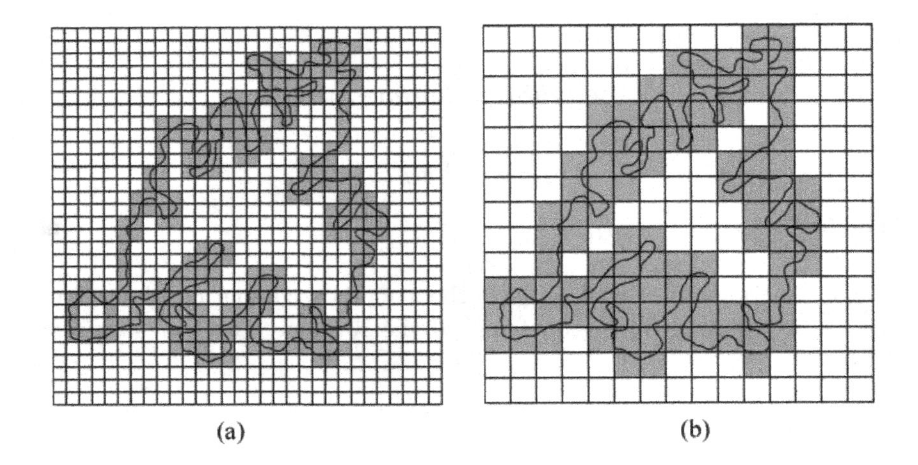

Fig. 6.16 Two box 'lengths' are respectively shown in (a) and (b). Boxes including the image are shaded.

of radius δ) required to cover the object. A more practical alternative is to superimpose a regular grid of pixels of length δ on the object and count the number of 'occupied' pixels on the curve, such as, Figure 6.16. Suppose that C is a non-self-intersecting curve, and $\delta > 0$. Let $M_\delta(C)$ be the maximum number of ordered sequence of points on curve C. Since the fractal dimensions D of non-self-intersecting curves are asymptotic values, we can derive their approximations based on the following expression:

$$D = \frac{log_{M_\delta(C)}}{-log_\delta},$$

when δ is set to be small enough.

In Figure 6.14, for each of the three non-self-intersecting curves, they have the same values of the fractal dimension in 360 subdivisible angles:

$$D_{0^o} = D_{60^0} = D_{320^0} = 1.05940657.$$

Therefore, the fractal dimensions of the non-self-intersecting curves are also called the rotation invariant signature (RIS) of themselves.

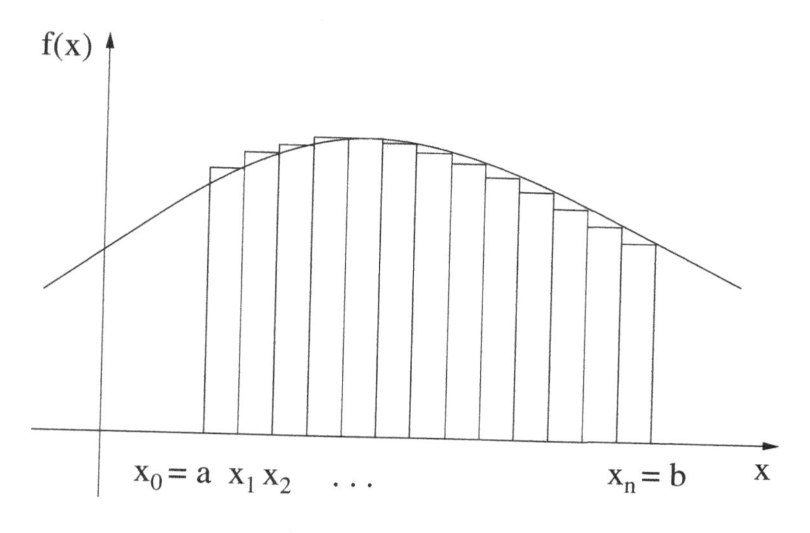

Fig. 6.17 Interpolating an integral as the sum of the areas of small rectangles.

6.4.1.2 *Extraction of Feature with Rotation Invariant Property*

Rotation invariant signature (RIS) is implied by the gird cell resolution of the computer model at a certain scale of the subdivisible angle. Therefore, the estimation of the RIS requires multiple resampling operations. If an alternative measure of subinterval density is required, the contour data may be examined. So that the vector of rotation invariant feature may be extracted. Since this scale of the subdivisible angle is arbitrarily defined by the preset value according to the resolution of the computer model, it may not be always appropriate for feature extraction. We allow the scale of the subdivisible angle used for sampling source data to vary according to the different subdivisible angles. Using an exact interpolator one can gather all available source data values. Thus, one would expect the fractal dimension of the non-self-intersecting curve to assort as the density of contours relative to the resolution of subdivisible angle increases. It can be done more efficiently using the interpolation method of composite trapezoidal rule [Rogers and Adams, 1990; Tiller, 1987] for estimating the set of rotation invariant features.

Interpolation curves are commonly used to restructure the shape of the curves. Using the composite trapezoidal interpolator, we can approximate the non-self-intersecting curve with a constant value in each subinterval and add the areas of the resulting rectangles, see in Figure 6.17. The smaller the subdivisions for the interval from a to b, the better the interpolation (up to a point). The method is called composite trapezoidal rule and an integral of $f(x)$ can be numerically approximated with the following summation

$$\int_a^b f(x)dx \approx \frac{h}{2}[f(a) + f(b) + 2\sum_{i=1}^{n-1} f(a+ih)]$$

where the interval from a to b is divided into n equal-width intervals:

$$\begin{aligned} h &= \frac{b-a}{n}; \\ x_i &= a + ih, \qquad i = 0, 1, 2, ...n. \end{aligned}$$

Actually, if the intervals get too small, the values of successive rectangular areas can get lost in the roundoff error. We repeat this procedure at each calculated interpolation until the difference between successive approximations is "small enough". Therefore, different interpolated curves have different fractal dimension according to the subdivisible angle respectively.

Rotation invariant signature (RIS) is defined as $D_{k,\Phi}$, which corresponds to its fractal dimension in the preset value of subdivisible angle. The vector of rotation invariant feature can be expressed by

$$(D_{k,360}, D_{k,720}, D_{k,1080}, \ldots, D_{k,\Phi})^T$$

where $k = 1, 2, 3, ..., 360, \quad \Phi = 360, 720, 1080, ..., 360k.$

In fact, the relationship between feature extraction and interpolation is not an arbitrary one. Both involve the characterization of as much useful source material as possible, from which a target is derived. The essential difference between interpolation and feature extraction is that the former involves extracting a target that is the same quantity as the source, whereas the latter involve a further transformation (for example, from elevation to morphometrics feature type). For the purposes of the RIS quality validation, a fractal curve is created using the interpolator of composite trapezoidal rule. For example, when the non-self-intersecting curve in Figure 6.8(c), is interpolated according to different subdivisible angle, the fractal and its contour representation are shown in Figure 6.18.

6.4.2 *Experimental Procedure and Results*

We outline the use of rotation invariant signatures (RIS) for representing the rotated pattern. It is composed of six major steps as follows:

Step 1 find the center of gravity and translate it to the center of the original pattern plane;
Step 2 correspond to the scale normalization of the pattern;
Step 3 the patterns are transformed by the central projection operation;
Step 4 extract the contour based on the CPT;
Step 5 obtain the curve to be a cyclic function according to the different starting points from which we unfold the contour;
Step 6 construct the vectors of the rotation invariant feature under the different subdivisible angles.

An example is illustrated in Figure 6.19. In this example, the Chinese character "Bei", in English means "north" is employed undergoing the CPT, unfolding and the measurement of the box computing dimension in 360 subdivisible angle. In Figure 6.19 these numbers of the RIS as follows:

$$\begin{aligned} D_{100,360} &= 1.03027896239, \\ D_{200,360} &= 1.03027896228, \\ D_{300,360} &= 1.03027896233. \end{aligned}$$

They indicate the values of the RIS for the Chinese character "Bei" with different rotation angles. It is obvious that the first ten digits of these three numbers are exactly same, the difference among these box computing dimensions is the last two digits. In particular, we only consider the first nine digits of the values of the rotation invariant signature (RIS) for extracting the feature vector with rotation invariant property.

The following example illustrates how to extract the feature with rotation invariant property for some two-dimensional objects. Figure 6.20(a) shows the input patterns with different sizes and orientations. Figure 6.20(b) shows the results obtained by size normalization. These patterns are transformed into the unique contour patterns by the CPT, Figure 6.20(c) extracts these closed contours respectively. Then the contours are unfolded into the non-self-intersecting curves. A cyclic function according to different

Table 6.2 Illustration of rotation invariant feature of the 2-D pattern.

k	Φ	**RIS** $(D_{k,\Phi})$	k	Φ	**RIS** $(D_{k,\Phi})$
1	360	1.05113971	21	7560	1.00251847
2	720	1.03007761	22	7920	1.00202960
3	1080	1.02247633	23	8280	1.00192026
4	1440	1.01918559	24	8460	1.00205697
5	1800	1.01606252	25	9000	1.00181250
6	2160	1.01359925	26	9360	1.00163637
7	2520	1.01138589	27	9720	1.00149610
8	2880	1.00949258	28	10080	1.00154478
9	3240	1.00900717	29	10440	1.00125622
10	3600	1.00717672	30	10800	1.00127388
11	3960	1.00600445	31	11160	1.00121481
12	4320	1.00547555	32	11520	1.00119125
13	4680	1.00483939	33	11880	1.00103719
14	5054	1.00451093	34	12240	1.00109535
15	5400	1.00404064	35	12600	1.00114393
16	5760	1.00364419	36	12960	1.00088956
17	6120	1.00328285	37	13320	1.00089665
18	6480	1.00303103	38	13680	1.00085428
19	6840	1.00279551	39	14040	1.00082888
20	7200	1.00237774	40	14400	1.00079550

rotations in the central projection space is shown in Figure 6.20(d). Moreover, the values of the RIS for extracting the feature vector with rotation invariant property are also shown in Table 6.2.

The proposed approach has been implemented in C++ and Matlab on a Sun workstation. A computer can be used to produce the perfect data sets, which are free from quantization noise and contains no ambiguous patterns. These sets can show that rotation invariant signature retain sufficient information for classification in the absence of noise, and these results can be used to assess the performance of this approach for invariant pattern recognition.

It remains now to demonstrate that the rotation invariant signature retains enough information to be usable for invariant pattern recognition. In order to demonstrate the feasibility of such a rotation invariant signature classifier a small simulation was performed. In the experiment, we have tested our new approach for three kinds of pattern in Figure 6.21.

Table 6.3 Classification rates for different patterns with three times interpolation using the RIS approach.

Input Pattern	# of Pattern	Classification Rate %
English Letter	10	80
Chinese Character	10	100
Logo	10	100

Table 6.4 Classification rates for 600 printed Chinese characters with different times of interpolation using the RIS approach.

Times of Interpolation	Classification Rate (%)
3	96.23
5	97.4
7	98.87

Three training sets were created containing four examples of each pattern, each differently rotated within an 128×128 input image. Three perfect test sets were created by computing rotated versions of the training data so that there was no quantization error. We make rotated patterns of $30^0, 60^0, \ldots$ and 330^0 for all of the test patterns. The final stage of the rotation invariant signature classifier was trained on this data, and the progress on the test sets monitored during training. The final classifier used the weighted Euclidean distance (WED), which is used to distinguish the different feature vectors. With this WED classifier, the smallest distance was considered the match. The results were shown in Table 6.3, which perform interpolation three times.

The simulation result indicates that the classification rate of Chinese character and Logo patterns are up to 100%. However, the classification rate of English letters is 80% on the test data, since the characters "b" and "d" were identical under reflection in the font used.

Moreover, we performed the classifier on the 600 printed Chinese character with the on-screen version of Black font using interpolation with different times for obtaining their feature vectors. We can see that the classification rate increases as the times of interpolation increases. The classification rates are shown in Table 6.4. The rotation invariant signatures are sufficient to characterize contours for many two-dimensional pattern recognition tasks. Results are presented demonstrating the utility of this technique for invariant pattern recognition.

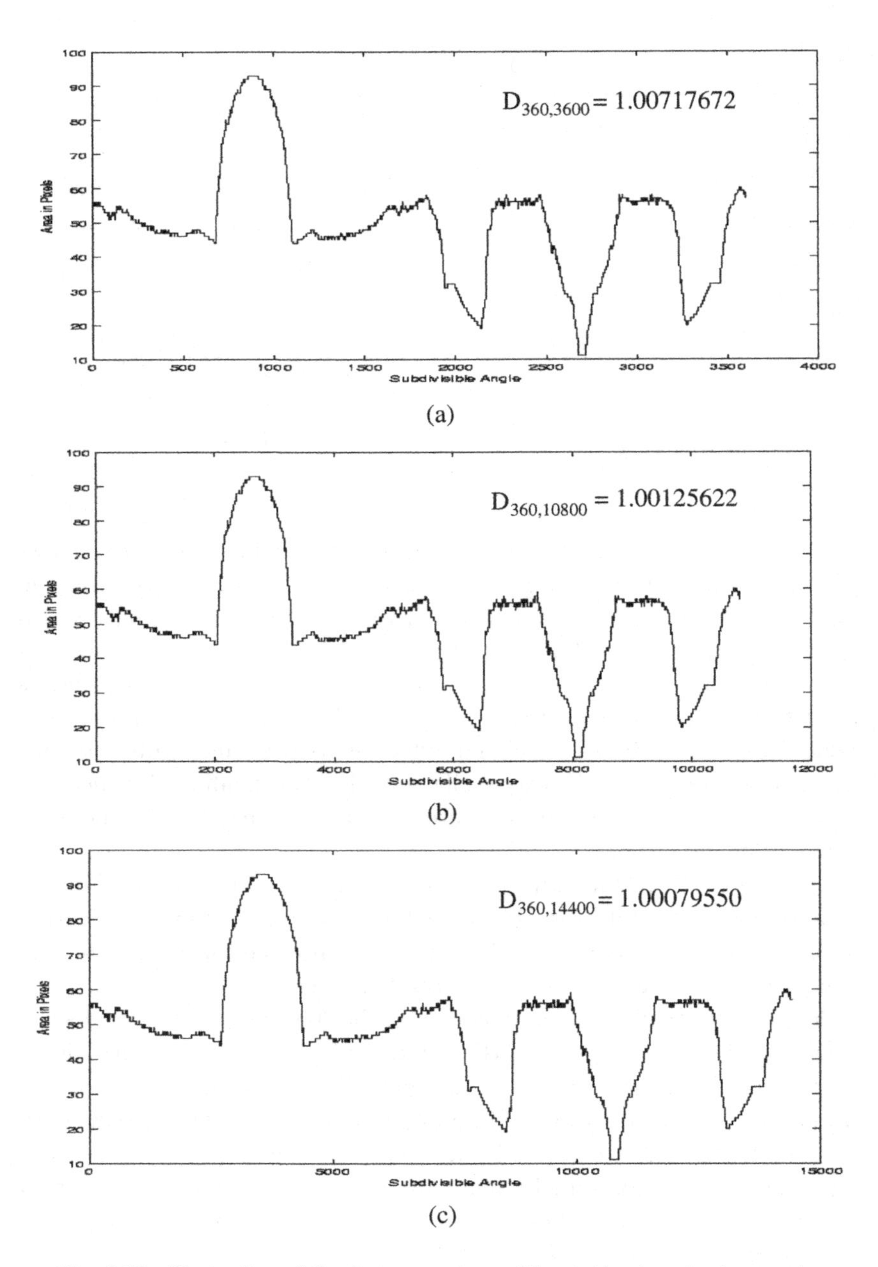

Fig. 6.18 Illustration of the feature vectors with rotation invariant property.

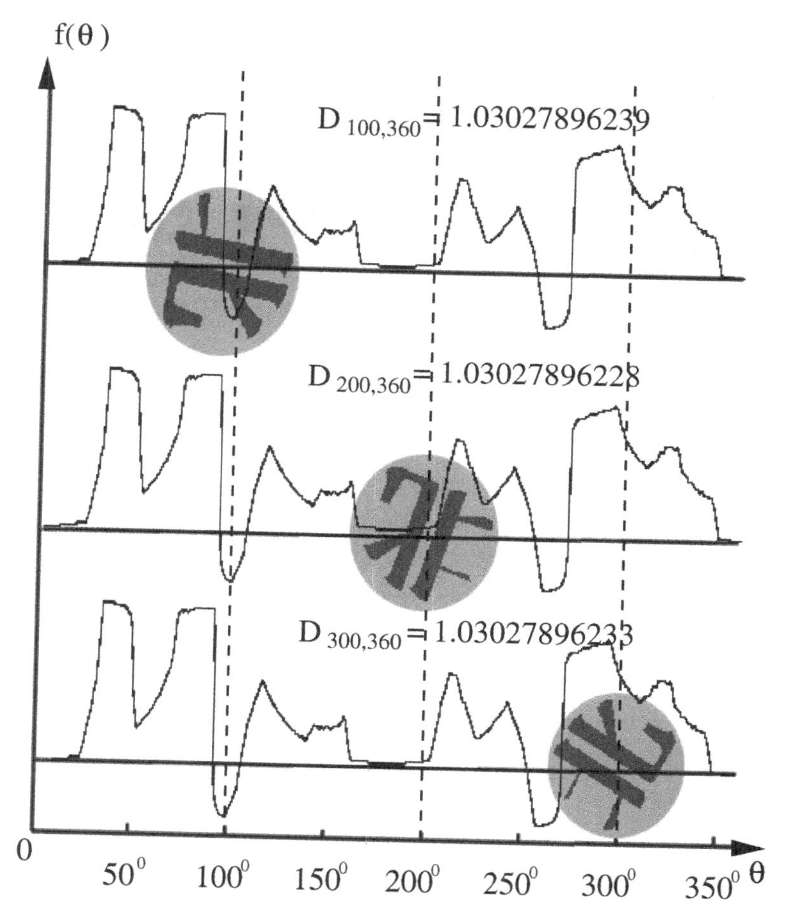

Fig. 6.19 Extracting rotation invariant feature for Chinese character.

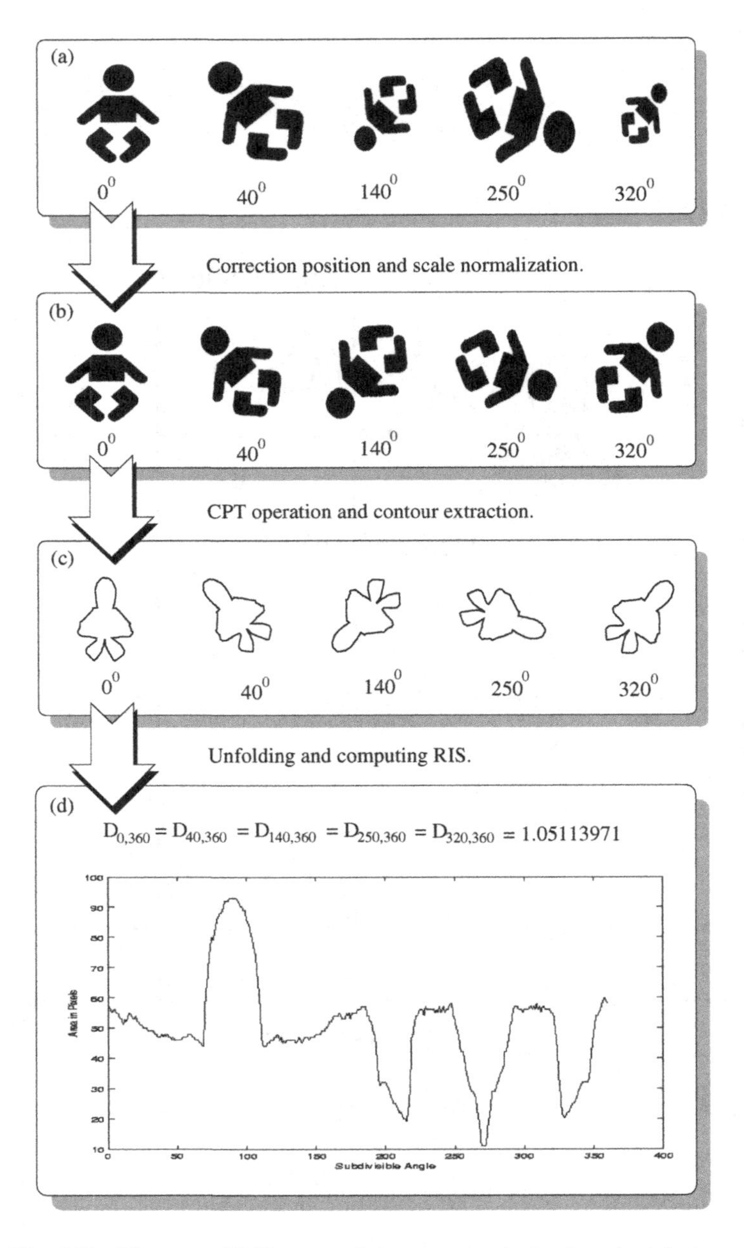

Fig. 6.20 Diagram of RIS approach to extract rotation invariant feature.

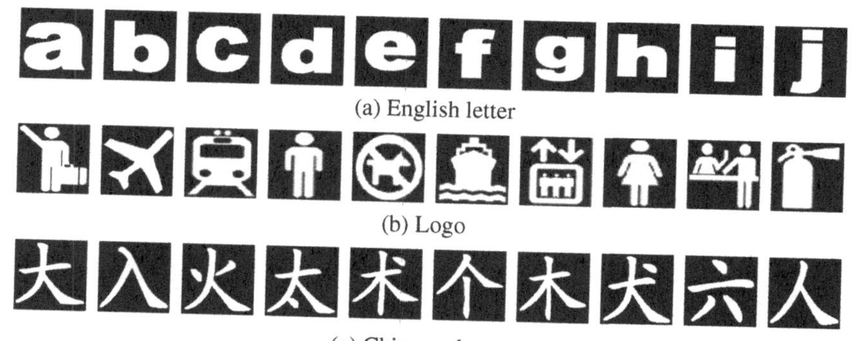

(a) English letter

(b) Logo

(c) Chinese character

Fig. 6.21 Some samples of pattern used for experiments.

Chapter 7

Wavelet-Based and Fractal-Based Methods for Script Identification

Documents printed in different languages need to be identified and processed in an international environment. This chapter proposes two methods to identify different scripts, namely:

- Wavelet-based method — script identification based on the features in the *frequency domain*, such as the energy features of the wavelet sub-images;
- Fractal-based method — script identification based on the features in the *time domain*, for example, the modified fractal signature (MFS).

(1) In wavelet-based approach, a document image is decomposed into several sub-images in the frequency domain by wavelet transform, and thereafter, two feature vectors are extracted from these sub-images, namely,

- WED - wavelet energy distribution, and
- WEDP - wavelet energy distribution proportion.

(2) In fractal-based method, a document image is divided into several non-overlapping sub-images in the time domain, each of which is then mapped onto a gray-level function. The δ parallel body technique, thereafter, is applied to compute the fractal signature.

Document images used in this chapter are in six languages, i.e., Chinese, English, Japanese, Korean, Russian, and Devanagari (a language used in Indian) with various fonts.

7.1 Introduction

In an international environment, increasing number of documents printed in various languages need to be identified and processed. In fact, different languages appear in books, newspapers, form-documents, letters, and many other documents. Especially, one document is often printed in two or even more languages. For instance, both English and Chinese are prevalent in publications in Hong Kong, Singapore, Taiwan, etc. An example of such a bilingual English-Chinese document can be found in Figure 7.1. Moreover,

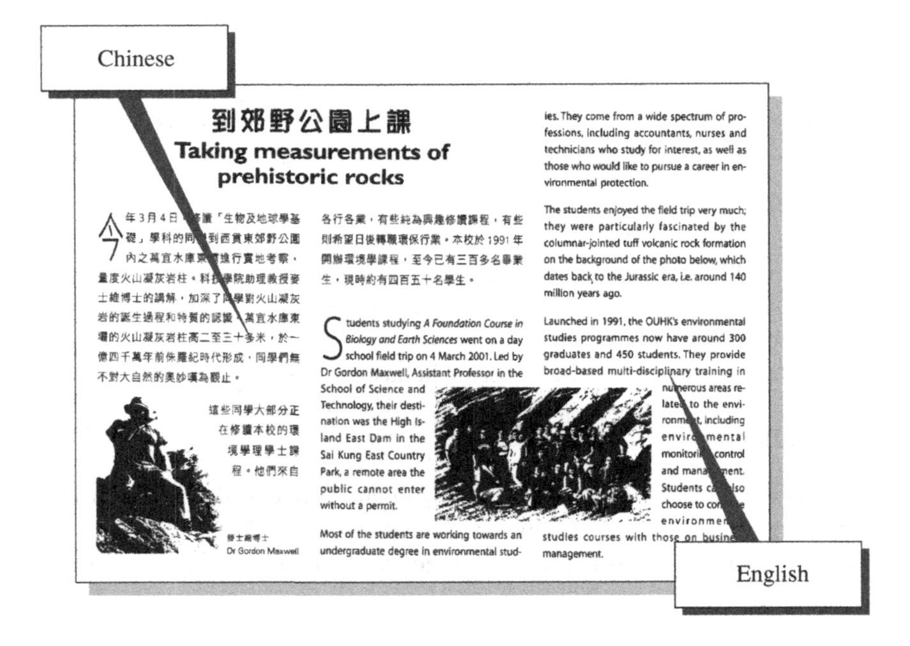

到郊野公園上課

Taking measurements of prehistoric rocks

今年3月4日修讀「生物及地球學基礎」學科的同學到西貢東郊野公園內之萬宜水庫東壩進行實地考察，量度火山凝灰岩柱。科技學院助理教授麥士維博士的講解，加深了同學對火山凝灰岩的誕生過程和特質的認識。萬宜水庫東壩的火山凝灰岩柱高二至三十多米，於一億四千萬年前侏羅紀時代形成，同學們無不對大自然的奧妙嘆為觀止。

這些同學大部分正在修讀本校的環境學理學士課程。他們來自各行各業，有些純為興趣修讀課程，有些則希望日後轉職環保行業。本校於1991年開辦環境學課程，至今已有三百多名畢業生，現時約有四百五十名學生。

麥士維博士
Dr Gordon Maxwell

Students studying *A Foundation Course in Biology and Earth Sciences* went on a day school field trip on 4 March 2001. Led by Dr Gordon Maxwell, Assistant Professor in the School of Science and Technology, their destination was the High Island East Dam in the Sai Kung East Country Park, a remote area the public cannot enter without a permit.

Most of the students are working towards an undergraduate degree in environmental studies. They come from a wide spectrum of professions, including accountants, nurses and technicians who study for interest, as well as those who would like to pursue a career in environmental protection.

The students enjoyed the field trip very much; they were particularly fascinated by the columnar-jointed tuff volcanic rock formation on the background of the photo below, which dates back to the Jurassic era, i.e. around 140 million years ago.

Launched in 1991, the OUHK's environmental studies programmes now have around 300 graduates and 450 students. They provide broad-based multi-disciplinary training in numerous areas related to the environment, including environmental monitoring, control and management. Students can also choose to combine environmental studies courses with those on business management.

Fig. 7.1 An example of the bilingual English-Chinese document.

a large quantity of postal matter has to be processed in an international post center. The receiver's addresses of these mails are written in various languages. Sorting machines with optical character recognition (OCR) techniques are employed in reading and processing the addresses of these mails. Although many OCR techniques have been developed, there is an important assumption in the most of existing OCR systems, that is, the language in the document to be processed must be known so that a proper

OCR algorithm can be chosen. Indeed, it has implied that the manual labour is inevitable in the identifications. When a large amount of multiple languages appear in the information system, the manual labour becomes infeasible, and it is necessary to identify languages automatically. For instance, 1.28 billion items of mails were processed in Hong Kong Airmail Center in 1997, and delivered to all over the world. In order to minimize the manual labour, an automatic machine that can identify the types of the language-images for input documents has to be developed.

In this chapter, we use term, "language-image" or "language" or "script" without any difference, because most of the existing methods including the proposed ones in this book are based on the image, not the language or script itself. On the other hand, it is vary confused with the terms "language" and "script". For instance, English is not a script. English is one of a large number of languages set in Roman or Latin script. Russian is not a script. Russian is one of a small number of languages set in Cyrillic script. Indian is not a script or language. India has many languages and approximately 10 scripts in common use.

Two techniques are often adopted in the identification of document image with different languages, i.e., statistics analysis and symbol matching. In reference [Ding et al., 1996], a united analysis method is used to identify statistical characteristics for oriental and European scripts. In article [Hochberg et al., 1997], templates of a script are generated based on the clusters of representative symbols from the set of training samples, and the comparison between the symbols in the test document image and these templates can then be made, so as to discover the script type with the best match. This strategy is able to recognize the document images with new languages or scripts by means of training learning. However, more than 40 symbols are required to be extracted from each document image in order to reach a high accuracy, and this identifier cannot overcome the differences in font types.

Because each document with a certain language has its distinguishable visual image, different document images in the same language can be treated as one type of texture. Thus, the identification of the language-images can be solved by the method of texture classification. However, identification of language-images will be more complicated than conventional texture classification because the documents with a certain language may have different contents, formats, and fonts.

Reference [Tan, 1998] involves the extraction of rotation invariant texture features and their application to identification of scripts. It is the extension of a popular technique, multi-channel Gabor filtering. In this way, the features are Fourier progressions of average energy along 16 channels (direction) in a single frequency band. The output energy in a channel is computed as the geometrical mean of two-dimensional convolutions between the original image and two Gabor filters respectively. Therefore, in order to extract these features from a document image, there are 32 two-dimensional convolutions between this image and Gabor filters. Obviously, the computational cost is still high even though the Fast Fourier Transform (FFT) is utilized. This technique reaches high accuracy of classification when the testing samples have similar formats to the training samples. However, document formatting undoubtedly affects the success of Gabor texture-based script identification. Unfortunately, Gabor texture-based script identification failed in processing Chinese document images with five different fonts.

The multi-resolution representation has led to increased interest in the method of texture analysis. Mental vision research has proven that images are processed in multiple scales by the human vision system. Wavelets can provide a convenient representation in multi-scale, and make the extraction of texture features easier. Thus, wavelets have been applied to texture classification in recent years [de Wouwer et al., 1999; de Wouwer et al., 1989]. The fractal dimension is also a powerful tool to extract the textural features, which can be utilized to identify document images with different scripts.

This chapter presents two methods to identify different scripts, namely: (1) Wavelet-based method — script identification based on energy features of the wavelet sub-images, (2) Fractal-based method — script identification based on a modified fractal signature (MFS). In the remaining sections of this chapter, we will provide more details of these two methods.

7.2 Wavelet-Based Approach

This section proposes a technique for automatic language-image identification based on multi-scale non-redundant wavelet texture analysis. It has less computational complexity, and more stability to handle the document image with a variety of formats and fonts. In this wavelet-based approach, a document image is first decomposed into four sub-images by wavelet transform. Further, the three detailed sub-images are employed to extract feature vectors, they are (1) wavelet energy distribution feature vector, and (2) wavelet energy distribution proportion feature vector. Finally, we use these feature vectors to identify the document images in different languages including Chinese, English, Japanese, Korean, Russian, and Devanagari (a language used in Indian). The process of this wavelet-based approach can be described in Figure 7.2.

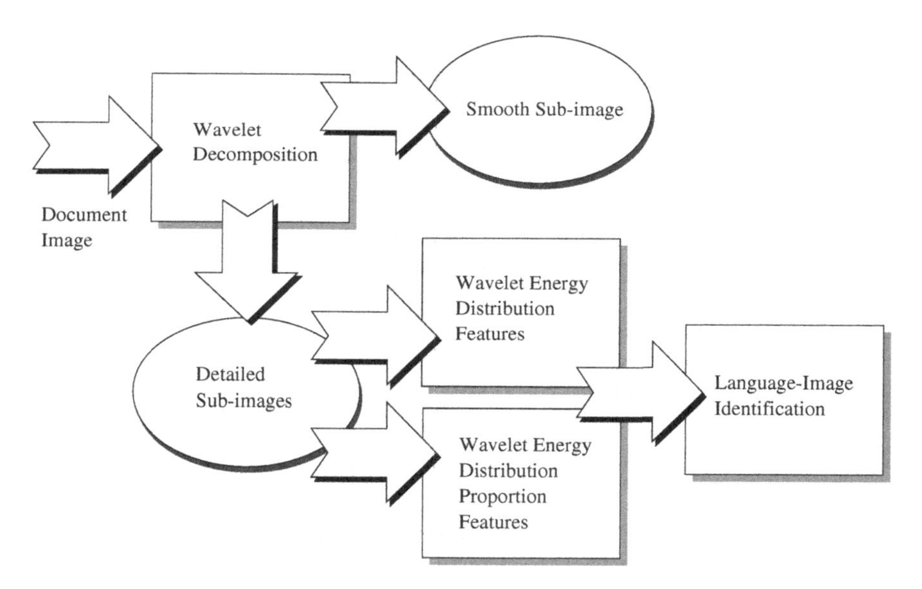

Fig. 7.2 Diagram of the wavelet-based approach to identification of various scripts in documents: A document image is first decomposed into four sub-images by wavelet transform. Further, the three detailed sub-images are employed to extract feature vectors, which can be used to identify the document images in different languages.

This section consists of two major parts:

- The basic idea of image decomposition by multi-scale wavelet transform,
- Extraction of features, where the wavelet energy distribution feature vector and wavelet energy distribution proportion feature vector are used.

7.2.1 *Image Decomposition by Multi-Scale Wavelet Transform*

Assume that point (x, y) is a pixel in an image. It has a gray function $f(x, y)$, which indicates the gray level at this pixel. Hence, the wavelet transform of function $f(x, y)$ is defined as [Mallat and Hwang, 1992]

$$Wf(x, y) = f(x, y) * \psi(x, y), \tag{7.1}$$

where, '$*$' stands for the two-dimensional convolution operator, and $\psi(x, y)$ is a two-dimensional wavelet function, which satisfies

$$c_\psi = (2\pi) \int_{-\infty}^{+\infty} \int_{-\infty}^{+\infty} (\hat{\psi}(\omega_x, \omega_y))^2 / (\omega_x^2 + \omega_y^2) d\omega_x d\omega_y < \infty. \tag{7.2}$$

In order to extract features from an image in different resolutions, the multi-scale wavelet function can be written explicitly as

$$\psi_s(x, y) = (1/s^2)\psi(x/s, y/s), \tag{7.3}$$

where, s is the scale, and the wavelet transform of $f(x, y)$ at scale s is,

$$\varpi_s f(x, y) = f(x, y) * \psi_s(x, y). \tag{7.4}$$

Furthermore, some constraints are forced into the 'mother' wavelet function so as to guarantee the transform to be non-redundant, complete, and form a multi-resolution representation for the original image. A well-known example is Daubechies wavelet, which is the orthonormal bases of compactly supported wavelet [Daubechies, 1992], from which the pagoda algorithm of wavelet decomposition can be drawn out:

In a certain scale, a one-dimensional filter is first used to convolute the rows of the input image, and one of every two columns is reserved. Then, another one-dimensional filter is applied to columns convolution on the image, and one of every two rows is reserved. Figure 7.3 shows this process. The complete description of multi-scale wavelet decomposition of an image can be found in [Mallat, 1989c].

A pair of filters, $\overline{G}$ and $\overline{H}$, are employed in decomposition [Daubechies, 1992; Mallat, 1989c], where $\overline{G}$ is the high-pass filter, and $\overline{H}$ is the low-pass filter. Hence, the input image is decomposed into 4 sub-images after

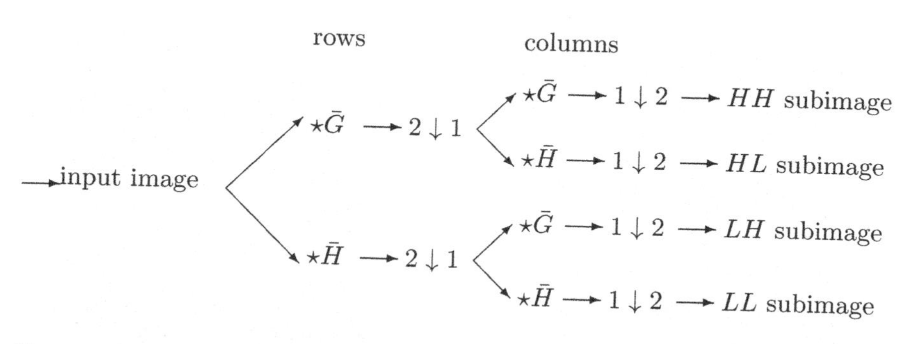

Fig. 7.3 The wavelet decomposition pyramidal algorithm at one scale; $\star X$: convolute (rows or columns) with filter X; $2 \downarrow 1$: keep one column out of two; $1 \downarrow 2$: keep one row out of two; $\bar{G}$ and $\bar{H}$ are 1-D mirror filters.

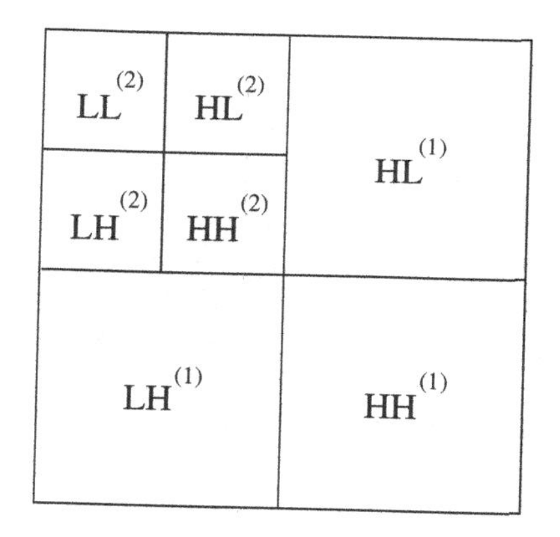

Fig. 7.4 Sub-images of wavelet decomposition at two scales.

wavelet decomposition: $LL^{(1)}$, $LH^{(1)}$, $HL^{(1)}$, and $HH^{(1)}$. Here, $LL^{(1)}$ is called the smooth sub-image, while $LH^{(1)}$, $HL^{(1)}$, and $HH^{(1)}$ are referred to as the detailed sub-images. Moreover, the area of the sum of the 4 sub-images is equal to the area of the input image. If we repeat the process of wavelet decomposition in sub-image $LL^{(1)}$, i.e. duplicate the scale, then we can obtain the sub-images, $LL^{(2)}$, $LH^{(2)}$, $HL^{(2)}$, and $HH^{(2)}$, as shown in Figure 7.4.

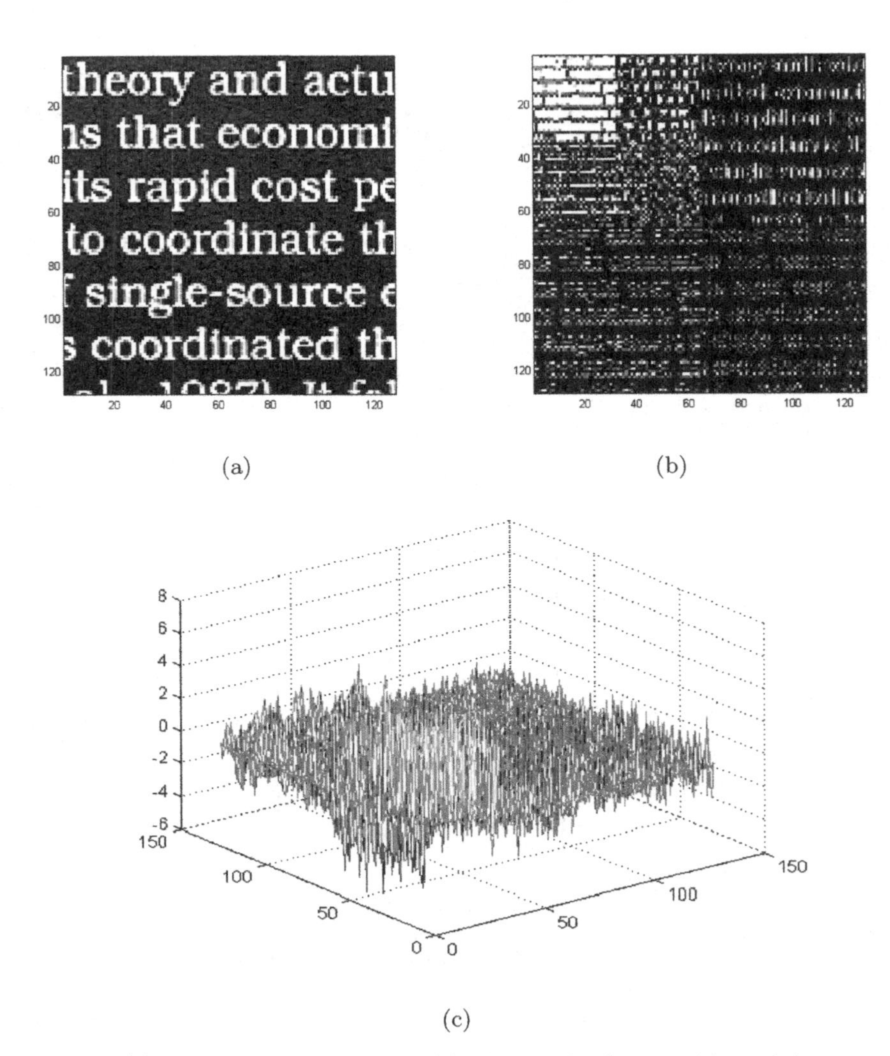

Fig. 7.5 (a) Original document image; (b) The wavelet decomposition sub-images at two scales; (c) 3-D display of the sub-images.

Suppose the image contains $N \times N$ sampling points, thus, the wavelet decomposition of this image can be computed with the following equations:

$$\begin{aligned} LL^{(k)}(m,n) &= [[LL^{(k-1)}_{rows} * \overline{H}]_{2\downarrow 1\ columns} * \overline{H}]_{1\downarrow 2} \\ m &= 1, \cdots, N/2^k;\ n = 1, \cdots, N/2^k \end{aligned} \tag{7.5}$$

$$LH^{(k)}(m,n) = [[LL_{rows}^{(k-1)} * \overline{H}]_{2 \downarrow 1 \ columns} * \overline{G}]_{1 \downarrow 2}$$
$$m = N/2^k + 1, \cdots, N/2^{k-1}; \ n = 1, \cdots, N/2^k \tag{7.6}$$

$$HL^{(k)}(m,n) = [[LL_{rows}^{(k-1)} * \overline{G}]_{2 \downarrow 1 \ columns} * \overline{H}]_{1 \downarrow 2}$$
$$m = 1, \cdots, N/2^k; \ n = N/2^k + 1, \cdots, N/2^{k-1} \tag{7.7}$$

$$HH^{(k)}(m,n) = [[LL_{rows}^{(k-1)} * \overline{G}]_{2 \downarrow 1 \ columns} * \overline{G}]_{1 \downarrow 2}$$
$$m = N/2^k + 1, \cdots, N/2^{k-1}; \ n = N/2^k + 1, \cdots, N/2^{k-1} \tag{7.8}$$

where, $s = 2^{k-1}$ $(k = 1, \cdots, \log_2 N)$, $LL^{(0)}$ is the original image, and $2 \downarrow 1$ $(1 \downarrow 2)$ indicates the sub-sampling (get 1 from 2) along rows (columns).

Figure 7.5 gives an example of wavelet decomposition of a document image. Figure 7.5(a) is an original document image. Figure 7.5(b) is the sub-images of the 2-D representation of the wavelet decomposition at two scales. Figure 7.5(c) is a 3-D display of these sub-images.

7.2.2 *Wavelet-Based Features*

In this method, an image is decomposed by wavelet transform into several sub-images, and thereafter, the average energies of the sub-images are computed. We can use them as a set of the features to identify different documents with different languages. Actually, two feature vectors are utilized, namely:

(1) wavelet energy distribution feature (F_d) vector, and
(2) wavelet energy distribution proportion feature (F_{dp}) vector.

They can be presented below.

7.2.2.1 *Average Energy of Document Image*

The average energy of an image of size $N \times N$ is defined by

$$energy\ f = \sum_{m=1}^{N}\sum_{n=1}^{N}\frac{f^2(m,n)}{N^2}. \tag{7.9}$$

Different language-images have different average energies. For instance, the average energies of six classes of language-images are presented in Figure 7.6, where document images with six languages, namely, Chinese, English, Japanese, Korean, Russian, and Devanagari (a language used in Indian), are analyzed. The six curves in this figure stand for the average energies of these six languages. Each language has four different document images, which are indicated by the numbers on the horizontal axis. From those, it is obvious that the different document images of the same language have different average energies. Thus, they must be normalized before we perform the wavelet decomposition. To normalize the energies, we have

$$f(m,n) \leftarrow \frac{f(m,n)}{(energy\ f)^{1/2}}, \qquad m,n = 1,\cdots,N. \tag{7.10}$$

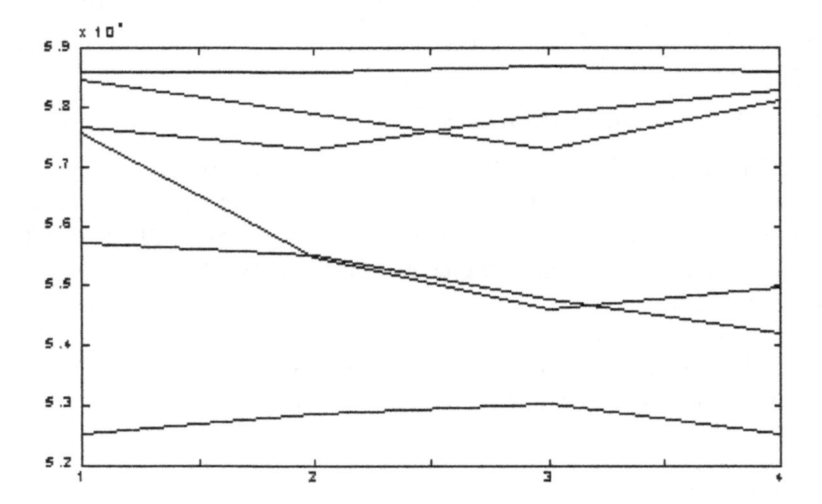

Fig. 7.6 The average energies of different document images: six curves stand for six languages (Chinese, English, Japanese, Korean, Russian, and Devanagari), each language has four different document images labeled by the numbers on the horizontal axis.

7.2.2.2 *Wavelet Energy Distribution Features (F_d)*

After the multi-scale wavelet decomposition, the original document image has been decomposed into detailed sub-images LH, HH, and HL. The average energy distribution of the detailed sub-images at k-th level of wavelet decomposition can be defined by

$$ELH^{(k)} = \sum_{m=(N/2^k)+1}^{N/2^{k-1}} \sum_{n=1}^{N/2^k} \frac{(LH^{(k)}(m,n))^2}{(N/2^k)^2}, \quad k = 1, \cdots, \log_2 N, \quad (7.11)$$

$$EHH^{(k)} = \sum_{m=(N/2^k)+1}^{N/2^{k-1}} \sum_{n=(N/2^k)+1}^{N/2^{k-1}} \frac{(HH^{(k)}(m,n))^2}{(N/2^k)^2}, \quad k = 1, \cdots, \log_2 N, \quad (7.12)$$

$$EHL^{(k)} = \sum_{m=1}^{N/2^k} \sum_{n=(N/2^k)+1}^{N/2^{k-1}} \frac{(HL^{(k)}(m,n))^2}{(N/2^k)^2}. \quad k = 1, \cdots, \log_2 N, \quad (7.13)$$

where, $ELH^{(k)}$, $EHH^{(k)}$, and $EHL^{(k)}$ ($k = 1, \cdots, \log_2 N$) respectively are multi-scale wavelet features of an image, which are called *wavelet energy distribution* features. They become the components of the wavelet energy distribution feature (F_d) vector:

$$F_d = \begin{pmatrix} EHL^{(k)} \\ EHH^{(k)} \\ ELH^{(k)} \end{pmatrix}, \quad k = 1, 2, ... \log_2 N. \quad (7.14)$$

Some examples of the wavelet energy distributions of LH sub-images in six languages, English, Chinese, Japanese, Korean, Russian and Devanagari, are illustrated in Figure 7.7.

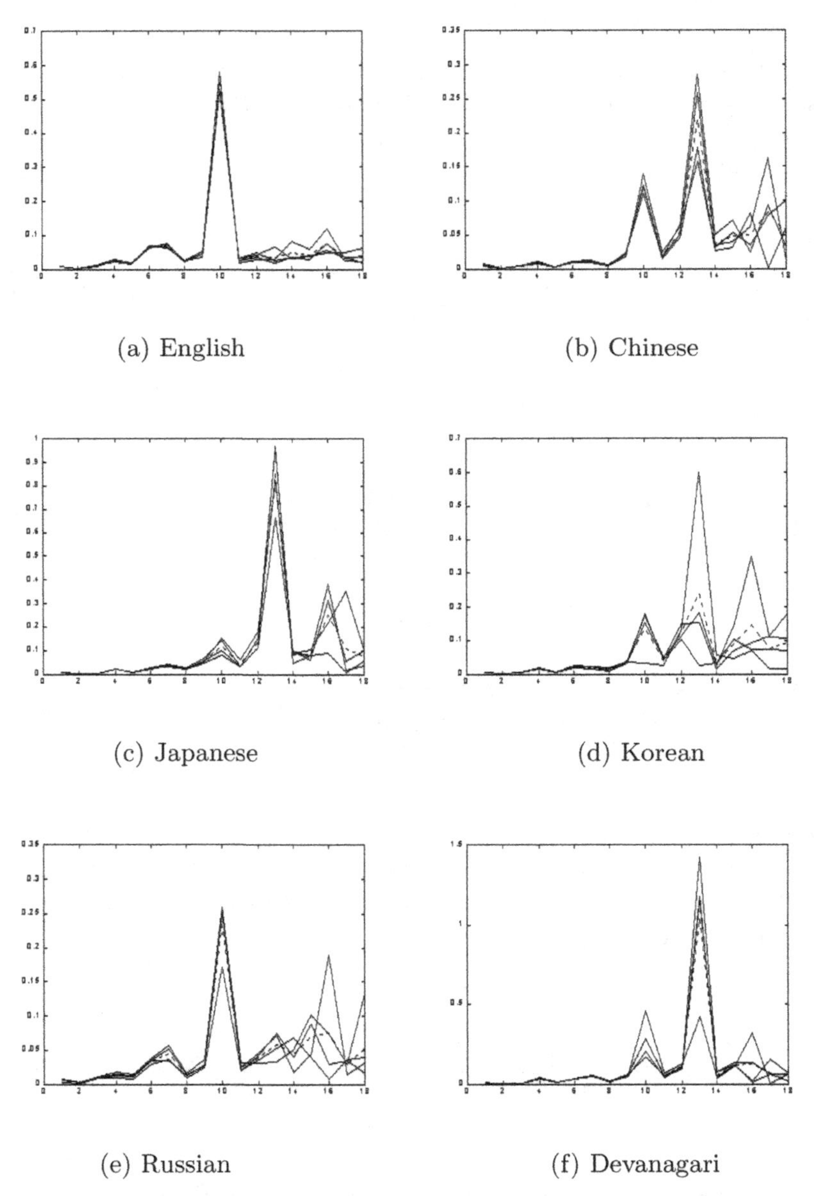

(a) English (b) Chinese

(c) Japanese (d) Korean

(e) Russian (f) Devanagari

Fig. 7.7 Examples of the wavalet energy distributions features F_d of the sub-images (LH) in (a) English, (b) Chinese, (c) Japanese, (d) Korean (e) Russian, and (f) Devanagari.

7.2.2.3 *Wavelet Energy Distribution Proportion Features* (F_{dp})

The wavelet energy distribution proportion features (F_{dp}) consist of three sets of features, namely:

$$F_{dp} = \begin{pmatrix} EPHL^{(k)} \\ EPHH^{(k)} \\ EPLH^{(k)} \end{pmatrix}, \quad k = 1, 2, \ldots \log_2 N, \tag{7.15}$$

where, $EPHL^{(k)}$, $EPHH^{(k)}$ and $EPLH^{(k)}$ stand for the energy distribution proportions of sub-image HL, HH and LH at the k-th level of the wavelet decomposition respectively.

Note the following facts: (1) Sub-image LH carries much more horizontal information than that of HL as well as HH. (2) Sub-image HL has more vertical features than those of LH and HH. (3) Sub-image HH contains more features in the direction of diagonal comparing with other two sub-images. In order to characterize the oriented property of the different language-images, we consider the ratio of the energy distribution of a specific detailed sub-image to that of all detailed sub-images at the same wavelet scale.

To illustrate the horizontal property of the different languages in document images, the ratio of the energy distribution of LH sub-image, (ELH), to those of all detailed sub-images at the same wavelet scale, (ELH + EHL + EHH), is defined as *energy distribution proportions* of sub-image LH, which is symbolized by EPLH. It can be computed by

$$EPLH^{(k)} = \frac{ELH^{(k)}}{ELH^{(k)} + EHH^{(k)} + EHL^{(k)}}, \quad k = 1, \cdots, \log_2 N. \tag{7.16}$$

Similarly, the ratio of the energy distribution of HH sub-image, (EHH), to those of all detailed sub-images at the k-th wavelet scale is defined as $EPHH^{(k)}$ - the energy distribution proportion of sub-image HH at the k-th level of the wavelet decomposition, and is represented by

$$EPHH^{(k)} = \frac{EHH^{(k)}}{ELH^{(k)} + EHH^{(k)} + EHL^{(k)}}, \quad k = 1, \cdots, \log_2 N, \tag{7.17}$$

which can characterize the diagonal property of the different languages in document images.

Using the similar way, we have $EPHL^{(k)}$ - the energy distribution proportion of sub-image HL at the k-th level of the wavelet decomposition,

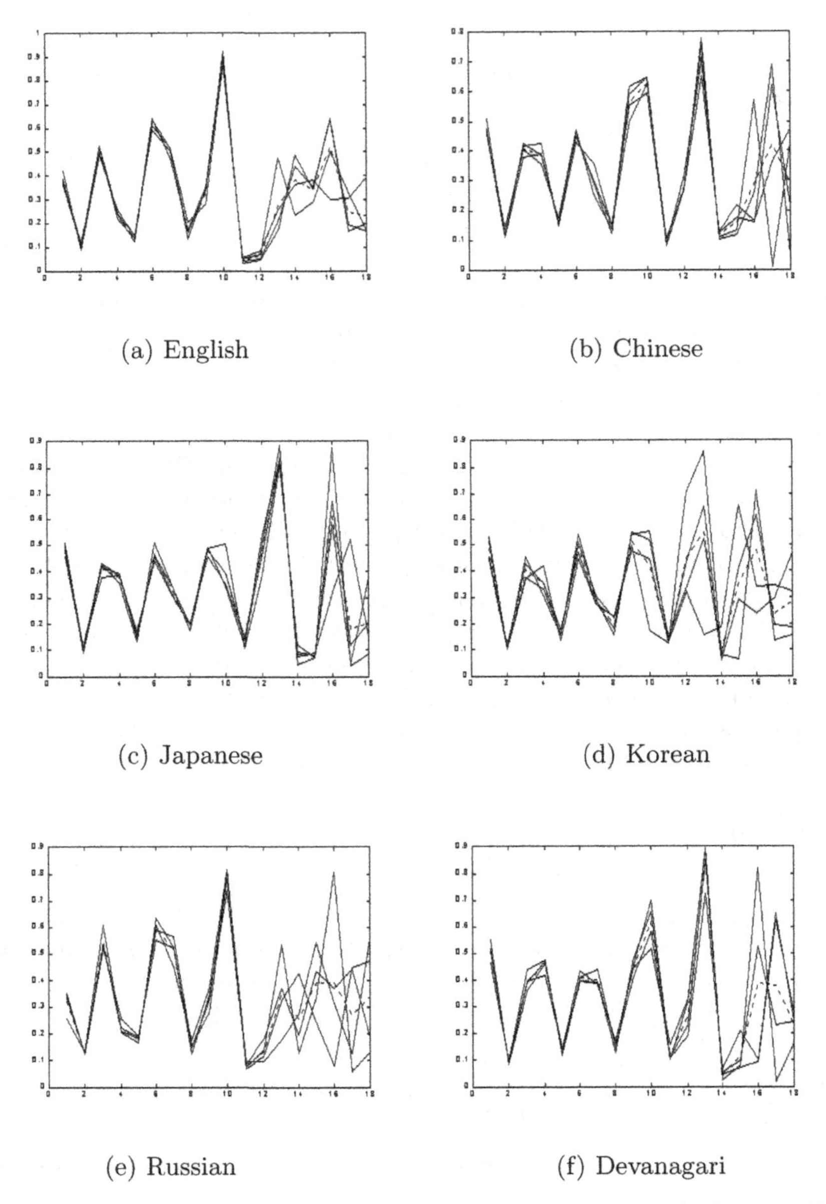

(a) English (b) Chinese

(c) Japanese (d) Korean

(e) Russian (f) Devanagari

Fig. 7.8 Examples of the wavalet energy distributions proportion features F_{dp} of the sub-images (LH) in (a) English, (b) Chinese, (c) Japanese, (d) Korean (e) Russian, and (f) Devanagari.

which is defined as

$$EPHL^{(k)} = \frac{EHL^{(k)}}{ELH^{(k)} + EHH^{(k)} + EHL^{(k)}}, \quad k = 1, \cdots, \log_2 N. \quad (7.18)$$

Some examples of the wavelet energy distribution proportions of LH sub-images in six languages, English, Chinese, Japanese, Korean, Russian and Devanagari, are illustrated in Figure 7.8, where each language has some samples.

The experiments have shown that the wavelet energy distribution proportion features (F_{dp}) such as $EPLH^{(k)}$, $EPHH^{(k)}$ and $EPHL^{(k)}$ for a document image are more stable.

7.2.3 *Experiments*

In our experiments, we choose six language-image samples (Chinese, English, Japanese, Korean, Russian, and Devanagari) with different fonts to demonstrate the potential of this wavelet-based technique.

In accordance with the requirements of the OCR techniques, each document is scanned on the grey level of $0 - 255$, and distorted by a bit of noise. In our experiments, we do not have any pre-processing to remove the noise. Some examples of the scanned document images are shown in Figure 7.9.

7.2.3.1 *Distance Functions*

To recognize different language-images using the wavelet-based features described in the previous sections, four distance functions are used in this work, namely, DEMW, DPMW, DEMWV and DPMWV. They are presented in this sub-section.

In order to identify a certain language-image from various types of language-images, we must first obtain the mean of multi-scale wavelet features corresponding to each language in the training images.

Let $averageF_d^j(m)$ and $averageF_{dp}^j(m)$ be the mean of the features vector for multi-scale wavelet energy distribution and multi-scale wavelet energy proportion associated with the mth element and the jth language in the image respectively. If there are c types of languages, a test sample will be classified to be the ith language when and only when one of the following distance functions reaches the minimum in the case of $j = i$:

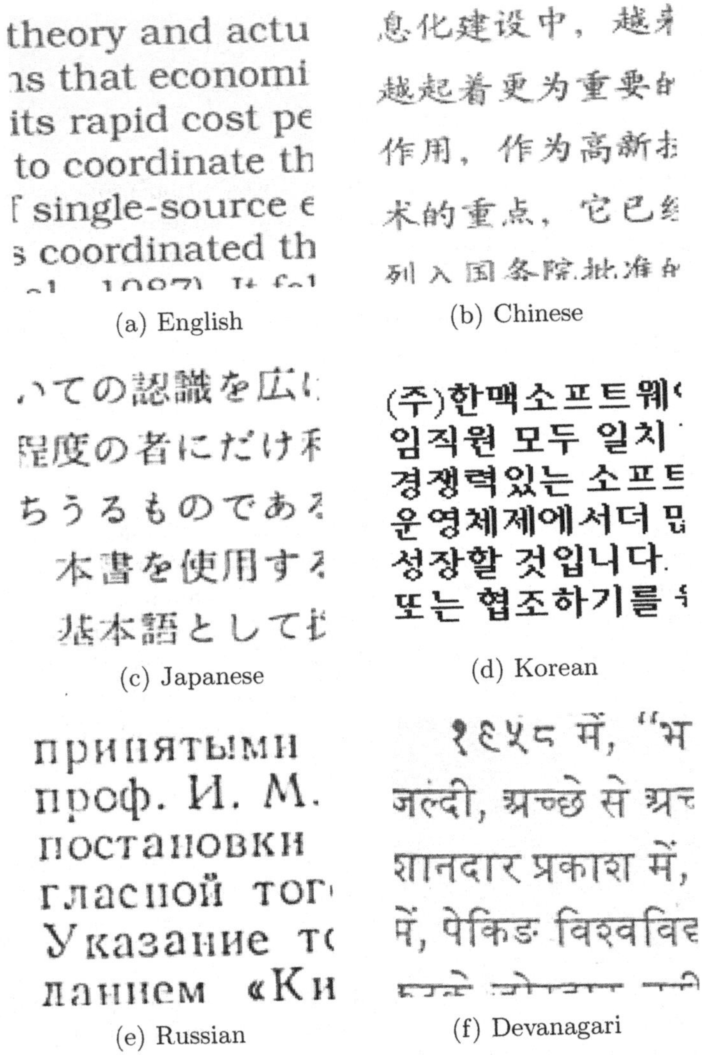

theory and actu
ns that economi
its rapid cost pe
to coordinate th
single-source e
s coordinated th

(a) English

息化建设中，越
越起着更为重要
作用，作为高新
术的重点，它已
列入国务院批准

(b) Chinese

いての認識を広
程度の者にだけ
ちうるものであ
本書を使用す
基本語として

(c) Japanese

(주)한맥소프트웨
임직원 모두 일치
경쟁력있는 소프트
운영체제에서더
성장할 것입니다.
또는 협조하기를

(d) Korean

принятыми
проф. И. М.
постановки
гласной тог
Указание т
ланием «Кн

(e) Russian

१९५८ में, "भ
जल्दी, अच्छे से अ
शानदार प्रकाश में,
में, पेकिङ विश्ववि

(f) Devanagari

Fig. 7.9 Examples of scanned document images in six languages.

$$\underset{j=i}{\text{MIN}}\ DEMW(j) = \underset{j=i}{\text{MIN}}\Big[\sum_{m=1}^{d\times 3}(F_d(m) - averageF_d^j(m))^2\Big], \qquad (7.19)$$
$$d = \log_2 N;\ j = 1, \cdots, c$$

or,

$$\underset{j=i}{\text{MIN}}\ DPMW(j) = \underset{j=i}{\text{MIN}}[\sum_{m=1}^{d\times 3}(F_{dp}(m) - averageF_{dp}^{j}(m))^2]. \qquad (7.20)$$
$$d = \log_2 N;\ j = 1, \cdots, c$$

Let $varianceF_d^j(m)$ and $varianceF_{dp}^j(m)$ be the variance of the feature vectors for multi-scale wavelet energy distribution and multi-scale wavelet energy proportion associated with the mth element and the jth language in the training image respectively. If there are c types of languages, a test sample will be classified to be the ith language when and only when one of the following distance functions reaches the minimum in the case of $j = i$:

$$\underset{j=i}{\text{MIN}}\ DEMWV(j) =$$
$$\underset{j=i}{\text{MIN}}\left[\sum_{m=1}^{d\times 3}\left((F_d(m) - averageF_d^j(m))^2/(varianceF_d^j(m))^2\right)\right], \qquad (7.21)$$
$$d = \log_2 N;\ j = 1, \cdots, c$$

or,

$$\underset{j=i}{\text{MIN}}\ DPMWV(j) =$$
$$\underset{j=i}{\text{MIN}}\left[\sum_{m=1}^{d\times 3}\left((F_{dp}(m) - averageF_{dp}^j(m))^2/(varianceF_{dp}^j(m))^2\right)\right]. \qquad (7.22)$$
$$d = \log_2 N;\ j = 1, \cdots, c$$

7.2.3.2 *Experimental Results*

In our experiments, all samples are scanned from some newspapers and books. To identify a certain language from a document image, which contains various types of languages, we must first obtain the mean and variance of wavelet texture features corresponding to each language from the training images. We choose some training samples for each language to accomplish the on-line training and identification. For example, in order to identify the languages from the books printed in various languages, users first perform the artificial identification based on the training samples selected in the books, then the computer extracts the mean and variance of wavelet-based features automatically from the training document images for each language, and finally the computer performs the automatic identification from samples in the books. The experiments show that a few training samples, for example, 10-20, will be enough to obtain the mean and variance of wavelet-basd features for one type of language.

Furthermore, experiments also display that the size of the selected scale must be suitable, since there is large variance in the texture features for multi-scale wavelet energy distribution (proportion) when the scale is selected too large.

Three experiments were conducted, and they are presented in this subsection.

Experiment 1

In this experiment, the test samples are chosen from the books, so that they have similar formats to the training samples. For example, both of the test samples and training ones have similar sizes, fonts, intervals between the characters, type formats, etc. Experiments show that the wavelet energy distribution features (F_d) and the wavelet energy distribution proportion feature (F_{dp}) of the test samples in a certain language-image are very close to those of the training samples. We classify the test samples in six different languages based on four distance functions expressed in Eqs. (7.19)–(7.22), and the classification accuracy for each is given in Table 7.1.

Table 7.1 Classification accuracy of test samples that have similar formats to the training samples.

Function	Scale	Accuracy of Classification			
		English	Chinese	Japanese	
DEMW	$2^0 - 2^2$	100%	100%	75%	
DPMW	$2^0 - 2^2$	100%	100%	75%	
DEMWV	$2^0 - 2^2$	100%	100%	100%	
DPMWV	$2^0 - 2^3$	100%	100%	50%	
		Korean	Russian	Indian	**Average**
		100%	75%	100%	92%
		75%	100%	75%	88%
		100%	100%	100%	100%
		100%	100%	100%	92%

According to the experimental results, we find that the distance function $DEMWV$ (Eq. 7.21) has the highest accuracy of classification for those test samples that have similar formats to the training samples.

Experiment 2

In the second experiment, the test samples were taken from those texts, so that their formats differ from that of the training samples. That implies that the test samples and the training ones have different formats, such as sizes, fonts, intervals between the characters, type formats, etc.

Table 7.2 Classification accuracy of test samples that different formats than the training samples.

Function	Scale	Accuracy of Classification			
		English	Chinese	Japanese	
DEMW	$2^0 - 2^2$	50%	50%	25%	
DPMW	$2^0 - 2^2$	75%	75%	50%	
DEMWV	$2^0 - 2^2$	75%	75%	50%	
DPMWV	$2^0 - 2^3$	100%	75%	50%	
		Korean	Russian	Indian	**Average**
		50%	50%	25%	42%
		100%	75%	75%	75%
		75%	50%	50%	63%
		75%	100%	100%	83%

Experiments show that the means of the wavelet energy distribution features (F_d) are different between the test and training samples in the same language. However, the means of their wavelet energy distribution proportion feature (F_{dp}) are similar. In this experiment, we also classify the test samples in six different languages based on four distance functions expressed in Eqs. (7.19)–(7.22), and the classification accuracy for each is given in Table 7.2.

The experimental results show that the distance function $DPMWV$ (Eq. (7.22)) has higher accuracy of classification for those test samples that have have the formats differ from the training samples.

Experiment 3

Finally, two sets of test samples, English and Chinese, with multi-font

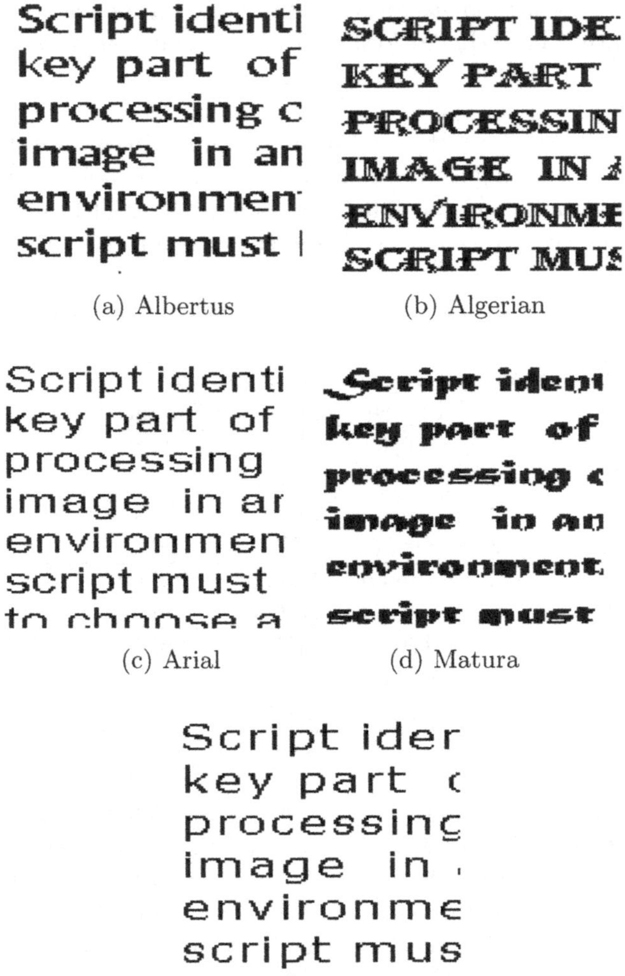

(a) Albertus

(b) Algerian

(c) Arial

(d) Matura

(e) Vordana

Fig. 7.10 Examples of scanned document images in five fonts of English.

are used. In the experiments, we choose five fonts of English, Albertus, Algerian, Arial, Matura, and Vordana as shown in Figure 7.10, and five fonts of Chinese, Song typeface, the traditional Song typeface, imitative Song-Dynasty-Style typeface, regular type, and tabular inscriptions of the

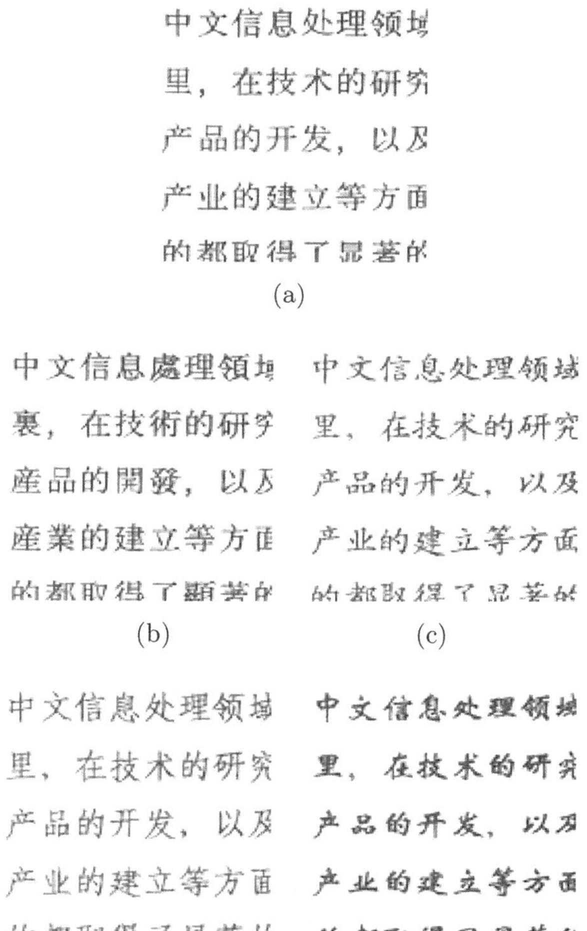

中文信息处理领域
里，在技术的研究
产品的开发，以及
产业的建立等方面
的都取得了显著的

(a)

中文信息處理領域 裏，在技術的研究 產品的開發，以及 產業的建立等方面 的都取得了顯著的	中文信息处理领域 里，在技术的研究 产品的开发，以及 产业的建立等方面 的都取得了显著的
(b)	(c)
中文信息处理领域 里，在技术的研究 产品的开发，以及 产业的建立等方面 的都取得了显著的	中文信息处理领域 里，在技术的研究 产品的开发，以及 产业的建立等方面 的都取得了显著的
(d)	(e)

Fig. 7.11 Examples of scanned document images in five fonts of Chinese. (a) Song typeface, (b) the traditional Song typeface, (c) imitative Song-Dynasty-Style typeface, (d) regular type, (e) tabular inscriptions of the Northern Dynasties.

Northern Dynasties, as shown in Figure 7.11. Although the traditional characters are seldom used in mainland China, they appear extensively in some oversea Chinese publications, as well as in the areas of Taiwan and Hong Kong. Therefore, in this experiment, both of the traditional and simple forms of Chinese characters are utilized.

We classify the test samples above into six types based on two distance functions (Eq. 7.20 and Eq. 7.22), and the accuracy of classification is about 90%.

Time Complexity

Because the non-redundant wavelet decomposition is employed, computational cost is remarkably reduced. In order to extract the multi-scale wavelet texture features from an image with $(N \times N)$ sampling points, the rank of computational cost is theoretically N^2. In fact, if the length of the filter is M ($M \ll N$), it will take $4MN^2$ multiplications when using the pagoda algorithm of wavelet decomposition with one-dimensional direct convolution, and the cost of energy computation is about N^2. On the other hand, since the sub-images in wavelet decomposition can replace the original image, storage cost is still about N^2.

7.3 Fractal-Based Approach

In fractal-based approach, the modified fractal signature (MFS) is used. In this way, a document image is mapped onto a gray-level function. Furthermore, this function can be mapped onto a gray-level surface, and from the area of such surface, the fractal dimension of the document image can be approximated. However, directly calculating the area of the gray-level surface of the document image is an obscure task. To simplify this computation, a special equivalent technique of the Minkowski dimension technique, which was mentioned in Chapter 2, is applied in this study, referred as δ Parallel Bodies. We analyze the box counting methodology from the standpoint of computing fractal features. Using the δ parallel body of the gray-level surface of the document image, we first thicken the gray-level surface, so that it becomes a three-dimensional parallel body. Then, we calculate the volume of that body, since the calculation of a volume is much easier than that of a gray-level surface. Direct computation of fractal dimension of a document image is not used in this method. Alternately, the volume of a δ parallel body is estimated to approximate the fractal dimension. The derails of this process can be found in Chapter 4 of this book.

7.3.1 *Algorithm*

Algorithm 7.1 (fractal signature)

Input: a page of document image;
Output: the fractal signature of the document;

Step-1 For x = 1 to X_{max} do
For y = 1 to Y_{max} do
F is mapped onto a gray-level function $g_k(x, y)$;
Step-2 For x = 1 to X_{max} do
For y = 1 to Y_{max} do
Substep-1 Initially, taking $\delta = 0$, the upper layer $u_0^k(x, y)$ and lower layer $b_0^k(x, y)$ of the blanket are chosen as the same as the gray-level function $g_k(x, y)$, namely:

$$u_0^k(x, y) = b_0^k(x, y) = g_k(x, y);$$

Substep-2 Taking $\delta = \delta_1$,
(a) $u_{\delta_1}(x, y)$ is computed according to the formula:

$$u_{\delta_1}(x, y) = max\left\{u_0(x, y) + 1, \quad \max_{|(i,j)-(x,y)|\leq 1} u_0(i, j)\right\};$$

(b) $b_{\delta_1}(x, y)$ is computed according to the formula:

$$b_{\delta_1}(x, y) = min\left\{b_0(x, y) - 1, \quad \min_{|(i,j)-(x,y)|\leq 1} b_0(i, j)\right\};$$

(c) The volume Vol_{δ_1} of the blanket is computed by the formula:

$$Vol_{\delta_1} = \sum_{x,y}(u_{\delta_1}(x, y) - b_{\delta_1}(x, y));$$

Substep-3 Taking $\delta = \delta_2$,
(a) $u_{\delta_2}(x, y)$ is computed according to

$$u_{\delta_2}(x, y) = max\left\{u_{\delta_1}(x, y) + 1, \quad \max_{|(i,j)-(x,y)|\leq 1} u_{\delta_1}(i, j)\right\};$$

(b) $b_{\delta_2}(x,y)$ is computed according to

$$b_{\delta_2}(x,y) = min\left\{b_{\delta_1}(x,y) - 1, \quad \min_{|(i,j)-(x,y)|\leq 1} b_{\delta_1}(i,j)\right\};$$

(c) The volume Vol_{δ_1} of the blanket is computed by

$$Vol_{\delta_2} = \sum_{x,y}(u_{\delta_2}(x,y) - b_{\delta_2}(x,y));$$

Step-3 The sub fractal signature A_δ^k is computed by the formula:

$$A_\delta^k = \frac{Vol_{\delta_2} - Vol_{\delta_1}}{2}.$$

Step-4 Combining sub fractal signatures A_δ^k, $k = 1, 2, ..., n$ into the whole fractal signature:

$$A_\delta = \bigcup_{k=1}^{n} A_\delta^k.$$

7.3.2 *Experiments*

By using the better output image and comparison with the fractal signature value, we can analyze the document image and distinguish the blocks as script, handwritten signature, graphic, text and so on. The main discrimination between Oriental and Euramerican scripts used in the document is their different complexities. The modified fractal signature (MFS) will be used to distinguish these two categories of scripts. For the document image that are fractal only within a certain range of scale this is a necessary precaution. The final estimate of fractal signature is obtained by linear regression over the range of k as a curve. Here, we have used the scale with k values from 1 to 16. The result of discrimination between Chinese and English in Figure 4.19 is shown in Figure 7.12. Figure 7.14 shows that the result is satisfacctory when our method is used to discriminate between Oriental and Euramerican scripts, with six samples of scripts (Chinese, English, Indian, Japanese, Korean and Russian) in Figure 7.13.

One example for three typical Oriental scripts (Chinese, Indian and Japanese) is shown in Figure 7.15. Comparing with the different MFS's curves between three categories of scripts shows that our experimental results are satisfactory.

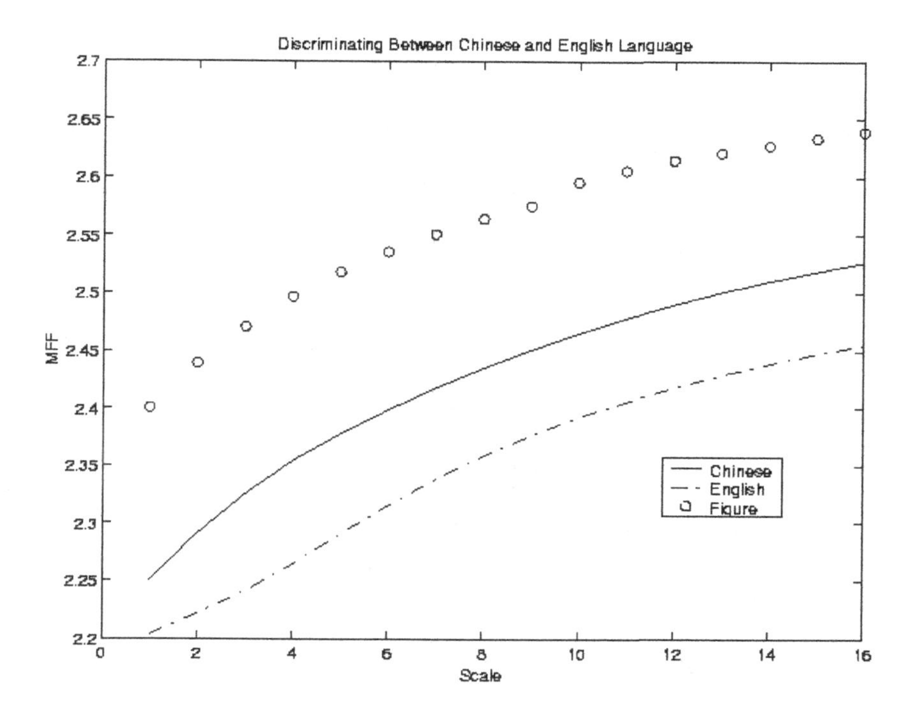

Fig. 7.12 An example of script identification between Chinese and English.

In our experimental database, there are 65 pages of Euramerican documents printed in 5 different scripts (English, Russian, French, German and Italian). For Oriental scripts, we have 65 pages of Chinese documents, 60 pages of Japanese documents, 40 pages of Indian documents and 43 pages of Korean documents. We use this method for all the documents images in our database. The main results are shown in Table 7.3. Moreover, Figure 7.16 clearly demonstrates the satisfactory results of discrimination between Oriental and Euramerican scripts.

Fig. 7.13 Samples of different Oriental and Euramerican scripts.

Table 7.3 Results of script classification.

Scripts	♯ Samples	Classification rate (%)
Chinese	65	92.55
Japanese	60	93.27
Korean	43	91.45
Indian	40	93.40
Euramerican	75	95.89

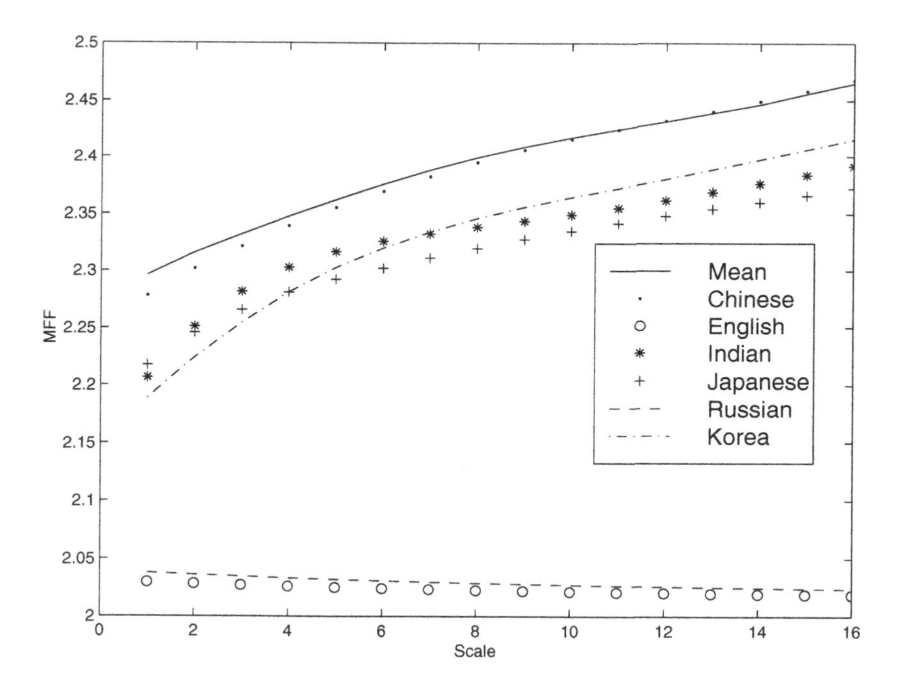

Fig. 7.14 An example of discrimination of Oriental and Euramerican scripts.

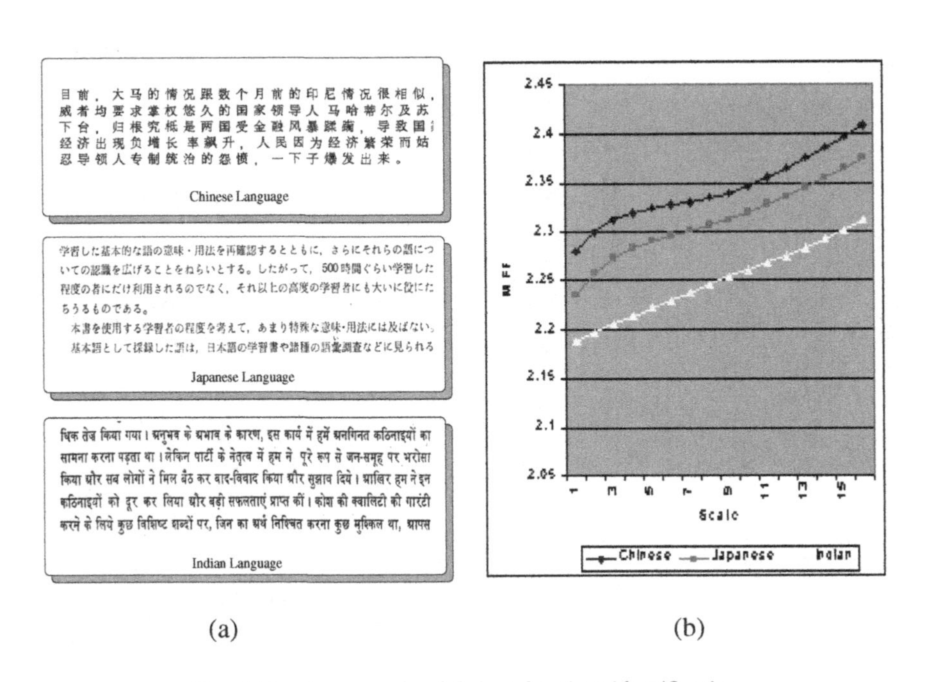

Fig. 7.15 An example of Oriental scripts identification.

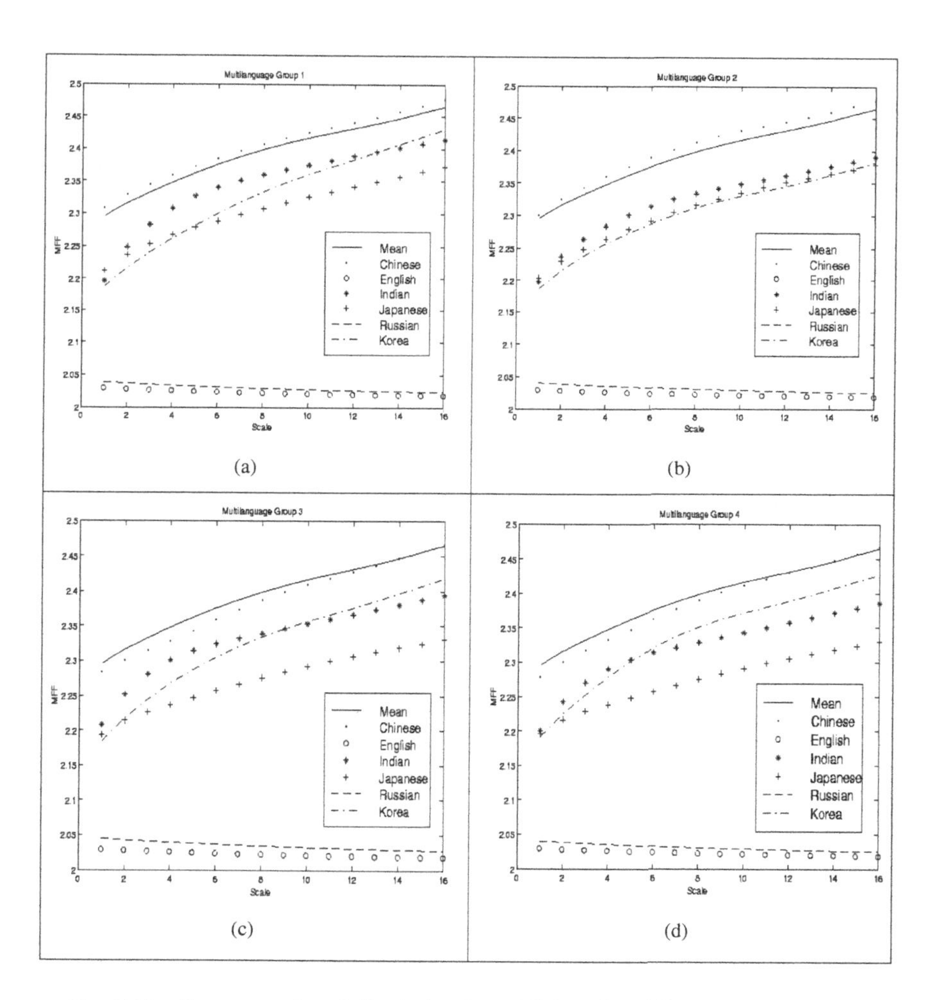

Fig. 7.16 Four examples of discrimination of Oriental and Euramerican scripts.

Chapter 8

Writer Identification Using Hidden Markov Model in Wavelet Domain (WD-HMM)

In this Chapter, we propose a method based on hidden Markov model in wavelet domain (WD-HMM) for writer identification. WD-HMM can handle the dependencies among wavelet coefficients of real world images with simple dependency structure and mixture distribution. In order to solve the non-Gaussianity problem of the images, WD-HMM considers each wavelet coefficient is a random variable and has a Gaussian Mixture distribution.

8.1 Introduction

Even in such a highly developed time, handwriting still plays a very important role in human society. Generally, handwriting is also regarded as a sign of the writer. A long history before, people has realized the importance of finding the true writer of one unknown handwriting document. In fact, writer identification based on the handwriting (there are many forms of handwritings, such as signatures, letters, notes, etc.) has a wide applicable field: to confirm the document authenticity in the financial sphere, to solve the expert problems in criminology, etc.

With the coming of computer era, most manual works can be well carried out by computer automatically. As a result, a great deal of time, efforts and money are saved. Naturally, the automatic writer identification comes into scientists' views. Nowadays, the automatic writer identification of handwriting is receiving growing interests from both academia and industry. More and more researchers and firms put money and energies on it. However, automatic writer identification is a complex problem involved in many science disciplines, including computer version, pattern recognition, image processing, machine learning, etc., and has many challenging items. Because of that, in spite of continuous efforts, writer identification is still a world challenging issue and far from being well solved.

Generally speaking, writer identification is a typical pattern recognition problem which consists of several steps: pre-processing, feature extraction, similarity measurement, performance evaluation. Among these steps, the feature extraction is the core one.

We begin this chapter with an introduction of the hidden markov model (abbreviated as HMM) and relevant statistical knowledge. Then we discuss the HMM in wavelet-domain (WD-HMM), which is a joint statistical model of the wavelet coefficients [Crouse et al., 1997].

8.2 Hidden Markov Model and Relative Statistical Knowledge

8.2.1 *Expectation Maximization (EM) Algorithm*

We can now extend our application of maximum likelihood techniques to permit the learning of parameters governing a distribution from training points, some of which have missing features. If we have uncorrupted data, we can use maximum likelihood, i.e., find $\hat{\theta}$ that maximizes the log-likelihood $L(\theta)$. The basic idea in the expectation maximization (EM) algorithm is to iteratively estimate the likelihood given the data that is present.

There are two main applications of the EM algorithm, namely:

- One is the observed data is uncompleted because of the limitations of the observation process.
- The second occurs when optimizing the likelihood function is analytically intractable but when the likelihood function can be simplified by assuming the existence of values for additional but missing (or hidden) parameters.

The later application is more common in the computational pattern recognition community [Simoncelli, 1997].

Let data X be the observed incomplete data, and Z be such the complete data that satisfies $Z = (X, Y)$, and specify a joint density function:

$$P(z|\theta) = P(x, y|\theta) = P(y|x, \theta)P(x|\theta) \tag{8.1}$$

With this new density function, a new likelihood function (called the complete-data likelihood) can be defined:

$$L(\Theta|Z) = L(\Theta|X, Y) = P(X, Y|\Theta) \tag{8.2}$$

The EM algorithm first finds the expected value of the complete-data log-likelihood function $\log P(X, Y|\Theta)$ with respect to the unknown data Y given the observed data X and the current parameter estimates. That is, we define

$$Q(\Theta, \Theta^{i-1}) = E[\log P(X, Y|\Theta)|X, \Theta^{i-1}] \tag{8.3}$$

where Θ^{i-1} are the current parameters estimates that we used to evaluate the expectation and Θ are the new parameters that we optimize to increase Q.

It must be noted that X and Θ^{i-1} are constants, Θ is the normal variable need to be adjusted, Y is a random variable governed by the distribution $P(y|X, \Theta^{i-1})$. Therefore, the right hand side of Eq. (8.3) can be rewritten as:

$$E[\log P(X, Y|\Theta)|X, \Theta^{i-1}] = \int_{y \in r} \log P(X, y|\Theta)P(y|X, \Theta^{i-1})dy \tag{8.4}$$

where r is the value space of y, and $P(y|X, \Theta^{i-1})$ is the marginal distribution of the unobserved data and is dependent on both the observed data X and on the current parameters Θ^{i-1}. In the best cases, this marginal distribution is a simple analytical expression of the assumed parameters Θ^{i-1} and perhaps the data. In the worst cases, this density might be very hard to obtain [Simoncelli, 1997]. In fact, sometimes, the density actually used is

$$\log P(y, X|\Theta^{i-1}) = P(y|X, \Theta^{i-1})P(X|\Theta^{i-1}) \tag{8.5}$$

The evaluation of this expectation $Q(\Theta, \Theta^{i-1})$ is called the E-setp of the algorithm. Maximizing the expectation computed in the E-step is called M-setp.

$$\Theta^i = \arg\max_{\Theta} Q(\Theta, \Theta^{i-1}) \tag{8.6}$$

The E-step and M-step are repeated as necessary. Sometimes, instead of maximizing $Q(\Theta, \Theta^{i-1})$, M-step is to find Θ^i such that

$$Q(\Theta^i, \Theta^{i-1}) > Q(\Theta, \Theta^{i-1}) \tag{8.7}$$

What is more, an important criteria of the EM iteration is that the EM iteration is guaranteed to converge. Some papers have concerned the converge problem [Jordan and Jacobs, 1996].

Let i be an iteration counter, and T be the convergence threshold, the EM iteration algorithm can be described as follows.

Algorithm (Expectation Maximization (EM))

1. **Initialization** Θ^0, $T, i = 0$
2. **do** $i \leftarrow i + 1$
3. **E step**: compute $Q(\Theta; \Theta^{i-1})$
4. **M step**: $\Theta^i \leftarrow \arg\max_{\Theta} Q(\Theta; \Theta^{i-1})$
5. **until** $Q(\Theta^i; \Theta^{i-1}) - Q(\Theta^{i-1}; \Theta^{i-2}) \leq T$
6. **return** $\hat{\Theta} \leftarrow \Theta^i$

8.2.2 *Gaussian Mixture Model (GMM) and Expectation Maximization (EM) for Gaussian Mixture Model (GMM)*

A phenomenon of wavelet coefficients is that most wavelet coefficients contain very little signal information and hence these coefficients have small, random value, whereas a few wavelet coefficients have large values that represent significant signal information. This distribution has a heavy tail and very sharp narrow peak comparing with the Gaussian density. Modelling the wavelet coefficients is to find a proper distribution function to describe the character of the wavelet coefficients. In general, two models can well approximate this character of wavelet coefficients, one is the Generalized Gaussian Density (GGD) model, the another is the Gaussian Mixture Model (GMM). The GMM can be well designed to connect with the WD-HMM. In fact, the distribution of the value of each wavelet coefficient in WD-HMM is assumed to satisfy a GMM distribution. The structure of GMM is presented in Figure 8.1. In GMM, we associate each variable W with a set of discrete hidden states $S = 0, 1, ..., M - 1$, which have probability mass function (pmf), $P_S(m)$. Given $S = m$, the probability density function (pdf) of the coefficient W is Gaussian function with mean μ_m and variance σ_m^2. An example of GMM is shown in Figure 8.2. In this model, M = 2, Mean=[-1 1], Cov=[1 0.5], Prior=[0.4 0.6].

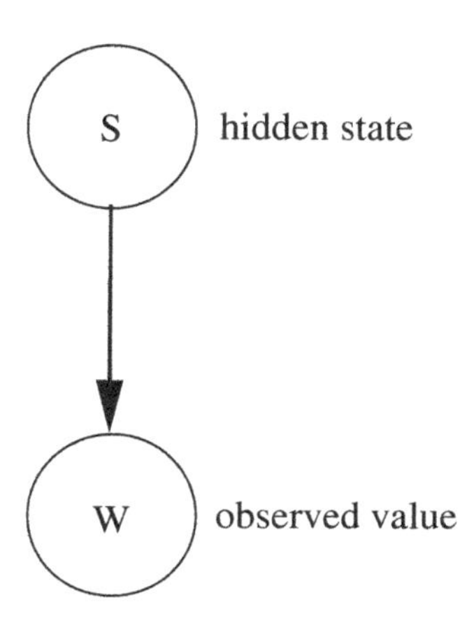

Fig. 8.1 An example of the bilingual English-Chinese document.

Generally, an M-state Gaussian mixture model for a variable W consists of:

(1) A pmf of state variable S with value m, $P_S(m)$, where $m \in \{1, 2, ..., M\}$.
(2) The Gaussian conditional pdfs $f_{W|S}(w|m)$.
(3) Finally, the pdf of ω is given as follows.

$$f_W(w) = \sum_{m=1}^{M} P_S(m) f_{W|S}(w|m) \tag{8.8}$$

where

$$f_{W|S}(w|m) = \frac{1}{2\pi\sigma_m^2} \exp \frac{-(w-\mu_m)^2}{\sqrt{2\sigma_m^2}} =: g(w|\mu_m, \sigma_m^2) \tag{8.9}$$

In signal and image processing using GMM, the value of wavelet coefficient ω is observed, but the value of the sate variable S is not. Therefore S is called hidden state value. Although each wavelet coefficient w is conditionally Gaussian given its state variable $S = s$, the wavelet coefficient has an overall non Gaussian density due to the randomness of S.

By increasing the number of states and allowing non-zero means, we can make the fit arbitrarily close to density function $f_W(w)$ with the desired fidelity. However, the accuracy of two-state, zero-mean GMM can meet our desires in most applications. In addition, compared to the multi-state, non-zero mean GMM, the two-state, zero-mean GMM is simpler, more

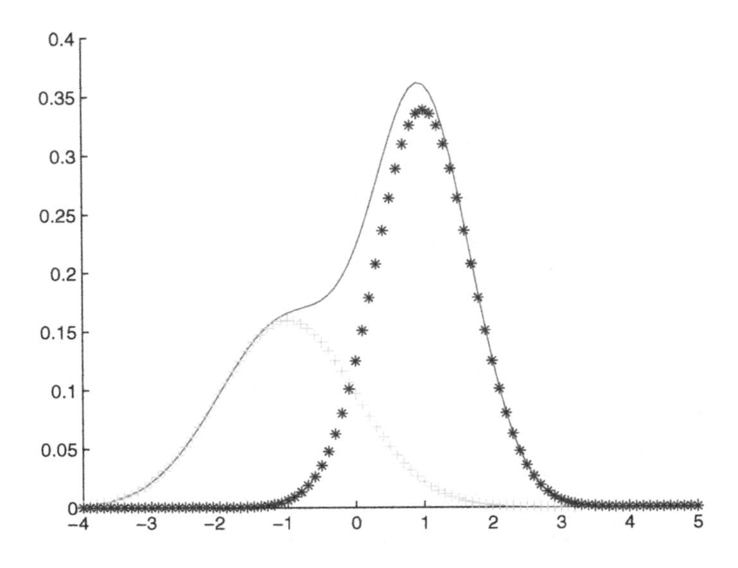

Fig. 8.2 An example of GMM. In this model, M = 2, Mean=[-1 1], Cov=[1 0.5], Prior=[0.4 0.6].

robust, and easier to be implemented, all of which are attractive features for practical applications.

EM algorithm is probably one of the most commonly used approaches to obtain the model parameters of GMM.

The expectation in GMM is given by [Simoncelli, 1997]:

$$\begin{aligned} Q(\Theta,\Theta^i) \quad &= E[\log L(\theta|X,Y)|X,\Theta^i] \\ &= \Sigma_{y\in\Upsilon} \log(L(\Theta|X,y))P(y|X,\Theta^i) \\ &= \Sigma_{y\in\Upsilon} \sum_{i=1}^{N} \log(a_{y_i}P_{y_i}(x_i|\theta_{y_i})) \prod_{j=1}^{N} P(y_j|x_j,\Theta^i) \\ &= \sum_{l=1}^{M}\sum_{i=1}^{N} \log(a_l P_l(x_i|\theta_l)) \sum_{y_1=1}^{M} \cdots \sum_{y_N=1}^{M} \delta_{l,y_i} \prod_{j=1}^{N} P(y_j|x_j,\Theta^i) \end{aligned} \tag{8.10}$$

where Υ is the value range of y. Since $\sum_{i=1}^{M} P(i|x_j, \theta^i) = 1$, we can get

$$\sum_{y_1=1}^{M} \cdots \sum_{y_N=1}^{M} \delta_{l,y_i} \prod_{j=1}^{N} P(y_j|x_j, \Theta^i) = \prod_{j=1,j\neq i}^{N} (\sum_{y_j=1}^{M} P(y_j|x_j, \Theta^i)) = P(l|x_i, \theta_i) \tag{8.11}$$

According to Eq. (8.11), the expectation can be simplified as:

$$\begin{aligned} Q(\Theta, \Theta^i) &= \sum_{l=1}^{M} \sum_{i=1}^{N} \log(a_l P_l(x_i|\theta_l)) P(l|x_i, \theta_i) \\ &= \sum_{l=1}^{M} \sum_{i=1}^{N} \log(a_l) P(l|x_i, \theta_i) + \sum_{l=1}^{M} \sum_{i=1}^{N} \log P_l(x_i|\theta_l)) P(l|x_i, \theta_i) \end{aligned} \tag{8.12}$$

In Gaussian Mixture Model, the model parameter can be set $\theta = (u, \Sigma)$.

The M-step is to maximize expectation after obtaining its expression in E-step. We can maximize the term containing a_l and θ independently since they are not related.

a_l is the answer to the following differential equation.

$$\frac{\partial}{\partial a_l} [\sum_{l=1}^{M} \sum_{i=1}^{N} \log(a_l) p(l|x_i, \Theta_i) + \lambda(\sum_{l} a_l - 1)] = 0 \tag{8.13}$$

we get

$$a_l^{i+1} = \frac{1}{N} \sum_{i=1}^{N} p(l|x_i, \Theta_i) \tag{8.14}$$

Taking the derivative of the right side of Eq. (8.12) with respect to u_l or $\sum_l$ and then setting it equal to zero, we can obtain

$$u_l^{i+1} = \frac{\sum_{i=1}^{N} x_i P(l|x_i, \Theta_i)}{\sum_{i=1}^{N} P(l|x_i, \Theta_i)} \tag{8.15}$$

$$\Sigma_l^{i+1} = \frac{\sum_{i=1}^{N} P(l|x_i, \theta_i)(x_i - u_l^{i+1})(x_i - u_l^{i+1})^T}{\sum_{i=1}^{N} P(l|x_i, \Theta_i)} \tag{8.16}$$

8.2.3 *Hidden Markov Model*

8.2.3.1 *Basic Frame of HMM*

Hidden Markov Model is a bi-random dynamic stochastic process, consisting of two joint stochastic processes, to describe the statistical properties of a random variable. One of this joint stochastic processes is hidden, unobserved markov random process and consequently is called hidden state process. The other is observed random process, of which each observed data is associated with the corresponding hidden state in hidden process.

The Markov random process consists of a finite sequence of states, each of which is associated with a probability distribution. The states are not independent of each other, and their relationship are governed by a set of probabilities named transition probabilities. Given a state, according to the corresponding probability distribution, the outcome (observation) is easily generated. The first order HMM is completely defined by the following model parameters [Warakagoda and Hogskole, 1998].

(1) The number of states M. Taking large value of M, the HMM model can fit a given random process more accurately, while at the same time, the model's complexity also greatly increases. Therefore, we call for a balance between accuracy and complexity.

(2) The number of observation N. That is, the sequence of observation $O = \{o_1, o_2, ..., o_N\}$.

(3) The collection of state transition probabilities $\Lambda = \epsilon_{m,r}^{t,t+1} = P\{S_{t+1} = r|S_t = m\}, \ r, m \in \{1, 2, ..., M\}$. Where S_t denotes the state at time t. $\epsilon_{m,r}$ is an conditional probability mass function, which satisfies

$$\epsilon_{m,r}^{t,t+1} \geq 0 \tag{8.17}$$

and

$$\sum_{r=1}^{M} \epsilon_{m,r}^{t,t+1} = 1 \tag{8.18}$$

In different HMM structures, Λ can be a vector (called transition chain) or matrix (called transition matrix).

(4) A probability density function in each of state f, $f_m(v) = P(o_t = v|S_t = m)$. Naturally, f_m satisfies the following statistical con-

strain:

$$f_m(v) \geq 0, \quad m \in \{1, 2, ..., M\}, v \in \Upsilon \tag{8.19}$$

where Υ is the value range of the observations.
And

$$\sum_{v=1}^{N} f_m(v) = 1 \tag{8.20}$$

Usually, the probability density function can be well characterized by a N-state Gaussian Mixture Model (N is the number of observation).

$$f_m(o_t) = \sum_{i=1}^{N} c_{m,i} g(o_t | \mu_{m,i}, \Sigma_{m,i}) \tag{8.21}$$

$c_{m,i}$ is a probability mass function satisfying

$$a_{m,i} \geq 0, \quad m \in \{1, 2, ..., M\}, i \in \{1, 2, ..., N\} \tag{8.22}$$

and

$$\sum_{i=1}^{N} a_{m,i} = 1 \tag{8.23}$$

In the later discussion of WD-HMM, we will find Eq. (8.21) implies that the different wavelet coefficients satisfy different GMM models. Sometimes, to reduce the WD-HMM's complexity, we assume that all wavelet coefficients satisfy the same GMM. If so, the Eq. (8.21) is simplified to

$$f_m(o_t) = a_m g(o_t | \mu_m, \Sigma_m) \tag{8.24}$$

(5) The initial state distribution, $\pi = \{\pi_i\}$, where

$$\pi_k = P\{S_1 = k\}, \quad k \in \{1, 2, ..., M\} \tag{8.25}$$

In short, a HMM model can be completely characterized by its parameter set

$$\Theta = \{\pi, \mu_{m,i}, \Sigma_{m,i}, a_{m,i}, \Lambda\}_{m=1,2,...,M; i=1,2,...,N} \tag{8.26}$$

In fact, for simplicity, we often use the notation Θ to represent the HMM model determined by it.

8.2.3.2 *Three Basic Problems for HMM*

For a HMM model, There are three canonical problems associated with it [Fan, 2001].

- **Model Training**: Given a sequence of observations

$$O = \{o_1, o_2, ..., o_k\},$$

determining the HMM parameters Θ that maximize $p\{O|\Theta\}$.
- **Model Evaluation**: Given a trained HMM Θ and a sequence of observations $O = \{o_1, o_2, ..., o_k\}$, determining the likelihood function $p\{O|\Theta\}$.
- **State Evaluation**: Given a trained HMM Θ and a sequence of observations $O = \{o_1, o_2, ..., o_k\}$, determining the most likely sequence of hidden states for the observed sequence.

Model evaluation can be viewed as a selection problem. For example, if there are serval existing HMM models, for a given observation sequence, the solution to this model evaluation help us to choose the HMM model that best characterize the given observation sequence.

State evaluation let us to uncover the "best" or "correct" state sequence. It should be clear that for all but the case of degenerated models where there is no "correct" state sequence to be found. Therefore, an optimality criterion ought to be set to solve this problem as best as possible. Generally, there are several reasonable optimality criterias and the choice of criteria is a function of the intended use for the uncovered state sequence. Typical uses might be to learn about the structure of the model, to find optimal state sequences, or to get average statistics of individual states, etc [Liao, 2004].

Model training can be considered to be an optimizing problem. That is, optimizing the model parameters to be best characterize the given observation sequence $O = \{o_1, o_2, ..., o_k\}$. In many pattern recognition applications, the trained model parameters are selected as the features to represent a pattern.

Forward or backward algorithm is the solution to the model evaluation problem, and Viterbi algorithm is the solution to the state evaluation problem. Please refer to [Liao, 2004] for details of these two algorithms. Beaum-Welch algorithm is an efficient solution to the Model Training problem, and we will discuss this algorithm in details in the later part of this chapter.

8.2.3.3 *Important Assumptions for HMM*

The theory of HMM is basically based on three assumptions [Fan, 2001].

- **Markov Assumption**: In the HMM, the next state only relies on the current state. The relationship between these two states can be described by the transition probability function.

$$\epsilon_{m,r} = p\{S_{t+1} = r | S_t = m\} \tag{8.27}$$

This is called the first order HMM. However, in many applications, the first order HMM is not enough to offer a satisfiable description for the states' relationship, and we need to use the high order HMM. In the kth order HMM, the transition probability function is given by:

$$\epsilon_{m_1,m_2,...,m_k,r} = p\{S_{t+1} = r | S_t = m_k, ..., S_{t-k+1} = m_1\} \tag{8.28}$$

That means the next state is dependent on the past k states.

- **Stationary Assumption**: The state transition probability function is time-shift invariant. That is, the transition probability function is independent of the actual time at which the transition takes place.

$$\begin{aligned} \epsilon_{m_1,m_2,...,m_k,r} &= P\{S_{t_1+1} = r | S_{t_1} = m_k, ..., S_{t_1-k+1} = m_1\} \\ &= P\{S_{t_2+1} = r | S_{t_2} = m_k, ..., S_{t_2-k+1} = m_1\} \end{aligned} \tag{8.29}$$

- **Output Independence Assumption**: It is assumed that the current observed value is only dependent on the current state value and independent on the previous observed values. Mathematically, by defining a sequence of observations,

$$O = \{o_1, o_2, ..., o_k\}$$

then in a HMM model Θ, this assumption is given by

$$P\{O | s_1, s_2, ..., s_k, \Theta\} = \prod_{t=1}^{k} P(o_t | s_t, \Theta) \tag{8.30}$$

8.3 Hidden Markov Models in Wavelet Domain

8.3.1 *GMM Model for a Single Wavelet Coefficient*

In [Mallat and Zhong, 1992; Mallat and Hwang, 1992; Orchard and Ramchandran, 1994], researchers found the wavelet coefficients satisfy two properties: clustering and persistence across scale. Clustering means if the value of one wavelet coefficient is large/samll, those wavelet coefficients nearby this coefficient have a large possibility to be also large/small. Persistence across scale means the large/small values of wavelet coefficients tend to propagate across scales.

For a natural image, most wavelet coefficients obtained by the DWT contain very little image information and therefore their magnitudes are small, random variables. On the contrary, a few wavelet coefficients contain most image information and hence have large values. As a result, we can regard each wavelet coefficient as a random variable which has two state values, "large", corresponding to a wavelet coefficient contributing most to the image energy, and "small", corresponding to a wavelet coefficient with little image energy. If we associate each wavelet coefficient's state with a probability density, we will find each wavelet coefficient state value satisfies the heavy-tailed and peaky non-Gaussian distribution that be well characterized by the 2-state GMM. Some researchers' works have proved the 2-state GMM for individual wavelet coefficient is simple but successful [Pesquet et al., 1996]. This 2-state GMM consists of

(1) $P_S(m)_{m\in\{1,2\}}$, the pmf of state value of the wavelet coefficient.
(2) the Gaussian condition pdfs

$$f_{W|S}(w|S=m) := g(w|\mu_m, \Sigma_m)_{m\in\{1,2\}}$$

Since the GMM can accurately approximate the pdf of a single wavelet coefficient, it seems logical to use joint GMM model to characterize the dependence relationship (joint pdf) of the entire wavelet transform.

8.3.2 *Independence Mixture Model*

The independence Mixture (IM) model is the simplest model which treats the wavelet coefficients independent to be the random variables satisfying independent GMMs at different scales. As a result, there is not exist any connection line between the nodes in IM graph. The probabilistic graph for IM model is illustrated in Figure 8.3. IM model generally assumes that the wavelet coefficients at the same scale have the same GMM model parameters $\{P_S(\cdot), \mu, \Sigma\}$. The IM can be completely represented by the following three parameters, i.e. $P_{S_j}(m)$, $\mu_{j,m}$ and $\Sigma_{j,m}$:

(1) $P_{S_j}(m)$: the pmf of state value of the wavelet coefficient node W_j at scale j , $j = 1, ..., J$, $m = 0, 1, ..., M - 1$.
(2) $\mu_{j,m}$ and $\Sigma_{j,m}$: the parameters of the Gaussian condition pdfs $f_{W_j|S}(w|S = m) := g(w|\mu_m, \Sigma_m)$

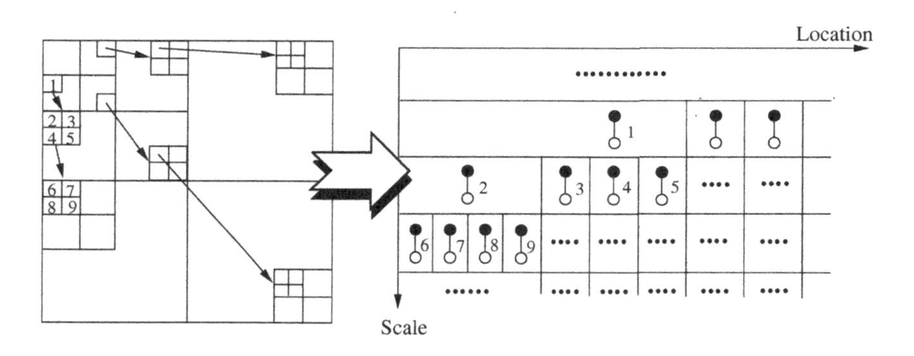

Fig. 8.3 Independence mixture model.

Because the discrete wavelet transform nearly decorrelates a wide variety of signals leading to wavelet coefficients are approximately independent, this model for the wavelet coefficients is intuitively plausible. In some applications, the IM is a substantial improvement over deterministic signal models that do not explicitly take the distribution of signal's wavelet values into account [Pesquet et al., 1996].

8.3.3 *WD-HMM and EM for WD-HMM*

Since a GMM can characterize the pdf of a single wavelet coefficient w, for an image after wavelet transform, another problem arises that how to characterize the joint statistics of the all wavelet coefficients at different subbands, at different scales. M.S.Crouse et al introduced that Markov dependencies between the hidden state variables, and offered a probabilistic graph for modeling the local dependencies between a set of wavelet coefficients [Crouse et al., 1997]. In this probabilistic graph, each random variable is associated with a node, and the dependencies between pairs of variables are represented by links connecting the corresponding nodes. Models of this type are commonly referred to as Hidden Markov Models in wavelet domain (WD-HMM).

Though the discrete wavelet transform can decorrelate an image into uncorrelated wavelet coefficients, it is widely understood that there are considerable amount of high-order dependencies among these wavelet coefficients. Clustering and persistence are two important properties to prove the existence of dependency between wavelet coefficients.

The persistence property of wavelet coefficients suggests an across-scale dependency between a wavelet coefficient ω_i at a coarse resolution and its corresponding coefficients at the next resolution, which are also called the children of ω_i (In Figure 8.5, tree nodes 2, 3, 4, 5 are children nodes of node 1. In other words, node 1 is the parent node of tree nodes 2, 3, 4, 5). The cluster property of wavelet coefficients reveals a strong inter-scale dependency between a wavelet coefficient ω_i and its neighbors within the same scale. The wavelet coefficients in Figure 8.4 clearly display these two properties.

Figure 8.5 is a vivid example to describe the dependency among wavelet coefficients. In Figure 8.5, each white node refers to the observed value of a continuous wavelet coefficient ω_i, each black node represents the mixture state value S_i for ω_i. Connecting the hidden state value nodes horizontally across in the same scale (dashed links) yields the Hidden Markov Chain (HMC) model. Connecting the hidden state value nodes vertically across scale (solid links) yields the Hidden Markov Tree (HMT) model.

Using an M-state Gaussian mixture model for each wavelet coefficient ω_i, the Hidden Makove Tree model can be completely defined by the following parameters.

(1) $P_{S_1}(m)$: the pmf of state value of the root node 1.
(2) $\epsilon_{mr}^{i,p(i)} = P_{S_i|S_{P(i)}}(m|r)$: the *parent* $\rightarrow$ *children* link between hidden states.
(3) μ_{im}, σ_{im}: the mean and variance of Gaussian pdf of wavelet coefficient ω_i given state $S_i = m$.

For simplicity, we usually assume $M = 2$ (that is, the state only can be in the states "small" and "large") because in this case the state value

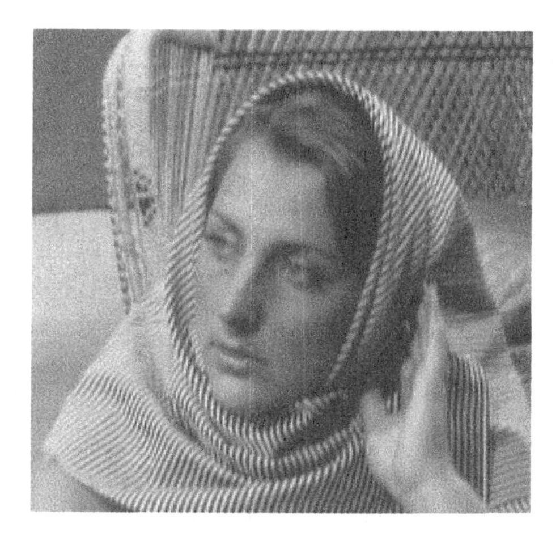

Fig. 8.4 "Woman" image and 3-scale DWT of "Woman" image with db2 wavelet. The gray value is the magnitude of the wavelet coefficients.

has a clear physical meaning. Wavelet coefficients with large value contain significant contributions of signal energy, wavelet coefficients with small value contain little signal energy. And since the wavelet coefficients are generated by the wavelet filters with zero sum, they can be considered to be zero-mean.

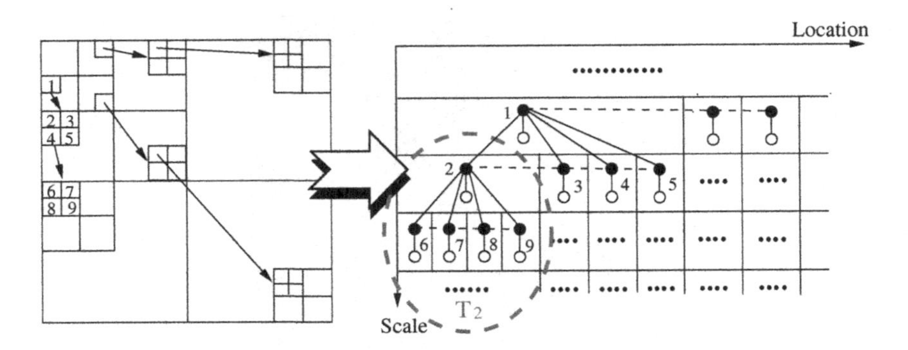

Fig. 8.5 The HMC and HMT models in wavelet domain.

8.4 Writer Identification Using WD-HMM

8.4.1 *The Whole Procedure*

The basic idea of using WD-HMT model to describe the feature of a handwriting image is to establish corresponding WD-HMT models for a handwriting image, and then the model parameter set θ can be regarded as the features of the handwriting. Figure 8.6 shows the whole procedure. By measuring the distance between the parameter set of query handwriting image and reference handwriting sorted in the database, we can obtain the matching results. The most important work is to training the WD-HMT model according to the input handwriting image.

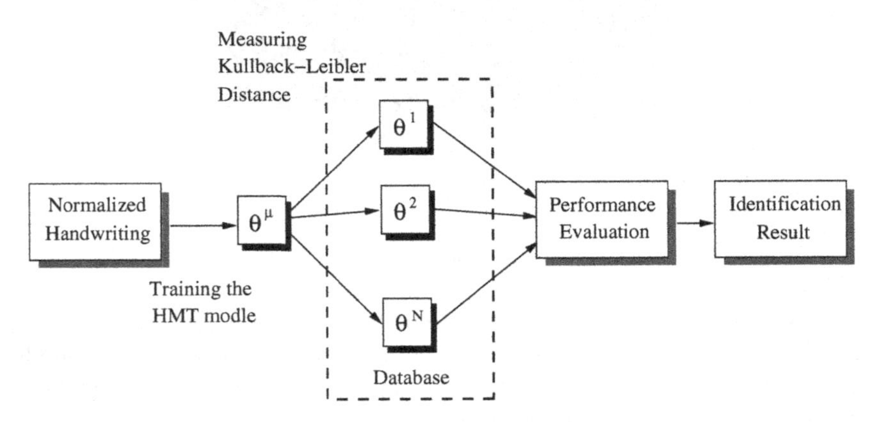

Fig. 8.6 The whole procedure of the algorithm.

8.4.2 *Feature Extraction*

The basic idea of using HMT model to describe the feature of a handwriting image is training a HMT model for the handwriting image, and then the HMT parameter set $\theta = \{P_{S_1}(m), \epsilon_{m,r}^{i,p(i)}, \mu_{im}, \sigma_{im}\}$ can be regarded as the features of the handwriting. The most important work is to train the HMT model according to the input handwriting image.

EM algorithm is a commonly used and efficient approach to reach the estimation of the WD-HMM model parameters. The EM algorithm for HMMs is often known as Baum-Welch algorithm [Crouse et al., 1997]. Its basic steps are given as follows:

(1) **Initialization:** Setting an initial model estimate Θ^0
(2) **E step (upward-downward algorithm):** Estimating $P\{S|W, \Theta^t\}$, the probability for the hidden state variables of the wavelet coefficients.
   (a) **Upward step:** Propagating the hidden state information up along the Hidden Markov Tree.
   (b) **Downward step:** Propagating the hidden state information down along the Hidden Markov Tree.
(3) **M step:** Updating $\Theta^{t+1} = \arg\max_\theta E[\ln f(W, S|\Theta)|W, \Theta^t]$, maximizing the expected likelihood function.
(4) **Iteration step:** Set $t = t + 1$. If not converged, then return to E step; else stop the whole procedure.

Generally, the HMT model is composed by multiple wavelet trees,

$$W = [W^1, ..., W^k, ..., W^K]$$

K is the total number of the wavelet trees. The observed wavelet coefficients in one wavelet tree are denoted as

$$W^k = [W_1^k, ..., W_i^k, ..., W_N^k]$$

and their hidden states are

$$S^k = [S_1^k, ..., S_i^k, ..., S_N^k]$$

where N is the total number of wavelet coefficients in one wavelet tree.

The parameters estimated by the Baum-Welch algorithm are given as follows.

$$P_{S_i}(m) = \frac{1}{K}\sum_{k=1}^{K} P(S_i^k = m|W^K, \Theta^t), \quad m = 1, ..., M. \tag{8.31}$$

$$\epsilon_{n,m}^{i,p(i)} = \frac{1}{KP_{S_i(m)}}\sum_{k=1}^{K} P(S_i^k = n, S_{p(i)}^k = m|W^K, \Theta^t), \quad m = 1, ..., M. \tag{8.32}$$

$$\mu_{i,m} = \frac{1}{KP_{S_i}(m)}\sum_{k=1}^{K} w_i^k P(S_i^k = m|W^K, \Theta^t), \quad m = 1, ..., M. \tag{8.33}$$

$$\sigma_{i,m}^2 = \frac{1}{KP_{S_i}(m)}\sum_{k=1}^{K} (w_i^k - \mu_{i,m})^2 P(S_i^k = m|W^K, \Theta^t), \quad m = 1, ..., M. \tag{8.34}$$

If we define $K = 1$, then Eqs. (8.31), (8.32), (8.33), (8.34) consist of the M-step of a HMT with single wavelet tree.

To prevent "overfitting", sometimes wavelet coefficients and their states within a certain subband are tied with a common set of density parameters. In this scenario, the M-step Eqs. (8.31), (8.32), (8.33), (8.34) can become simpler. Please refer to [Crouse et al., 1997] for details.

8.4.3 *Similarity Measurement*

After obtaining the parameter set of HMT model θ, we can believe that a handwriting image is completely represented by its corresponding parameter set of HMT model. In other words, the similarity between two handwriting images can be considered to be the similarity between the two corresponding HMT parameter sets.

Inspired by other researchers' works on statistical framework, we adopt the Kullback-Leibler distance (KLD) to measure the similarity between two HMT models.

KLD between two pmfs is given by [Cover and Thomas, 1991],

$$D(\omega||\bar{\omega}) = \sum_i \omega_i \log \frac{\omega_i}{\bar{\omega_i}} \tag{8.35}$$

KLD between two pdfs is defined as follows:

$$D(P(X;\theta_i)||P(X;\theta_j)) = \int P(x;\theta_i) log \frac{P(x;\theta_i)}{P(x;\theta_j)} \quad (8.36)$$

In HMT model, the probability function is very complex which can be viewed as a mixture of large number of pdfs, and we do not have a simple, direct expression for the KLD. Monte-Carlo method is a traditional approximation of the KLD, while it needs high computational cost [Juang and Rabiner, 1985]. To save the computational cost, we decide to compute the upper bound for the KLD instead of using the Monte-Carlo method [Do, 2003].

Lemma 8.1 *[Cover and Thomas, 1991]. The KLD between two mixture densities $\sum_i \omega_i f_i$ and $\sum_i \bar{\omega}_i \bar{f}_i$ is upper bounded by*

$$D(\sum_i \omega_i f_i || \sum_i \bar{\omega}_i \bar{f}_i) \leq D(\omega||\bar{\omega}) + \sum_i \omega_i D(f_i||\bar{f}_i) \quad (8.37)$$

with equality if and only if $(\omega_i f_i)/(\bar{\omega}_i \bar{f}_i) = const.$

Proof

$$\begin{aligned} D(\sum_i \omega_i f_i || \sum_i \bar{\omega}_i \bar{f}_i) &= \int (\sum_i \omega_i f_i) \log \frac{\sum_i \omega_i f_i}{\sum_i \bar{\omega}_i \bar{f}_i} \\ &\leq \sum_i \omega_i f_i \log \frac{\omega_i f_i}{\bar{\omega}_i \bar{f}_i} \\ &= \sum_i \omega_i \log \frac{\omega_i}{\bar{\omega}_i} + \sum_i \omega_i \int f_i \log \frac{f_i}{\bar{f}_i} \\ &= D(\omega||\bar{\omega}) + \sum_i \omega_i D(f_i||\bar{f}_i) \quad (8.38) \end{aligned}$$

Proof ends. ■

For a tree node i, its conditional probability density of its observation value given its state is m, which is defined as:

$$P(O_i = o|S_i = m) = b^i_m(o) \quad (8.39)$$

Generally, $b^i_m(o)$ is a Gaussian function.

Then, for a tree node i, define β^i_m is the conditional probability density of the observation value of the subtree of the node i given its state is m, $p(i)$ is the parent of the tree node i, $C(i)$ is the set of children node of tree node i, T_i is the subtree of all nodes with root at i (in Figure 8.5, the tree inside the circle is T_2), and O_{T_i} is the wavelet values (observed value) of this subtree. Here, for simplicity, we only consider the case that the number of mixture state $M = 2$. Based on the Hidden Markov Chain rule, we can get

$$\beta_m^i(O_{T_i}) = b_m^i(o_i) \prod_{c \in C(i)} \sum_{n=1}^{2} \epsilon_{n,m}^{c,i} \beta_n^c(O_{T_c}) \tag{8.40}$$

For a leaf node i without any children node,

$$\beta_m^i(O_{T_i}) = b_m^i(o_i) \tag{8.41}$$

For the root node 1, the probability of the whole observation tree is defined as

$$P(O_{T_1} = o_{T_1}|\theta) = \sum_{m=1}^{2} P_{S_1}(m)\beta_m^1(o_{T_1}) \tag{8.42}$$

Then, based on t lemma 8.1, KLD between two HMTs is upper bounded as

$$D(P(O_{T_1} = o_{T_1}|\theta)|P(\bar{O}_{T_1} = \bar{o}_{T_1}|\bar{\theta})) \leq D(P_{S_1}||\bar{P}_{S_1}) + \sum_{m=1}^{2} P_{S_1}(m)D(\beta_m^1||\bar{\beta}_m^1) \tag{8.43}$$

According to the chain rule of KLD for independent data set and lemma 8.1, $D(\beta_m^1||\bar{\beta}_m^1)$ can be calculated in the following way.

$$\begin{aligned} D(\beta_m^1||\bar{\beta}_m^1) &= D(b_m^1||\bar{b}_m^1) + \sum_{c \in C(1)} D(\sum_{n=1}^{2} \epsilon_{n,m}^{c,1}\beta_n^c || \sum_{n=1}^{2} \bar{\epsilon}_{n,m}^{c,1}\bar{\beta}_n^c) \\ &\leq D(b_m^1||\bar{b}_m^1) + \sum_{c \in C(1)} [D(\epsilon_m^1||\bar{\epsilon}_m^1) + \sum_{n=1}^{2} \epsilon_{n,m}^{c,1} D(\beta_n^c||\bar{\beta}_n^c)] \end{aligned} \tag{8.44}$$

This induction can be iteratively operated till the leaf node. For leaf node i of the HMT, the KLD can be written as

$$D(\beta_m^i||\bar{\beta}_m^i) = D(b_m^i||\bar{b}_m^i) \tag{8.45}$$

$D(b^i_m||\bar{b}^i_m)$ is a KLD between two Gaussian pdfs, and $D(\epsilon^1_m||\bar{\epsilon}^1_m)$ is a KLD between two pmfs. The following expression for KLD between two d-dimensional Gaussian pdfs is used [Singer and Warmuth, 1998].

$$D(N(.;\mu,\Sigma)||D(N(.;\bar{\mu},\bar{\Sigma}) = \frac{1}{2}[\log\frac{|\bar{\Sigma}|}{\Sigma} - d + trace(\bar{\Sigma}^{-1}\Sigma) + (\mu-\bar{\mu})^T\bar{\Sigma}^{-1}(\mu-\bar{\mu})] \tag{8.46}$$

As we mentioned above, the pdf of wavelet coefficients is a Gaussian with zero-mean, thus, Eq. (8.46) can be simplified as

$$D(N(.;0,\Sigma)||D(N(.;0,\bar{\Sigma}) = \frac{1}{2}[\log\frac{|\bar{\Sigma}|}{\Sigma} - d + trace(\bar{\Sigma}^{-1}\Sigma)] \tag{8.47}$$

In sum, the KLD of two Hidden Markvo Trees can be expressed by Eq. (8.43), and then $D(\beta^1_m||\bar{\beta}^1_m)$ in Eq. (8.43) can be calculated by using Eq. (8.44) iteratively till leaf node. The KLD between two leaf nodes is acquired by Eq. (8.45).

8.4.4 *Performance Evaluation*

The evaluation criteria used in the method is similar to which used in the wavelet-based GGD method. That is, identification result is to find out the top N handwriting images which are most similar to the query handwriting image and does not consider the problem of setting the threshold value.

8.5 Experiments

In our experiments, we make a comparison between the HMT method with the 2-D Gabor method not only on the identification accuracy, but also on the calculational time.

In the HMT method, we decompose the handwriting image via the discrete wavelet transform (DWT), and the used wavelets are Daubechies orthogonal wavelets.

Figure 8.7 shows a graph illustrating this comparison of HMT and Gabor on retrieval performance. Obviously, the identification rate changes at varied number of top matches considered.

An example of writer identification using our new approach is shown in Figure 8.8. The training handwriting written by the same writer with the query handwriting is ranked at the top 1, this is the very successful

Table 8.1 Writer Identification Rate 1 (%).

Number of Top Matches	HMT	Gabor, f = 16	Gabor, f = 16, 32	Gabor, f = 16, 32, 64, 128
1	48	18	22	48
2	60	30	42	60
3	64	42	48	66
4	70	44	52	68
5	72	48	58	70
6	74	56	68	76
7	80	66	72	80
8	80	74	78	82
9	86	78	80	86
10	88	84	86	90

identification. The KLD values between the query handwriting and top matches are much less than those between the query handwriting and undermost matches. In this example, the KLD values of the top 9 matches are $\{1.75\times10^4, 3.87\times10^4, 4.70\times10^4, 6.12\times10^4, 6.34\times10^4\ 6.64\times10^4, 7.11\times10^4, 1.00\times10^4, 1.05\times10^4\}$, the KLD values of the undermost 5 matches are $\{1.99\times10^6, 2.00\times10^6, 2.08\times10^6, 2.13\times10^6, 2.19\times10^6\}$. Our experiments also record the calculational time of HMT and Gabor method. The record of average elapsed time is given in Table 8.2 and Figure 8.9.

Table 8.2 Average Elapsed Time 2 for Writer Identification (Second).

Number of Top Matches	HMT	Gabor, f = 16	Gabor, f = 16, 32	Gabor, f = 16, 32, 64, 128
Elapsed time	50.65	52.89	106.79	213.26

From Table 8.1, it is clear that in Gabor method, the more frequencies are combined, the higher identification rate is achieved, unfortunately at the same time, the elapsed time also increases greatly. The identification rate of the Gabor model combing four frequencies $f = 16, 32, 64, 128$ is nearly same to that of HMT method, while its cost time is 4 times of that used in HMT model. The elapsed time of the Gabor model with $f = 16$ is equal to that of HMT, however its identification rate is much lower than

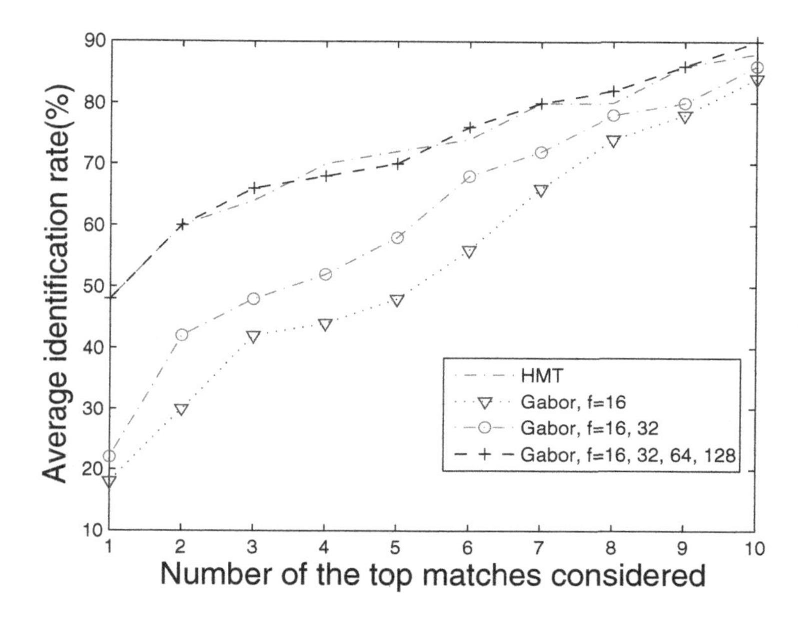

Fig. 8.7 Identification rate according to the number of top matches considered.

that of HMT. Comprehensively, the HMT method exceeds the Gabor model on both identification performance and the computational efficiency. The improvement of computational efficiency of our method is very valuable in the practical identification applications which may involve in thousand of handwritings.

To increase the writing samples for one writer in order to enhance the persuasion of the performance evaluation, we divide one normalized handwriting image of 512×512 pixels into 4 non-overlapped subimages of 256×256 pixels. In chapter 3, we have described the process of image division, here we ignore the details about that. The evaluation criterion keeps the same as that used in the previous chapter. The identification rates of HMT and Gabor model with different frequency combinations are offered in Table 8.3.

The average elapsed time of HMT and Gabor model in this experiments is offered in Table 8.4 and Figure 8.5. Combining the data in Table 8.3 and Table 8.4, HMT model still exceeds the Gabor model in this experiment.

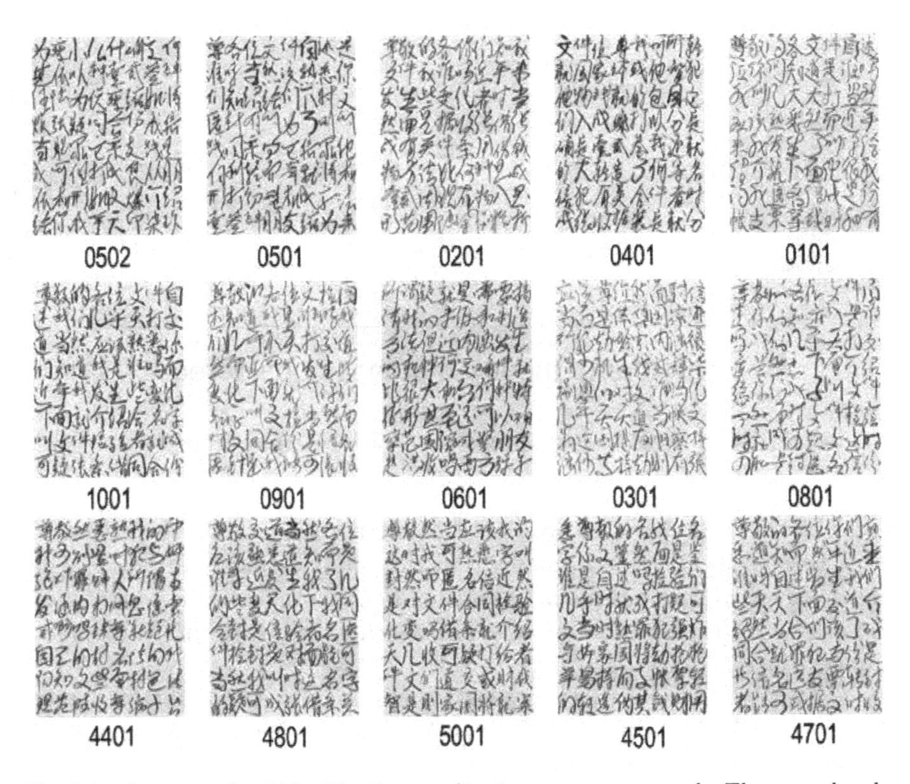

Fig. 8.8 One example of identification result using our new approach. The query handwriting "0502" is on the top left corner. The upper two rows (except for the query handwriting "0502") are the top 9 matches, the last row is the undermost 5 matches. The handwriting matches are ranked from left to right, from top to bottom.

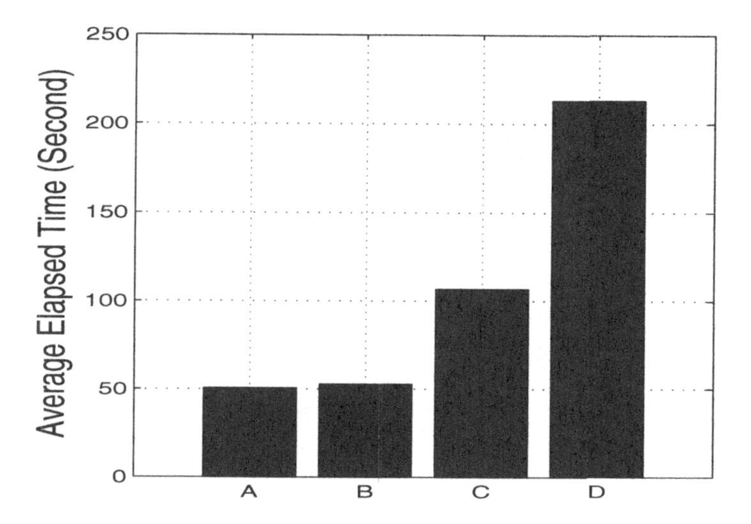

Fig. 8.9 Average elapsed time of different methods. In X-axis, A refers to HMT method, B refers to Gabor with $f = 16$, C refers to Gabor with $f = 16, 32$, D refers to Gabor with $f = 16, 32, 64, 128$.

Table 8.3 Writer Identification Rate 2 (%).

Number of Top Matches	HMT	Gabor, f = 16	Gabor, f = 16, 32	Gabor, f = 16, 32, 64, 128
1	58.50	19.75	29.75	55.25
2	51.25	18.86	26.75	52.50
3	48.33	19.16	25.08	48.41
4	43.56	17.68	23.75	42.93
5	39.40	17.25	21.80	40.25
6	36.29	15.87	21.12	38.00
7	34.96	15.53	20.14	36.92
11	40.89	19.60	25.82	40.21
21	51.96	31.35	40.57	50.46
31	59.64	40.82	51.17	64.25
41	66.14	49.96	59.07	70.50
51	70.89	57.71	64.89	74.00

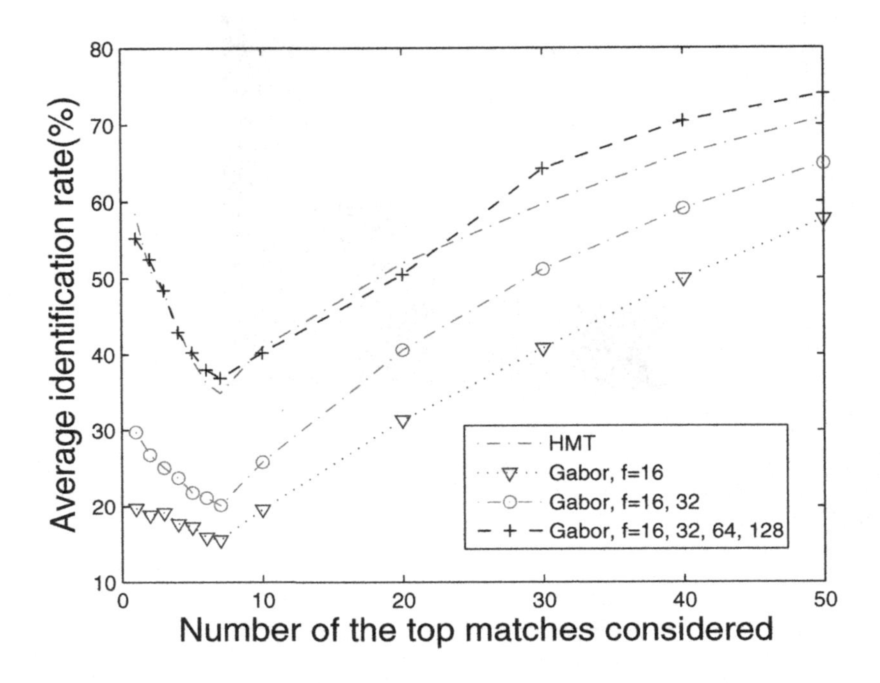

Fig. 8.10 Identification rate according to the number of top matches considered.

Table 8.4 Average Elapsed Time 2 for Writer Identification (Second).

Number of Top Matches	HMT	Gabor, f = 16	Gabor, f = 16, 32	Gabor, f = 16, 32, 64, 128
Elapsed time	10.67	4.53	8.96	17.37

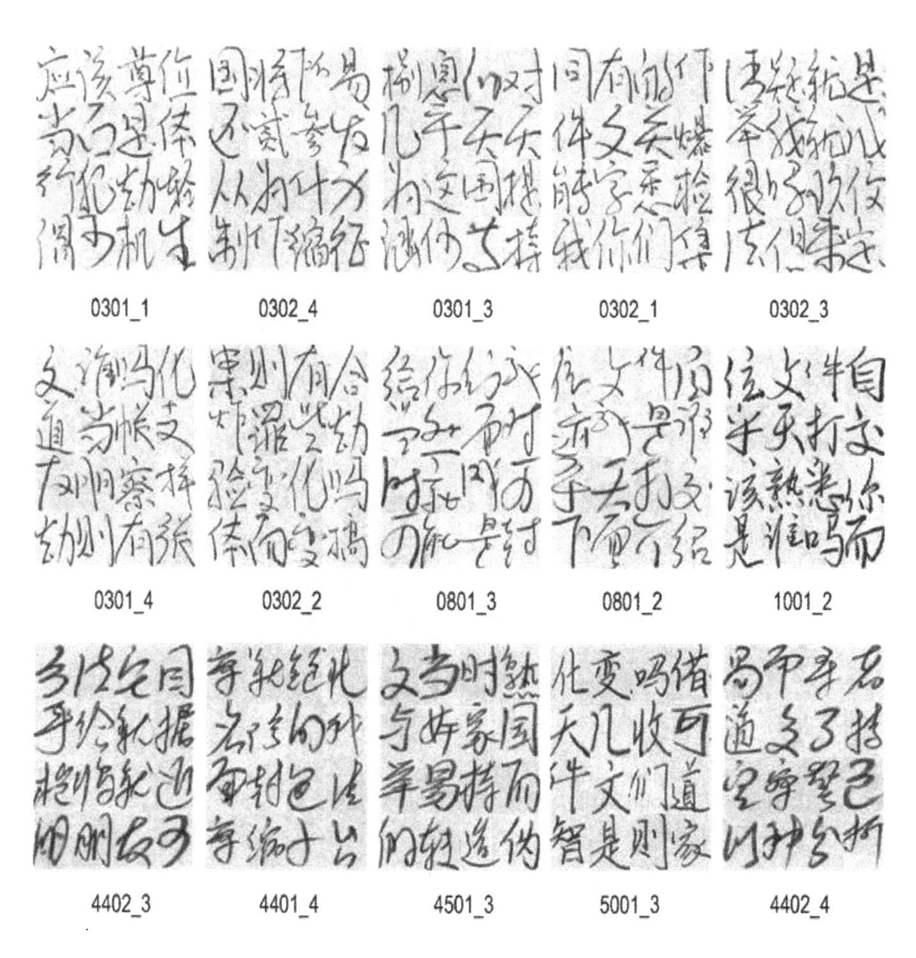

Fig. 8.11 An example of identification result using our new approach. The query handwriting "0301_1" is on the top left corner. The upper two rows are the top 9 matches, the last row is the undermost 5 matches. The KLD values of the top 9 matches are $\{6.13 \times 10^3, 7.22 \times 10^3, 8.07 \times 10^3, 8.44 \times 10^3, 8.59 \times 10^3\ 8.62 \times 10^3, 9.08 \times 10^3, 1.15 \times 10^4, 1.18 \times 10^4\}$, the KLD values of the undermost 5 matches are $\{7.15 \times 10^5, 7.22 \times 10^5, 7.32 \times 10^5, 7.53 \times 10^5, 7.97 \times 10^5\}$.

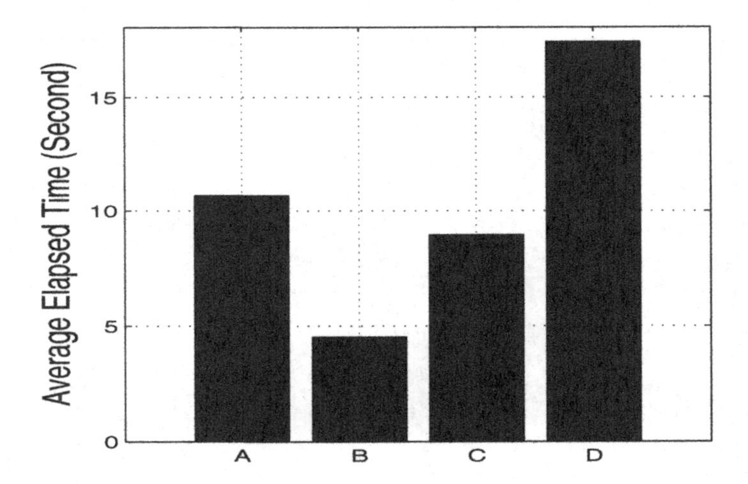

Fig. 8.12 Average elapsed time of different methods. In X-axis, A refers to HMT method, B refers to Gabor with $f = 16$, C refers to Gabor with $f = 16, 32$, D refers to Gabor with $f = 16, 32, 64, 128$.

Bibliography

Abele, L., Wahl, F., and Scheri, W. (1981). Procedures for an automatic segmentation of text graphic and halftone regions in document. In Proc. of 2nd Scandinavian Conference on Image Analysis, pages 177–182.

Abu-Mostafa, Y. S. and Psaltis, D. (1984a). Image normalization by complex moments. *IEEE Transactions on Pattern Analysis and Machine Intelligence*, 7(1):46–55.

Abu-Mostafa, Y. S. and Psaltis, D. (1984b). Recognitive aspects of moment invariants. *IEEE Transactions on Pattern Analysis and Machine Intelligence*, 6(6):698–706.

Ahmed, P. and Suen, C. Y. (1982). Segmentation of unconstrained handwritten postal zipcodes. In *Proc. of 6th International Conference on Pattern Recognition*, pages 545–547.

Ahmed, P. and Suen, C. Y. (1987). Computer recognition of totally unconstrained handwritten zipcodes. *International Journal of Pattern Recognition and Artificial Intelligence*, 1(1):1–9.

Akiyama, T. and Hagita, N. (1990). Automated entry system for printed documents. *Pattern Recognition*, 23(11):1141–1154.

arbter, K., Snyder, W. E., Burkhardt, H., and Hirzinger, G. (1990). Application of affine-invariant fourier descriptors to recognition of 3-d objects. *IEEE Transactions on Pattern Analysis and Machine Intelligence*, 12(6):640–647.

Ascher, R. N., Koppelman, G. M., Miller, M. J., Nagy, G., and Jr., G. L. S. (1971). An interactive system for reading unformatted printed text. *IEEE Transactions on Computers*, 20(12):1527–1543.

Ayache, N. and Faugeras, O. D. (1986). Hyper: a new approach for the recognition and positioning of two-dimensional objects. *IEEE Transactions on Pattern Analysis and Machine Intelligence*, 8(1):44–54.

Bagdanov, A. and Worring, M. (2002). Granulometric analysis of document images. In *Proc. of 16th International Conference on Pattern Recognition*, pages 468–471.

Baldoni, M., Baroglio, C., Cavagnino, D., and Saitta, L. (1998). *Towards automatic fractal feature extraction for image recognition*, pages 356–373. Kluwer Academic Publishers, in huan liu and hiroshi motoda, editors, feature extraction, construction and selection: a data mining perspective edition.

Barnsley, M. F. and Sloan, A. D. (1988). A better way to compress images. *Byte*, (January):215–223.

Bartneck, N. (1988). Knowledge based address block finding using hybrid knowledge representation schemes. In *Proc. of 3rd USPS Advanced Technology Conference*, pages 249–263.

Barton, C. C. and Larsen, E. (1985). Fractal geometry of two-dimensional fracture networks at yucca mountain, southwestern nevada. In *Proc. of the International Sysmposium on Fundamentals of Rock Joints*, pages 77–84, Bjorkliden, Sweden.

Bergman, A., Bracha, E., Mulgaonkar, P. G., and Shaham, T. (1988). Advanced research in address block location. In *Proc. of 3rd USPS Advanced Technology Conference*, pages 218–232.

Bhatnagar, G. and Raman, B. (2010). Distributed multiresolution discrete fourier transform and its application to watermarking. *International Journal of Wavelets, Multiresolution and Information Processing*, 8(2):225–241.

Bixler, J. P. (1988). Tracking text in mixed-mode document. In *Proc. of ACM Conference on Document Processing Systems*, pages 177–185.

Bourgeois, F. L., Emptoz, H., Trinh, E., and Duong, J. (2001). Networking digital document image. In *Proc. of The Sixth International Conference on Document Analysis and Recognition (ICDAR'01)*, pages 379–383.

Bunke, H., Wang, P. S. P., and Baird, H. S. (1994). *Document Image Analysis.* World Scientific Publishing Co. Pte, Ltd., Singapor.

Casey, R. G., Ferguson, D. R., Mohiuddin, K. M., and Walach, E. (1992). An intelligent forms processing system. *Machine Vision and Applications*, 5(3):143–155.

Cass, T. A. (1988). A robust parallel implementation of 2-d model-based recognition. In *Proc. of Computer Vision Pattern Recognition Conference*, pages 879–884, Ann Arbor, MI.

Cesar, M. and Shinghal, R. (1990). An algorithm for segmentation handwritten postal codes. *Man Machine Studiess*, 33:63–80.

Ciardiello, G., Degrandi, M. T., Poccotelli, M. P., Scafuro, G., and Spada, M. R. (1988). An experimental system for office document handling and text recognition. In *Proc. of 9th International Conference on Pattern Recognition*, pages 739–743.

Combettes, P. L. and Pesquet, J. C. (2004). Wavelet-constrained image restoration. *International Journal of Wavelets, Multiresolution and Information Processing*, 2(4):371–389.

Cover, T. M. and Thomas, J. A. (1991). *Elements of Information Theory.* Wiley Inter-Science, New York.

Crouse, M., Nowak, R., and Baraniuk, R. (1997). Wavelet-based signal processing using hidden markov model. *IEEE Transactions on Signal Processing*, Special Issue on Wavelets and Filter banks:886–902.

Daubechies, I. (1992). *Ten Lectures on Wavelets.* Capital City Press, Montpelier, Vemont.

Daugman, J. (2003). Demodulation by complex-valued wavelets for stochastic pattern recognition. *International Journal of Wavelets, Multiresolution and Information Processing*, 1(1):1–317.

Davis, P. J. (1977). Plane regions determined by complex moments. *Journal of Approximation Theory*, 19:148–153.

Davis, R. H. and Lyall, J. (1986). Recognition of handwritten characters - a review. *Image and Vision Computing*, 4(4):208–218.

de Wouwer, G. V., Scheunders, P., and Dyck, D. V. (1999). Statistical texture characterization from discrete wavelet representations. *IEEE Transactions on Image Processing*, 8(4):592–598.

de Wouwer, G. V., Scheunders, P., Livens, S., and Dyck, D. V. (1989). Wavelet correlation signatures for color texture characterization. *Pattern Recognition*, 32(3):443–451.

Dejey and Rajesh, R. S. (2010). An improved wavelet domain digital watermarking for image protection. *International Journal of Wavelets, Multiresolution and Information Processing*, 8(1):19–31.

Demjanenko, V., Shin, Y. C., Sridhar, R., Palumbo, P., and Srihari, S. (1990). Real-time connected component analysis for address block location. In *Proc. of 4th USPS Advanced Technology Conference*, pages 1059–1071.

Dengel, A. (1990). Document image analysis - expectation-driven text recognition. In *Proc. of International Conference on Syntactic and Structural Pattern Recognition (SSPR'90)*, pages 78–87.

Dengel, A. and Barth, G. (1988). Document description and analysis by cuts. In *RIAO Conference*, MIT, USA.

Ding, J., Lam, L., and Suen, C. Y. (1996). Classification of oriental and European scripts by using characteristic feature. In *Proc. of International Conference on Document Analysis*, pages 1023–1027.

Dinstein, I., Landau, G. M., and Guy, G. (1991). Parallel (pram erew) algorithms for contour-based 2-d shape recognition. *Pattern Recognition*, 24(10):929–942.

Do, M. N. (2003). Fast approximation of kullback-leibler distance for dependence trees and hidden markov models. *IEEE Signal Processing Letters*, 10:115–118.

Doster, W. (1984). Different states of a document's content on its way from the gutenbergian world to the electronic world. In *Proc. of 7th International Conference on Pattern Recognition*, pages 872–874.

Downton, A. C. and Leedham, C. G. (1990). Preprocessing and presorting of envelope images for automatic sorting using ocr. *Pattern Recognition*, 23(3/4):347–362.

Dudani, S., Breeding, K., and McGee, R. (1972). Aircraft identification by moment invariants. *IEEE Transactions on Computers*, 26:39–45.

Ehler, M. and Koch, K. (2010). The construction of multiwavelet bi-frames and applications to variational image denoising. *International Journal of Wavelets, Multiresolution and Information Processing*, 8(2):431–455.

El-Khamy, S. E., Hadhoud, M. M., Dessouky, M. I., Salam, B. M., and El-Samie, F. E. A. (2006). Wavelet fusion: A tool to break the limits on lmmse image super-resolution. *International Journal of Wavelets, Multiresolution and Information Processing*, 4(1):105–118.

Esposito, F., Malerba, D., Semeraro, G., Annese, E., and Scafuro, G. (1990). An experimental page layout recognition system for office document automatic classification: an integrated approach for inductive generalization. In *Proc. of 10th International Conference on Pattern Recognition*, pages 557–562.

Falconer, K. L. (1985). *The Geometry of Fractal Sets*. Cambridge University Press.

Falconer, K. L. (1990). *Fractal Geometry: Mathematical Foundation and Applications*. Wiley, New York.

Fan, G. (2001). *Wavelet Domain Statistical Image Modeling and Processsing*. PhD Thesis, Institutt for Teleteknikk.

Fenrich, R. (1991). Segmentation of automatically located handwritten words. In *Proc. of 3rd International Workshop on Frontiers in Handwriting Recognition*, pages 33–44, Chateau de Bonas, France.

Fisher, J. L., Hinds, S. C., and D'Amato, D. P. (1990). A rule-based system for document image segmentation. In *Proc. of 10th International Conference on Pattern Recognition*, pages 567–572.

Fletcher, L. A. and Kasturi, R. (1988). A robust algorithm for text string separation from mixed text/graphics images. *IEEE Transactions on Pattern Analysis and Machine Intelligence*, 10(6):910–918.

Fujisawa, H. and Nakano, Y. (1990). A top-down approach for the analysis of document images. In *Proc. of International Workshops on Structural and Syntactic Pattern Recognition (SSPR'90)*, pages 113–122.

Fukumi, M., Omatu, S., Takeda, F., and Kosak, T. (1992). Rotation-invariant neural pattern recognition system with application to coin recognition. *IEEE Transactions on Neural Networks*, 3(2):272–279.

Giannakis, G. B. and Tsatsanis, M. K. (1990). Signal detection and classification using matched filtering and higher order statistics. *IEEE Transactions on Acoustics, Speech*, 38:1284–1296.

Gilbarg, D. and Trudinger, N. S. (1977). *Elliptic Partial Differential Equations of Second Order*. Springer, Berlin ; New York.

Govindan, V. K. and Shivaprasad, A. P. (1990). Character recognition - a review. *Pattern Recognition*, 23(7):671–683.

Granlund, G. H. (1972). Fourier preprocessing for hand print character recognition. *IEEE Transactions on Computers*, 21:195–201.

Haley, G. M. and Manjunath, B. S. (1999). Rotation-invariant texture classification using a gomplete space-frequency mode. *IEEE Transactions on Image Processing*, 8(2):255–269.

He, D. C. and Wang, L. (1991). Texture features based on texture spectrum. *Pattern Recognition*, 24(5):391–399.

He, D. C. and Wang, L. (1992). Detecting texture edges from image. *Pattern Recognition*, 25(6):595–600.

Higashino, J., Fujisawa, H., Nakano, Y., and Ejiri, M. (1986). A knowledge-based segmentation method for document understanding. In *Proc. of 8th International Conference on Pattern Recognition*, pages 745–748.

Hilderbrandt, T. H. and Liu, W. (1993). Optical recognition of handwritten chinese characters: advances since 1980. *Pattern Recognition*, 26(2):205–225.

Hinds, S. C., Fisher, J. L., and D'Amato, D. P. (1990). A document skew detection method using run-length encoding and the hough transform. In *Proc. of 10th International Conference on Pattern Recognition*, pages 464–468.

Hochberg, J., Kelly, P., Thomas, T., and *et al* (1997). Automatic script identification from document images using cluster-based templates. *IEEE Transactions on Pattern Analysis and Machine Intelligence*, 19(2):176–181.

Hose, M. and Hoshino, Y. (1985). Segmentation method of document images by two-dimensional fourier transformation. *System and Computers in Japan*, 16(3):38–47.

Hu, J. Y. and Bagga, A. (2003). Identifying story and preview images in news web pages. In *Proc. of 7th International Conference on Document Analysis and Recognition*, pages 640–644.

Hu, M. K. (1961). Visual pattern recognition by moment invariants. In *Proc. of the IRE*, pages Vol–49, 1428.

Hu, M. K. (1962). Visual pattern recognition by moment invariants. *IRE Transactions on Information Theory*, 8(2):179—187.

ICDAR'09 (2009). In *Proc. of 10th International Conference on Document Analysis and Recognition.*

ICDAR'91 (1991). In *Proc. of First International Conference on Document Analysis and Recognition.*

Impedovo, S., Ottaviano, L., and Occhinegro, S. (1991). Optical character recognition - a survey. *International Journal of Pattern Recognition and Artificial Intelligence*, 5(1):1–24.

Inagaki, K., Kato, T., hiroshima, T., and Sakai, T. (1984). Macsym: A hierarchical parallel image processing system for event-driven pattern understanding of documents. *Pattern Recognition*, 17(1):85–108.

Ingold, R. and Armangil, D. (1991). A top-down document analysis method for logical structure recognition. In *Proc. of First International Conference on Document Analysis and Recognition*, pages 41–49, Saint-Malo, France.

Ishitani, Y. (1993). Document skew detection based on local region complexity. In *Proc. of Second International Conference on Document Analysis and Recognition*, pages 49–52, Tsukuba Science City, Japan.

Ishitani, Y. (2003). Document transformation system from papers to xml data based on pivot xml document method. In *Proc. of 7th International Conference on Document Analysis and Recognition*, pages 250–255.

ISO (1989). *8613: Information Processing-Text and Office Systems-Office, Document Architecture (ODA) and Interchange Format.* International Organization for Standardization.

Iwaki, O., Kida, H., and Arakawa, H. (1985). A character/graphic segmentation method using neighbourhood line density. *Transactions of the Institute of Electronics and Communication Engineers of Japan*, J68D(4):821–828.

Iwaki, O., Kida, H., and Arakawa, H. (1987). A segmentation method based on office document hierarchical structure. In *Proc. of IEEE International Conference on Systems, Man and Cybernetics*, pages 759–763, Alexandria, VA, USA.

Jacob, T., Rao, K. R., and Kim, D. N. (2010). Image mirroring and rotation in the wavelet domain. *International Journal of Wavelets, Multiresolution and Information Processing*, 8(1):61–69.

Jain, A. K. and Bhattacharjee, S. K. (1992a). Address block location on envelopes using gabor filters: supervised method. In *Proc. of 11th International Conference on Pattern Recognition*, pages 264–266.

Jain, A. K. and Bhattacharjee, S. K. (1992b). Text segmentation using gabor filters for automatic document processing. *Machine Vision and Applications*, 5(3):169–184.

Jain, A. K. and Namboodiri, A. M. (2003). Indexing and retrieval of on-line handwritten documents. In *Proc. of 7th International Conference on Document Analysis and Recognition*, pages 655–659.

Jain, P. and Merchant, S. N. (2004). Wavelet-based multiresolution histogram for fast image retrieval. *International Journal of Wavelets, Multiresolution and Information Processing*, 2(1):59–73.

Johnston, E. G. (1974). Short note: printed text discrimination. *Computer Graphics and Image Processing*, 3(1):83–89.

Jordan, M. and Jacobs, R. (1996). Convergence results for the em appraoch to mixtures of experts architecutres. *Neural Networks*, 8:1409–1431.

Juang, B. H. and Rabiner, L. R. (1985). A probabilistic distance measure for hidden markov models. *AT&T Tech. Journal*, pages 391–408.

Kanai, J., Krishnamoorthy, M. S., and Spencer, T. (1986). Algorithms for manipulating nested block represented images. In *Advance Printing of Paper Summaries, SPSE's 26th Fall Symposium*, pages 190–193, Arlington, Virginia.

Karatzas, D. and Antonacopoulos, A. (2003). Two approaches for text segmentation in web images. In *Proc. of 7th International Conference on Document Analysis and Recognition*, pages 131–136.

Khotanzad, A. and Hong, Y. H. (1990). Invariant image recognition by zernike moments. *IEEE Transactions on Pattern Analysis and Machine Intelligence*, 11(4):489–497.

Kimura, F. and Shridhar, M. (1991). Recognition of connected numerals. In *Proc. of the First International Conference on Document Analysis and Recognition*, pages 731–739, Saint-Malo, France.

Kouzani, A. Z. and Ong, S. H. (2003). Lighting-effects classification in facial images using wavelet packets transform. *International Journal of Wavelets, Multiresolution and Information Processing*, 1(2):199–215.

Kreich, J., Luhn, A., and Maderlechner, G. (1988). Knowledge based interpretation of scanned business letters. In *Proc. of IAPR Workshop on Computer Vision*, pages 417–420.

Krishnamoorthy, M., Nagy, G., Seth, S., and Viswanathan, M. (1993). Syntactic segmentation and labeling of digitized pages from technical journal. *IEEE Transactions on Pattern Analysis and Machine Intelligence*, 15(7):737–747.

Ksantini, R., Ziou, D., Dubeau, F., and Harinarayan, P. (2006). Image retrieval based on region separation and multiresolution analysis. *International Journal of Wavelets, Multiresolution and Information Processing*, 4(1):147–175.

Kubo, M., Aghbari, Z., and Makinouchi, A. (2003). Content-based image retrieval technique using wavelet-based shift and brightness invariant edge feature. *International Journal of Wavelets, Multiresolution and Information Processing*, 1(2):163–178.

Kubota, K., Iwaki, O., and Arakawa, H. (1983). Image segmentation techniques for document processing. In *Proc. of 1983 International Conference on Text Processing with a Large Character Set*, pages 73–78.

Kubota, K., Iwaki, O., and Arakawa, H. (1984). Document understanding system. In *Proc. of 7th International Conference on Pattern Recognition*, pages 612–614.

Kumar, S. and Kumar, D. K. (2005). Visual hand gestures classification using wavelet transforms and moment based features. *International Journal of Wavelets, Multiresolution and Information Processing*, 3(1):79–101.

Kumar, S., Kumar, D. K., Sharma, A., and McLachlan, N. (2003). Visual hand gestures classification using wavelet transforms. *International Journal of Wavelets, Multiresolution and Information Processing*, 1(4):373–392.

Kunte, R. S. and Samuel, R. D. S. (2007). Wavelet descriptors for recognition of basic symbols in printed kannada text. *International Journal of Wavelets, Multiresolution and Information Processing*, 5(2):351–367.

Lau, K. K. and Leung, C. H. (1994). Layout analysis and segmentation of chinese newspaper articles. *Computer Processing of Chinese and Oriental Languages*, 8(8):97–114.

Lee, S. W., Kim, C. H., Ma, H., and Tang, Y. Y. (1996). Multiresolution recognition of unconstrained handwritten numerals with wavelet transform and multilayer cluster neural network. *Pattern Recognition*, 29(12):1953–1961.

Lee, S. W. and Kim, K. C. (1994). Address block location on handwritten korean envelope by the merging and splitting method. *Pattern Recognition*, 27(12):1641–1651.

Lenz, R. (1990). Agroup theoretical methods in image processing. *Lecture Notes in Computer Science*, 413.

Li, H. (2006). Wavelet-based weighted average and human vision system image fusion. *International Journal of Wavelets, Multiresolution and Information Processing*, 4(1):97–103.

Li, H. F., Jayakumar, R., and Youssef, M. (1989). Parallel algorithms for recognizing handwritten characters using shape features. *Pattern Recognition*, 22(6):641–652.

Li, S. (2008). Multisensor remote sensing image fusion using stationary wavelet transform: Effects of basis and decomposition level. *International Journal of Wavelets, Multiresolution and Information Processing*, 6(1):37–50.

Liang, K. H., Chang, F., Tan, T. M., and Hwang, W. L. (1999). Multiresolution Hadamard representation and its application to document image analysis. In *Proc. of The Second International Conference on Multimodel Interface (ICMI'99)*, pages V1–V6, Hong Kong.

Liang, K. H. and Tjahjadi, T. (2006). Adaptive scale fixing for multiscale texture segmentation. *IEEE Transactions on Image Processing*, 15(1):249–256.

Liao, Z. (2004). *Image Denoising Using Wavelet Domain Hidden Markov Model.* PhD Thesis, Hong Kong Baptist Univeristy.

Liao, Z. and Tang, Y. Y. (2005). Signal denoising using wavelets and block hidden markov model. *International Journal of Pattern Recognition and Artificial Intelligence*, 19(5):681–700.

Long, R. (1995). *High-Dimensional Wavelet Analysis.* World Books Co. Pte. Ltd, Beijing, China.

Ma, H., Zhou, J., and Tang, Y. Y. (2002). Order statistic filter (osf): A novel approach to document analysis. *International Journal of Pattern Recognition and Artificial Intelligence*, 16(5):551–571.

Makino, H. (1983). Representation and segmentation of document images. In *Proc. of IEEE Computer Society International Conference on Pattern Recognition and Image Processing*, pages 291–296.

Malik, J. and Perona, P. (1990). Preattentive texture discrimination with early vision mechanisms. *Journal Opt. Soc. Amer. A.*, 7(5):923–932.

Mallat, S. and Zhong, S. (1992). Characterization of signals from multiscale edges. *IEEE TTransactions on Pattern Analysis and Machine Intelligence*, 14(7):710–732.

Mallat, S. G. (1989a). Multifrequency channel decompositions of images and wavelet models. *IEEE Transactions on Acoust. Speech Signal Process.*, 37(12):2091–2110.

Mallat, S. G. (1989b). Multiresolution approximations and wavelet orthonormal bases of $l^2(r)$. *Transactions on Amer. Math. Soc.*, 315:69–87.

Mallat, S. G. (1989c). A theory for multiresolution signal decomposition: the wavelet representation. *IEEE Transactions on Pattern Analysis and Machine Intelligence*, 11(7):674–693.

Mallat, S. G. and Hwang, W. L. (1992). Singularity detection and processing with wavelet. *IEEE Transactions on Information Theory*, 38(2):616–642.

Mandelbrot, B. B. (1983). *The Fractal Geometry of Nature.* Freeman, New York.

Mantas, J. (1986). An overview of character recognition methodologies. *Pattern Recognition*, 19(6):425–430.

Mao, J. C. and Jain, A. K. (1992). Texture classification and segmentation using multiresolution simultaneous autoregressive models. *Pattern Recognition*, 25(2):173–188.

Maragos, P. and Sun, F. K. (1993). Measuring the fractal dimension of signals: morphological covers and iterative optimization. *IEEE Transactions on Signal Processing*, 41(1):108–121.

Masuda, I., Hagita, N., Akiyama, T., Takahashi, T., and Naitoo, S. (1985). Approach to smart document reader system. In *Proc. of 1985 IEEE Computer Society Conference on Computer Vision and Pattern Recognition (CVPR'85)*, pages 550–557.

Meyer, Y. (1990). *Ondelettes et operateurs, I, II.* Hermann, Paris.

Mitchell, B. T. and Gillies, A. M. (1989). A model-based computer vision system for recognizing handwritten zip codes. *Machine Vision and Applications*, 2:231–243.

Moghaddam, H. A., Khajoie, T., Rouhi, A. H., and Tarzjan, M. S. (2005). Wavelet correlogram: A new approach for image indexing and retrieval. *Pattern Recognition.*, 38(12):2506–2518.

Mori, S., Suen, C. Y., and Yamamoto, K. (1992). Historical review of ocr research and development. *Proceeding of the IEEE*, 80(7):1029–1058.

Mori, S., Yamamoto, K., and Yasuda, M. (1984). Research on machine recognition of handprinted characters. *IEEE Transactions on Pattern Analysis and Machine Intelligence*, 6(4):386–405.

Mukherjee, S., Yang, G. Z., Tan, W. F., and Ramakrishnan, I. V. (2003). Automatic discovery of semantic structures in html document. In *Proc. of 7th International Conference on Document Analysis and Recognition*, pages 245–249.

Muneeswaran, K., Ganesan, L., Arumugam, S., and Harinarayan, P. (2005). A novel approach combing gabor wavelet transforms and moments for texture segmentation. *International Journal of Wavelets, Multiresolution and Information Processing*, 3(4):559–572.

Munkres, J. R. (1975). *Topology, A First Course.* Prentice-Hall, Inc., New Jersey.

Murtagh, F. and Starck, J. L. (1998). Pattern clustering based on noise modeling in wavelet space. *Pattern Recognition*, 31:47–855.

Nagy, G. (1968). A preliminary investigation of techniques for the automated reading of unformatted text. *Comm, ACM*, 11(7):480–487.

Nagy, G. (1990). Towards a structured-document-image utility. In *Proc. of International Workshop on Structural and Syntactic Pattern Recognition (SSPR'90)*, pages 293–309.

Nagy, G., Kanai, J., and Krishnamoorthy, M. (1988). Two complementary techniques for digitized document analysis. In *Proc. of ACM Conference on Document Processing Systems*, pages 169–176.

Nagy, G., Seth, S. C., and Stoddard, S. D. (1993). *Document analysis with an expert system*, pages 149–159. Elsevier Science Publishers B. V. (North-Holland), in e. s. gelsema and l. n. kanal (editors), pattern recognition practice ii edition.

Nakano, Y., Fujisawa, H., Kunisaki, O., Okada, K., and Hananoi, T. (1986). A document understanding system incorporating with character recognition. In *Proc. of 8th International Conference on Pattern Recognition*, pages 801–803.

Nakano, Y., Shima, Y., Fujisawa, H., and Fujiwara, J. H. M. (1990). An algorithm for the skew normalization of document image. In *Proc. of 10th International Conference on Pattern Recognition*, pages V–2, 8–13.

Nguyen, P. T. and Quinqueton, J. (1988). Space filling curves and texture analysis. In *Proc. of 9th International Conference on Pattern Recognition*, pages 81–84, Toronto, Canada.

Niyogi, D. and Srihari, S. N. (1986). A rule-based system for document understanding. In *Proc. of International Conference on Artificial Intelligence (AAAI'86)*, pages 789–793.

O'Gorman, L. (1993). The document spectrum for structural page layout analysis. *IEEE Transactions on Pattern Analysis and Machine Intelligence*, 15(11):1162–1173.

O'Gorman, L. and Kasturi, R. (1995). *(Eds.) Document Image Analysis*. IEEE Computer Society Press, New York.

Orchard, M. T. and Ramchandran, K. (1994). An investigation of wavelet-based image coding using an entropy constrained quantization framework. In *Proc. of International Conference on Data Compression*, pages 341–350, Snowbird, Utah.

Otterloo, P. J. V. (1991). *Contour Oriented Approach to Shape Analysis*. Pretice Hall, New York.

Parrish, E. A. (1989). A foreword to knowledge and data engineering. *IEEE Transactions on Knowledge and Data Engineering*, 1(1):5–7.

Pavlidis, T. (1982). *Algorithm for Graphics and Image Processing*. Computer Science Press, Maryland.

Peleg, S., Naor, J., Hartley, R., and Avnir, D. (1984). Multiple resolution texture analysis and classification. *IEEE Transactions on Pattern Analysis and Machine Intelligence*, 6(4):518–523.

Pentland, A. (1984). Fractal-based description of nature scenes. *IEEE Transactions on Pattern Analysis and Machine Intelligence*, 6(6):661–674.

Persoon, E. and Fu, K. S. (1977). Shape discrimination using fourier descriptors. *IEEE Transactions on Systems, Man, and Cybernetics*, 7(3):170–179.

Pesquet, J. C., Krim, H., and Hamman, E. (1996). Bayesian approach to best basis selection. In *Proc. of IEEE International Conference on Acoust., Speech, Signal*, pages 2634–2637, Atlanta, USA.

Pitas, I. and Kotropoulos, C. (1992). A texture-based approach to the segmentation of sesmic image. *Pattern Recognition*, 25(9):929–945.

Ramachandran, S. and Kashi, R. (2003). An architecture for ink annotation on web documents. In *Proc. of of The 7th International Conference on Document Analysis and Recognition*, pages 256–260.

Ramamoorthy, C. V. and Wah, B. W. (1989). Knowledge and data engineering. *IEEE Transactions on Knowledge and Data Engineering*, 1(1):9–15.

Rastogi, A. and Srihari, S. N. (1986). *Recognizing textual blocks in document images using the Hough transform.* TR 86-01, Dept. of Computer Science, SUNY Buffalo, NY.

Reiss, T. H. (1991). The revised fundamental theorem of moment invariants. *IEEE Transactions on Pattern Analysis and Machine Intelligence*, 13(8):830–834.

Rogers, D. F. and Adams, J. A. (1990). *Mathematical elements for computer graphics.* 2nd Ed, MaGraw-Hill, New York.

Rudin, W. (1974). *Real and Complex Analysis.* McGraw-Hill Book Company, 2nd edition.

Sabourin, M. (1992). Optical character recognition by a neural network. *Neural Networks*, 5(5):843–852.

Shankar, B. U., Meher, S. K., and Chosh, A. (2007). Neuro-wavelet classifier for multispectral remote sensing images. *International Journal of Wavelets, Multiresolution and Information Processing*, 5(4):589–611.

Sharnia, A., Kumart, D. K., and Kumar, S. (2004). Wavelet directional histograms of the spatio-temporal templates of human gestures. *International Journal of Wavelets, Multiresolution and Information Processing*, 2(3):283–298.

Shi, S., Zhang, Y., and Hu, Y. (2010). A wavelet-based image edge detection and estimation method with adaptive scale selection. *International Journal of Wavelets, Multiresolution and Information Processing*, 8(2):385–405.

Shridhar, M. and Badreldin, A. (1986). Recognition of isolated and simply connected handwritten numerals. *Pattern Recognition*, 19(1):1–12.

Simoncelli, E. P. (1997). *A Gentle Tutorial on the EM Algorithm and Its Application to Parameter Estimation for Gaussian Mixture and Hidden Markov Modelsl.* Technical Report, University of Berkeley.

Singer, Y. and Warmuth, M. K. (1998). Batch and on-line parameter estimation of gaussian mixtures based on the joint entropy. In *Advances in Neural Information Processing Systems 11 (NIPS'98)*, pages 578–584.

Spitz, A. (2002). Progress in document reconstruction. In *Proc. of 16th International Conference on Pattern Recognition*, pages 464–467.

Spitz, A. L. and Dengel, A. (1995). *(Eds.) Document Analysis Systems.* World Scientific Publishing Co. Pte, Ltd., Singapor.

Srihari, S. N. and Govindaraju, V. (1989). Analysis of textual images using the hough transform. *Machine Vision and Application*, 2:141–153.

Srihari, S. N., Wang, C. H., Palumbo, P. W., and Hull, J. J. (1987). Recognizing address blocks on mail pieces: specialized tools and problem-solving architecture. *AI Magazine*, 8(4):25–40.

Suen, C. Y., Berthod, M., and Mori, S. (1980). Automatic recognition of hand-printed characters - the state of the art. *Proceeding of the IEEE*, 68(4):469–487.

Tan, T. N. (1998). Rotation invariant texture feature and their use in automatic script identification. *IEEE Transactions on Pattern Analysis and Machine Intelligence*, 20(7):751–756.

Tang, Y. Y. (2009). *Wavelets Theory Approach to Pattern Recognition.* World Scientific Publishing Co. Pte, Ltd., Singapor.

Tang, Y. Y., Cheng, X., Tao, L., and Suen, C. Y. (1993a). Parallel regional projection transformation (rpt) and vlsi implementation. *Pattern Recognition*, 26(4):627–641.

Tang, Y. Y., Cheriet, M., Liu, J., Said, J. N., and Suen, C. Y. (1999). *Document analysis and recognition by computers*, pages 579–612. World Scientific Publishing Co. Pte, Ltd., Singapor, in c. h. chen, l. f. pau and p. s. p. wang, editors, handbook of pattern recognition and computer vision (2nd edition) edition.

Tang, Y. Y., Li, B., Ma, H., and Liu, J. (1998a). Ring-projection-wavelet-fractal signatures: A novel approach to feature extraction. *IEEE Transactions on Circuits and Systems II*, 45(8):1130–1134.

Tang, Y. Y., Ma, H., B.Li, and Liu, J. (1996). Character recognition based on Doubechies wavelet. In *Proc. 1st Int. Conf. on Multimodal Interface*, pages 215–220, Beijing, China.

Tang, Y. Y., Ma, H., Liu, J., Li, B., and Xi, D. (1997a). Multiresolution analysis in extraction of reference lines from documents with graylevel background. *IEEE Transactions on Pattern Analysis and Machine Intelligence*, 19(8):921–926.

Tang, Y. Y., Ma, H., Xi, D., Cheng, Y., and Suen, C. Y. (1995a). A new approach to document analysis based on modified fractal signature. In *Proc. of 3-nd International Conference on Document Analysis and Recognition*, pages 567–570, Montreal, Canada.

Tang, Y. Y., Ma, H., Xi, D., Mao, X., and Suen, C. Y. (1997b). Modified fractal signature (mfs): A new approach to document analysis for automatic knowledge acquisition. *IEEE Transactions on Knowledge and Data Engineering*, 9(5):747–762.

Tang, Y. Y., Suen, C. Y., and Yan, C. D. (1991a). Chinese form pre-processing for automatic data entry. In *Proc. of International Conference on Computer Processing of Chinese and Oriental Languages*, pages 313–318.

Tang, Y. Y., Suen, C. Y., and Yan, C. D. (1994). Document processing for automatic knowledge acquisition. *IEEE Transactions on Knowledge and Data Engineering*, 6(1):3–21.

Tang, Y. Y., Yan, C. D., Cheriet, M., and Suen, C. Y. (1991b). Document analysis and understanding: a brief survey. In *Proc. of First International Conference on Document Analysis and Recognition*, pages 17–31, Saint-Malo, France.

Tang, Y. Y., Yan, C. D., Cheriet, M., and Suen, C. Y. (1993b). *Automatic analysis and understanding of documents*, pages 625–654. World Scientific Publishing Co. Pte, Ltd., Singapor, in patrick s.p. wang, c.h. chen and l.f. pau, editors, handbook of pattern recognition and computer vision edition.

Tang, Y. Y., Yan, C. D., Cheriet, M., and Suen, C. Y. (1995b). Financial document processing based on staff line and description language. *IEEE Transactions on Systems, Man, and Cybernetics*, 25(5):738–754.

Tang, Y. Y., Yang, F., and Liu, J. M. (2001). Basic processes of chinese character based on cubic b-spline wavelet transform. *IEEE Transactions on Pattern Analysis and Machine Intelligence*, 23(12):1443–1448.

Tang, Y. Y., Yang, L. H., and Feng, L. (1998b). Contour detection of handwriting by modular-angle-separated wavelets. In *Proc. of 6-th International Workshop on Frontiers of Handwriting Recognition (IWFHR-VI)*, pages 357–366.

Tang, Y. Y., Yang, L. H., Liu, J. M., and Ma, H. (2000). *Wavelets Theory and Its Application to Pattern Recognition.* World Scientific Publishing Co. Pte, Ltd., Singapor.

Tang, Y. Y. and You, X. G. (2003). Skeletonization of ribbon-like shapes based on a new wavelet function. *IEEE Transactions on Pattern Analysis and Machine Intelligence*, 25:1118–1133.

Tieng, Q. M. and Boles, W. W. (1997). Wavelet-based affine invariant representation: A tool for recognizing planar objects in 3D space. *IEEE Transactions on Pattern Analysis and Machine Intelligence*, 19(6):846–857.

Tiller, W. (1987). Rational b-spline for curve and surface representation. *IEEE Computer Graphics and Applications*, 3:61–69.

Toyoda, J., Noguchi, Y., and Nishimura, Y. (1982). Study of extracting japanese newspaper article. In *Proc. of 6th International Conference on Pattern Recognition*, pages 1113–1115.

Tsuji, Y. (1988). Document image analysis for generating syntactic structure description. In *Proc. of 9th International Conference on Pattern Recognition*, pages 744–747.

Tsujimoto, S. and Asada, H. (1990). Understanding multi-articled documents. In *Proc. of 10th International Conference on Pattern Recognition*, pages 551–556.

Tucker, L. W., Feyman, C. R., and Fritzsche, D. M. (1988). Object recognition using the connection machine. In *Proc. of Computer Vision Pattern Recognition Conference*, pages 871–878, Ann Arbor, MI.

Turney, J. L., Mudge, T. N., and Volz, R. A. (1985). Recognizing partially occluded parts. *IEEE Transactions on Pattern Analysis and Machine Intelligence*, 7:410–421.

Viswanathan, M. (1990). Analysis of scanned documents - a syntactic approach. In *Proc. of International Conference on Syntactic and Structural Pattern Recognition (SSPR'90)*, pages 450–556.

Wahl, F., Abele, L., and Scheri, W. (1981). Merkmale fuer die segmentation von dokumenten zur automatischen textverarbeitung. In *Proc. of 4th DAGM-*

Symposium.

Wang, D. and Srihari, S. N. (1989). Classification of newspaper image blocks using texture analysis. In *Proc. of Conference on Computer Vision, Graphics and Image Processing (CVGIP)*, pages 327–352.

Warakagoda, N. D. and Hogskole, N. T. (1998). *A Hybrid ANN-HMM ASR System with NN based Adaptive Preprocessing*. Master Thesis, University of Delaware.

Watanabe, S. (1985). *Pattern Recognition: Human and Mechanical*. Wiley-Interscience Publication, New York.

Wiskott, L. (1999). Segmentation from motion: combining gabor-and mallat-wavelets to overcome the aperture and correspondence problems. *Pattern Recognition*, 32(12):1751–1766.

Wong, K. Y., Casey, R. G., , and Wahl, F. M. (1982). Document analysis system. *IBM Journal of Research Develop*, 26(6):647–656.

Wu, A. Y., Bhaskar, S. K., and Rosenfeld, A. (1989). Parallel processing of region boundaries. *Pattern Recognition*, 22(2):165–172.

Wunsch, P. and Laine, A. F. (1995). Wavelet descriptors for multiresolution recognition of handprinted characters. *Pattern Recognition*, 28(8):1237–1249.

Yan, C. D., Tang, Y. Y., and Suen, C. Y. (1991). Form understanding system based on form description language. In *Proc. of First International Conference on Document Analysis and Recognition*, pages 283–293, Saint-Malo, France.

Yang, L. H., Bui, T. D., and Suen, C. Y. (2003). Image recognition based on nonlinear wavelet approximation. *International Journal of Wavelets, Multiresolution and Information Processing*, 1(2):151–161.

Yeh, P. S., Antoy, S., Litcher, A., and Rosenfeld, A. (1987). Address location on envelopes. *Pattern Recognition*, 20(2):213–227.

Yin, H. and Liu, H. (2010). A bregman iterative regularization method for wavelet-based image deblurring. *International Journal of Wavelets, Multiresolution and Information Processing*, 8(2):485–499.

You, X., Chen, Q., Fang, B., and Tang, Y. Y. (2006). Thinning character using modulus minima of wavelet transform. *International Journal of Pattern Recognition and Artificial Intelligence*, 20(3):361–376.

You, X. and Tang, Y. Y. (2007). Wavelet-based approach to character skeleton. *IEEE Transactions on Image Processing*, 16:1220–1231.

Yu, C. L., Tang, Y. Y., and Suen, C. Y. (1995). Document skew detection based on the fractal and least squares method. In *Proc. of 3-nd International Conference on Document Analysis and Recognition*, pages 1149–1152, Montreal, Canada.

Zahn, C. T. and Roskies, R. Z. (1972). Fourier descriptors for plain closed curves. *IEEE Transactions on Computers*, 21:269–281.

Zhang, G. (1986). Lecture of functional analysis. *Beijing University Press*, 1.

Zhang, T., Fan, Q., and Gao, Q. (2010). Wavelet characterization of hardy space h^1 and its application in variational image decomposition. *International Journal of Wavelets, Multiresolution and Information Processing*, 8(1):71–87.

Zhu, X. Y. and Yin, X. X. (2002). A new textual/non-textual classifier for document skew correction. In *Proc. of 16th International Conference on Pattern Recognition*, pages 480–482.

Index